INTRODUCTION TO MASS SPECTROMETRY

Instrumentation and Techniques

Introduction to

MASS SPECTROMETRY

INSTRUMENTATION AND TECHNIQUES

JOHN ROBOZ

Air Reduction Company, Inc.
Murray Hill, New Jersey

INTERSCIENCE PUBLISHERS

A DIVISION OF JOHN WILEY & SONS, NEW YORK • LONDON • SYDNEY • TORONTO

Library of Congress Catalog Number 67-13960
SBN 470 72840

PRINTED IN THE UNITED STATES OF AMERICA

To My Wife, Julia

Preface

Mass spectrometry is probably the most comprehensive and versatile of all instrumental methods of analysis. Almost any material can be analyzed with a suitable mass spectrometer, and more information is obtained per microgram of sample than with any other analytical technique. A mass spectrometric analysis provides both identification of the materials present and the concentration of each. In addition, systematic interpretation of the spectra provides a detailed picture of the ionization process which, in turn, may be utilized in the elucidation of molecular structure.

The science, or perhaps art, of mass spectrometry is now half a century old. It has been successfully applied to a wide variety of problems, ranging from basic research of matter through trace impurity analysis and empirical formula determination to routine quality control, in many areas of physics, chemistry, biochemistry, geology, and in virtually every industry. The development of analytical techniques and interpretation procedures best suited to a particular field of application, together with the availability of an extraordinary diversity of reliable commercial mass spectrometers, have resulted in a considerable compartmentalization of the subject. This is evidenced by the current surge of books devoted entirely to applications, e.g., interpretation of mass spectra of organic compounds, trace analysis of inorganic solids, space research, etc. Such works are, of course, inevitable and essential for serious research in any specific field.

The objectives of this book are: (*1*) to present an introduction to the theory, design, and operation of the various types of mass spectrometers; (*2*) to provide assistance in the selection of commercial mass spectrometer systems and components by discussing required instrument performance in many fields of applications; (*3*) to acquaint the reader with the general experimental techniques employed in mass spectrometry, and illustrate these methods in typical applications.

For the future mass spectroscopist this book is intended to serve as a comprehensive introduction to the subject, a *primer* to specialized works, and a source of further references. For those who would like to obtain information about how mass spectrometry can help in their own research problems, the text provides all the essentials needed. The book will be useful for courses in mass spectrometry, advanced instrumental analysis, physical chemistry, and vacuum techniques. It should be a helpful reference for organic chemists attending the increasingly popular short courses on the instrumental methods of structure determination.

The reader is not expected to have any prior knowledge of mass spectrometry. A background in science equivalent to a college degree in chemistry or physics is, however, assumed. The treatment is elementary, derivations are omitted, and the emphasis is on the basic principles and considerations involved in designs and applications. Special effort was made to define and thoroughly explain the terms and expressions of mass spectrometry which are so casually used in both publications and discussions, and a comprehension of which is so very essential.

Although the discussion on ion optics is short and non-analytical, the chapters on mass analyzers and the types of mass spectrometers are likely to be found more difficult than the rest of the text. Since beginners in mass spectrometry are usually also neophytes in high vacuum techniques, the chapter on this subject is relatively long. The objective of the chapter on commercial instruments is to acquaint the reader with the currently available basic instrument types and accessories from which a mass spectrometer system best suited for a particular problem may be assembled. Individual instruments are not described in detail since specifications change continuously and current brochures are readily available from the manufacturers.

In the chapters on applications the emphasis is on general experimental techniques that can be applied to a variety of problems. Whenever possible, limitations, experimental errors, inherent accuracy are discussed, and further application possibilities indicated. Interpretation of organic mass spectra is not discussed in detail. There are many recent texts devoted to this subject and the purpose of this book is to provide background for such works.

The number of references is kept relatively small. Lists of classified references are now readily available and an entire chapter deals with information and data on mass spectrometry. Those references marked with * are intended as suggested reading material to be studied not only for their specific content but also to introduce the reader to the language and style of original research.

Illustrations and tables adapted or reproduced from previously published material are identified and acknowledged either by a reference number or by direct citation in the caption. Wherever it appeared appropriate, permission of the copyright holder has been obtained. I wish to express my appreciation to the various publishers for their courtesy, and to all authors whose work is referred to.

I am greatly indebted to the manufacturers of mass spectrometers (Table 8-1) who have so generously furnished schematic diagrams, photographs, and technical information of their instruments.

Throughout the preparation of the manuscript, I have had the assistance of many people. Special acknowledgments are due to W. Brubaker, Teledyne Co., D. Damoth, Bendix Corp., W. Downer, Consolidated Electrodynamics Corp., F. Gollob, Gollob Analytical Services, R. Hein, Finnigan Instruments Corp., G. Junk, Iowa State University, H. Liebl, Max Planck Institute, Wm. McFadden, International Flavors & Fragrances, Inc., S. Meyerson, American Oil Co., D. Rhum, Air Reduction Co., and J. Sullivan, Institute of Exploratory Research, U.S. Army, who read various parts of the manuscript. I profited materially from their suggestions. As I did not always heed the good advice offered, I am alone responsible for the shortcomings of this book.

I am obliged to Dr. Gordon Arquette, Director of Research & Engineering, Air Reduction Company, for his permission to publish this work.

I am indebted to Thomas Polanyi and Robert Wallace, friends and former associates, for stimulating my interest in mass spectrometry.

Finally, I would like to express very special thanks to my wife Julia, for her encouragement and understanding, and for having dispensed the necessary prodding.

JOHN ROBOZ

September, 1968

Contents

PART II.   APPLIED MASS SPECTROMETRY

INSTRUMENTATION

Chapter One

Introduction

I. TERMINOLOGY

A. Ions, Mass, Abundance

A mass spectrometer is an apparatus which produces a beam of gaseous ions from a sample, sorts out the resulting mixture of ions according to their mass-to-charge ratios, and provides output signals which are measures of the relative abundance of each ionic species present. Mass spectrometers are usually classified on the basis of how the mass separation is accomplished, but they all can be described as ion optical devices which separate ions according to their mass-to-charge ratios by utilizing electric and/or magnetic force fields.

The mass spectrum of a sample reveals in a graphical (pictorial) or tabular form the measured mass-to-charge ratios of the separated ions and their corresponding intensities. The knowledge of the mass-to-charge ratios of the ions enables one to determine WHAT is present, while the measured ion intensities answer the question of HOW MUCH is present. In addition, systematic interpretation of the mass spectra provides a detailed picture of the ionization process which, in turn, may be utilized in the elucidation of molecular structures.

Ions are positively or negatively charged atoms, groups of atoms, or molecules. The process whereby an electrically neutral atom or molecule becomes electrically charged, due to losing or gaining one or more of its extranuclear electrons, is called ionization. Although both positive and negative ions can be studied by mass spectrometry, the majority of instruments are used to investigate positive ions because in most ion sources they are produced in larger number ($\sim 10^3$) than negative ions. In this text "ions" denotes positive ions unless otherwise specified.

There is a minimum amount of energy, characterized by the "ionization potential," that must be provided in order for ion formation to occur. The *first* ionization potential of an atom or molecule is defined as the energy

3

input required to remove (to infinite distance) a valence electron from the highest occupied atomic or molecular orbital of the neutral particle to form the corresponding atomic or molecular ion, also in its ground state. When only one electron is removed the ion is called an atomic or molecular ion; often the term "parent ion" is used. The formation of parent ions may be considered as ionization without cleavage. The numerical magnitude of the ionization potential is influenced by such factors as the charge upon the nucleus, the atomic or molecular radius, the shielding effect of the inner electronic shells, and the extent to which the most loosely bound electrons penetrate the cloud of electronic charge of the inner shells. The latter effect results in the fact that, other factors being equal, an s electron is harder to remove than a p electron, etc. First ionization potentials are in the 5–15 eV range for most elements and in the 8–12 eV range for most organic molecules and radicals. Numerical values are given in Appendix II. To remove a second, third, etc. electron, additional energy is needed (2nd, 3rd, . . . ionization potentials). When excess energy is available, fragmentation of the molecules may also occur during the process of ionization. There are many important types of such "fragment ions" (Chapter 9). The minimum energy required for the appearance of a particular fragment ion in the mass spectrum is designated by the term "appearance potential."

The mass of a body is generally understood to be a measure of the quantity of matter the body contains and it is assumed to be inherent and inalienable. It is difficult, however, to define precisely what is meant by the "amount of matter" in an object. Clearly, one cannot use size or volume, and also "weight" is unsatisfactory since the weight of the same body varies from place to place depending on gravity. In physics, mass is defined (Newton's second law) as a measure of inertia, i.e., a property of a body that determines the acceleration it will acquire when acted upon by a given force. In mass spectrometry the term "mass" has several connotations which require qualifications.

It will become clear in the discussion of mass analyzers that mass spectrometers measure only mass-to-charge ratios. The *mass-to-charge ratio*, m/ne, is the ratio of the mass, m, of the ion (atomic, molecular, fragment) to the number, n, of electric charges, e, lost or gained during ionization. When equations are derived in mass spectrometry, m means mass measured in kg and e means charge measured in coulombs. Atoms and molecules cannot, of course, be weighed directly (an oxygen atom weighs 2.6×10^{-26} kg) and for most ordinary purposes one is more interested in the relative weights of atoms and molecules than their absolute weights. A more convenient unit for the measurement of the masses of nuclides is the *atomic mass unit* (Section B in this chapter), u or amu, which is defined so that the mass of a ^{12}C atom is exactly $12u$. Final results of derivations are normally converted to include mass in atomic mass units, M. The conversion equation is $M = m/1.6604 \times 10^{-27}$. Since the *mass number,** A, of an atom is the sum of the number of protons (atomic number, Z) and the number of neutrons (N) in the nucleus, and because of the approximate equality of the proton and neutron masses ($1.007277u$ and $1.008665u$, respectively) and the relative insignificance of that of the electron ($5.48 \times 10^{-4}u$), the mass number gives a useful rough figure for the atomic mass, e.g., $^1H = 1.007825u$.

When the terms "mass" or "m/e value" ("M/e value") are used in everyday mass spectrometry, m (M) refers to the sum of the mass numbers of the atoms composing the particular ion and e gives the number of electrons lost during ionization. Although the majority of ions in a mass spectrometer are singly charged, atoms and certain molecules may lose more than one electron without disintegration. A doubly charged ion exhibits an apparent mass one-half that of the corresponding singly charged ion. For example, doubly charged carbon dioxide (CO_2^{2+}) appears at m/e 22, triply charged argon (Ar^{3+}) at m/e 13.3.

In the mass spectrum of carbon dioxide shown in Figure 1-1 the abscissa is an m/e scale and the ordinate shows relative ion abundances. The terms "ion abundance" or "ion intensity" refer to the number of ions detected in each separated ion beam. A very early method of detecting ions involved actual counting with the aid of a screen made of a metal, such as zinc sulfide, that fluoresces when bombarded by ions. Ion abundances are now measured by either electrical or photographic techniques. The former measures the intensity of the ion current caused by the ions arriving at a collector; the latter utilizes the fact that the darkening of certain photographic emulsions is proportional to the number of impinging ions. Although the actual number of ions detected may be determined with both methods, the knowledge of relative ion intensities is adequate in many applications. The ordinate of Figure 1-1, for example, is measured in recorder chart divisions. Such *peak heights* can be directly compared after multiplication by the proper scale attenuation factor. For example, the relative intensity of the $m/e = 28$ peak in Figure 1-1 is approximately three times that of the $m/e = 22$ peak. The displacement which is read on the

---

* The following designations are conventional for a chemical element, X,

$$^{\text{Mass number}}_{\text{Atomic number}} X ^{\text{State of ionization}}_{\text{Number of atoms of element in a molecule}}$$

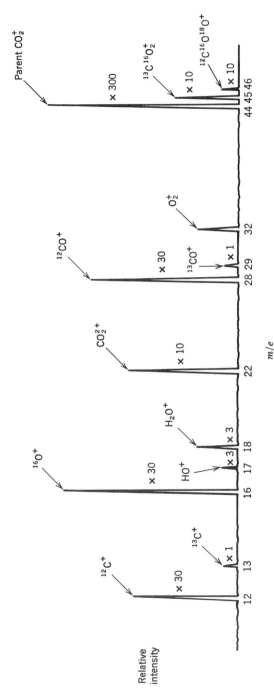

Figure 1-1. Mass spectrum of carbon dioxide. Relative peak intensities are plotted against m/e.

potentiometric recorder is a voltage which is proportional to the ion current caused by the ions arriving at the collector. If possible, information should also be given about the numerical relationship between measured chart divisions and ion current, e.g., in Figure 1-1 the peak height at $m/e = 13$ corresponds to 5×10^{-14} A. If the spectrum shown in Figure 1-1 had been taken with photographic detection, the ordinate would represent relative intensities obtained by densitometric determination of the darkening of the mass spectral *lines* on the photoplate. In this case the ordinate may be given in terms of optical density, transmission percent, or some other convenient quantity which is proportional to the number of ions hitting the photoplate. Similarly to electrical detection, information should be given, if possible, about the numerical relation between measured darkening and numbers of ions or equivalent ion current.

In the past, instruments primarily designed for precise m/e determinations were called mass *spectrographs* because they employed photographic methods for ion detection. Instruments used for ion abundance measurements were designated as mass *spectrometers* on account of their electrical means of detection. The term mass *spectroscope* has often been used in general discussions. During the last few years, with the development of modern instruments with interchangeable and/or simultaneous electrical and photoplate detection, these distinctions have become less meaningful. In current usage, instruments and techniques are customarily described by the generic term "mass spectrometer" and "mass spectrometric," respectively. However, data obtained by the use of ion-sensitive photoplates is still described as "mass spectrographic." Scientists working with mass spectroscopes are often called mass spectroscopists or mass spectrometrists.

B. Isotopes, Mass Scales

Isotopes are atoms whose nuclei contain the same number of protons but a different number of neutrons. Isotopes thus have the same atomic number but they have different atomic masses and therefore different mass numbers. The word *isotope* (Greek, *equal places*) was first suggested by Soddy (1913) to describe radioactive elements occupying identical positions in the periodic table. Since isotopes have the same number and arrangement of orbital electrons, and since chemical properties are determined by extranuclear electrons, isotopes are chemically indistinguishable, except for certain small kinetic effects.

Appendix I lists the nuclidic masses of all stable isotopes of the elements, together with their relative abundances. An examination of the data permits several conclusions and generalizations to be made: (*1*) There are only about 20 elements which are monoisotopic, the most important being Be, F, Na, Al, P, Mn, Co, As, I, Au, and Bi. (*2*) No stable nucleus, except $^1_1$H, has a mass less than twice its charge, i.e., contains fewer neutrons than protons. (*3*) Elements with odd nuclear charge never have more than two stable isotopes. (*4*) Elements with even nuclear charge exist in several isotopic forms, e.g., tin has 10 stable isotopes, xenon has 9. (*5*) Elements with even nuclear charges are more abundant and more stable than those with odd charges. In the course of mass spectrometric practice one encounters practically all stable isotopes at one time or another, and a "by heart" knowledge of the most common isotopes and their approximate abundances is not only impressive but also a time saver. When mass number is not indicated in a formula, the most abundant isotopes of the element are assumed to be present.

As mentioned already, masses of nuclides are conveniently measured in atomic mass units (*u* or amu). By definition, one atomic mass unit is equal to 1/12 the mass of the most abundant isotope of carbon. A ^{12}C atom thus has a mass exactly $12.0000u$. Using Avogadro's number ($N_A = 6.02252 \times 10^{26}$ kmole^{-1}) one obtains

$$1 \text{ amu} = 1/N_A = 1.66043 \times 10^{-27} \text{ kg} \qquad (1\text{-}1)$$

The atomic mass, or exact mass, of a nuclide is the weight of its atoms in amu, including the extranuclear electrons. The atomic mass thus defined is slightly different from the mass number, A, in every instance except for the carbon-12 standard. The difference is the *mass defect*. Formerly the magnitude of the mass defect was considered a measure of the stability of a nucleus. It is now replaced by the more precise concept of the binding energy (Section 13-I-B). The energy equivalent of 1 amu is 931.4 MeV or 1.49×10^{-3} erg.

Prior to 1961 there were two different scales for atomic masses. In the *physical* scale the mass of the most abundant oxygen isotope was set equal to precisely 16.000000 amu. In the *chemical* scale the average mass of the three oxygen isotopes, according to their natural abundances, was set equal to precisely 16.000000 amu. The conversion factor was

$$\frac{\text{Physical scale}}{\text{Chemical scale}} = \frac{16 \text{ exactly}}{15.99560} = 1.000275 \qquad (1\text{-}2)$$

This situation was the consequence of the discovery of the oxygen isotopes (1929). Physicists converted to the ^{16}O scale while chemists continued to use the old scale to save existing data compilations. In 1960–1961 the International Union of Pure and Applied Physics (IUPAP) and the

International Union of Pure and Applied Chemistry (IUPAC) accepted a new scale based on carbon-12. Conversion factors are

$$\frac{\text{Physical scale}}{^{12}\text{C scale}} = \frac{12.00382}{12 \text{ exactly}} = 1.000318 \qquad (1\text{-}3)$$

$$\frac{\text{Chemical scale}}{^{12}\text{C scale}} = \frac{12.00052}{12 \text{ exactly}} = 1.000043 \qquad (1\text{-}4)$$

The change to the new unified scale amounts to only 43 ppm on the chemical scale and 318 ppm on the physical scale. It is, as E. A. Guggenheim remarked, "a tremendous triumph of reasonableness over confusion." The change is trivial in most chemical calculations but highly significant in precise mass measurements. All nuclidic masses and atomic weights in Appendix I are given in the unified carbon-12 scale. Atomic weights are averages of the isotopic masses of the elements in proportion to their natural abundances, and they are dimensionless.

A few related concepts should also be reviewed at this point. The *mole*, the chemist's expression of the "amount," is the quantity of substance that contains the same number of molecules (or ions, or atoms, or electrons, as the case may be) as there are atoms in exactly 12 g of the pure carbon nuclide ^{12}C. Avogadro's number is the number of units in a mole.

Isobars are atomic species which have the same mass number but differ in atomic number, e.g., $^{40}_{19}\text{K}$ (39.9640) and $^{40}_{20}\text{Ca}$ (39.9626). In high-resolution mass spectrometry, ions of the same nominal mass but of different elemental composition are often called isobaric. For example, the molecular ions of ethane ($C_2H_6^+$) and formaldehyde (CH_2O^+) and a fragment ion of ethylamine (CH_4N^+) all have the same nominal mass of 30 while their exact masses are 30.0469, 30.0105, and 30.0344, respectively. When a mixture of these compounds is analyzed in a mass spectrometer with adequate resolution, three separate peaks are obtained in the m/e 30 region. In nuclear chemistry, nuclei identical in mass and atomic number are termed *isomeric*. They differ only in energy content and the more energetic nucleus normally decays, accompanied by γ-ray or photon emission, to the less energetic. In organic chemistry isomerism refers to the existence of more than one compound with the same molecular formula; e.g., C_2H_6O stands both for ethyl alcohol and for dimethyl ether. Mass spectrometry plays an important role in isomer analysis and many examples will be discussed.

C. Mass Spectrometer Systems

To best understand the operation of mass spectrometers, they should be considered as *systems*, i.e., integrated assemblies of interacting elements designed to carry out cooperatively a function which, in the present case, is the determination of the m/e ratios together with their relative intensities of the various species in a sample. Such a system consists of a number of elements, and each must be treated separately by specifying the *input* it must receive, the *process* it must perform, and the *output* it generates.

Every mass spectrometer, regardless of its special design, comprises four functional elements: (*1*) the *source*, where a beam of ions, representative of the sample under investigation, is generated; (*2*) the *analyzer*, in which the separation is effected either in space or in time; (*3*) the *detector*, where the resolved ions are detected and their intensity is measured; and (*4*) the *vacuum system*, which provides the "environment" for all these processes. The sometimes quite elaborate *sample inlet* systems and the various *recorders* may be considered as subsystems of the vacuum system and the detector, respectively. A diagram showing the main features of a mass spectrometer system is shown in Figure 1-2.

The most important method to produce ions is based on collisions between neutral atoms or molecules and energetic ($\sim$70 eV for reasons discussed later) electrons at low pressure (10^{-4} to 10^{-6} torr) in the *electron bombardment sources*. Almost all organic mass spectrometry is carried out with such electron impact sources. Two other types of sources offer certain unique advantages in the study of organic compounds: (*1*) ionization is achieved by strong electrical fields in the *field emission* source and the spectra are quite simple; (*2*) ultraviolet light of sufficient wavelength is employed in *photoionization* sources, and the process of ionization can be studied under controlled conditions. Inorganic solids are usually analyzed either in *thermal ionization sources*, where ionization takes place during evaporation from a heated metal surface, or in various *vacuum discharge* sources, where ions are formed in sparks or arcs.

Mass analyzers are classified into two broad groups. The *static* type analyzers utilize the momentum dispersion properties of magnetic fields and the energy dispersion properties of electric fields in such a manner

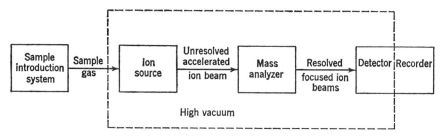

Figure 1-2.   Diagram of mass spectrometer system.

that instrumental parameters are kept constant in time, except for certain relatively slow changes needed for recording the spectrum; there are many possible field combinations. In *dynamic* type instruments the time dependence of a parameter of the system is the basis of mass separation. Differences in the time of flight of ions in evacuated tubes, the time dispersion properties of the radiofrequency field, and other properties are utilized in addition to those mentioned for the static types.

Ion detectors are based either on the "conversion" of the individual separated ion beams into a proportional electron current which can be amplified and recorded, or on the simultaneous collection of all ion beams using ion-sensitive photoemulsions.

Part I of this text is devoted to the description of the operation of the various subsystems and to the principles underlying the combination of these subsystems.

D. Measures of Performance

Mass spectrometers answer the basic questions of WHAT is present and HOW MUCH is present by determining ionic masses and abundances. Accordingly, the two most important performance characteristics are instrument resolution and sensitivity. These concepts are discussed and illustrated from many points of view as the text progresses; however, it appears desirable to present here a brief description of their meaning.

The *resolving power* or *resolution* of a mass spectrometer is a measure of its ability to separate and identify ions of slightly different masses. Figure 1-3a shows a *summation peak*. Here the resolution of the instrument is inadequate to separate the two neighboring masses. The result of the overlap is a summation output between the adjacent masses. The resultant peak is larger than each peak would normally be alone since the contribution of the overlapped portions is additive. Such a summation peak is the one at $m/e = 28$ in Figure 1-1. It is marked as a CO^+ peak but a small fraction of the peak height is caused by N_2^+ present in the instrument background. It is noted that the complex nature of this peak cannot be detected by simple visual observations; the peak appears similar to the others in the spectrum. Much more will be said about the $m/e = 28$ peak later.

Two peaks are considered "just resolved" when one can ascertain that two peaks are present (Fig. 1-3b). In qualitative analysis it is often adequate merely to establish the presence of an ionic species, without much regard to the accuracy with which abundances could be determined. In quantitative analysis, contributions from neighboring masses should be as small as possible. The required resolution thus depends not only on the accuracy with which the mass is to be determined but also on the relative quantities of the two masses present.

Resolution can be calculated on the basis of ion optics (Section 2-VI-B) or it can be measured. The ratio of theoretical to measured resolution may serve as a measure of the quality of instrument design and manufacture. For the practicing mass spectrometrist only the experimentally determined resolution is significant.

Resolution is usually defined in terms of the *largest mass at which a given criterion is met.* There are several criteria in common use today, leading to a considerable amount of confusion. Perhaps the most popular is the "valley" definition which expresses resolution in terms of the highest mass at which two adjacent peaks of equal height, differing in mass by one unit,

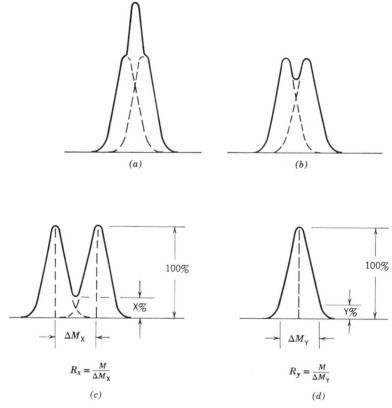

$$R_x = \frac{M}{\Delta M_X}$$

(c)

$$R_y = \frac{M}{\Delta M_Y}$$

(d)

Figure 1-3. (a) Summation peak. (b) Two peaks "just detectable." (c) "Percent valley" definition of resolution. (d) "Peak width" definition of resolution.

exhibit a valley between the peaks not greater than a certain percentage, such as 2 or 10%, of the peak height (Fig. 1-3c). To be precise, one should take the average of the two masses, but this has relatively little significance in this definition of resolution. Unless otherwise noted, resolution values in this text refer to the valley definition employing 10% levels, i.e., each peak contributes 5% to the valley.

In practice, resolution must often be determined using an isolated peak in the mass spectrum. Resolution, R, may be determined from the formula $R = M/\Delta M$, where M is the mass of the peak in atomic mass units and ΔM is the width of the peak at 5% peak height level, also in atomic mass units (Fig. 1-3d). This definition implies singly charged ions; for multiply charged ions the number of charges must be considered. The 5% *peak width* definition is technically equivalent to the 10% valley definition provided the peak is symmetrical and the system linear in the range between the 5 and 10% levels of the peak.

The resolution of the instrument yielding the spectrum of Figure 1-1 is about 1 part in 130. This resolution is adequate to separate the isotopes of xenon as shown in Figure 1-4a. At this resolving power the isotopes of mercury (m/e 196–204) would not be seen at "unit" resolution, but adequate details would be available to make an identification on the basis of isotopic abundances. It is obvious that an instrument with a resolving power of only 130 can be used for only a limited number of applications.

In most cases the minimum resolving power required is such that the *molecular ion* could be distinguished from adjacent peaks. For example, when the sample has a nominal mass of 500, the value of $M/\Delta M$ should be at least 500. Such a resolution would give the molecular weight of the

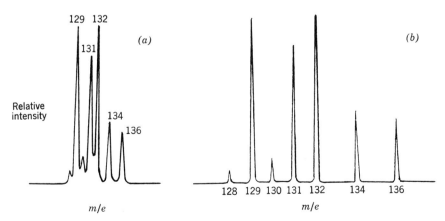

Figure 1-4.   Mass spectrum of xenon isotopes: (*a*) resolution 130: (*b*) resolution 500.

compound in question as a whole number. It would also give valuable information on fragments, again in whole number units. For example, a peak 15 mass units below the molecular ion peak indicates the presence of a fragment ion which originated from the parent ion by the loss of a methyl group. Those peaks for which the resolution is more than adequate will, of course, be well separated, e.g., the isotopes of xenon at a resolution of 500 (Fig. 1-4*b*).

The magnitude of separation varies in different applications. In quantitative analysis, contributions from neighboring masses should be as small as possible, while for qualitative identifications it is often adequate merely to establish the presence of an ionic species. It is emphasized that the principal objective of *high-resolution* instruments is not to separate and determine adjacent unit masses at the numerical value of the resolving power, but to separate peaks only millimass units apart at much lower nominal masses. For example, a resolving power of 2500 is required to separate the N_2^+ peak (mass = 28.006148) from the CO^+ peak (mass = 27.994915), even though the nominal mass is only 28. As another example, consider the $^{14}N^+-^{12}CH_2^+$ and the $^{12}C_{20}H_{36}^{14}N^+-^{12}C_{21}H_{38}^+$ *doublets*. The exact mass difference in both cases is 0.013. In the first case the nominal mass is 14, and the required resolution is about 1000. In the second case, the nominal mass is 290, and the required resolution for mass separation is about 22,000.

The need for even higher resolution in organic mass spectrometry becomes obvious when one considers that for any nominal mass number many possible combinations of carbon, oxygen, and nitrogen atoms exist. Moreover, the number of possible combinations increases rapidly with mass (Section 8-II). In modern high resolution instruments for organic structure elucidation work, a nominal resolution of 30,000 or higher is provided.

Since the maximum attainable resolving power of a mass spectrometer depends on basic optical design, which once selected cannot be altered, much care must be exercised in instrument selection. In addition, the cost of mass spectrometers is almost a direct function of the resolution desired. The various methods of expressing resolution are further discussed in Section 2-VI, while resolution requirements in various applications are detailed in Section 8-II.

Sensitivity, the second important performance characteristics, is dependent upon the detector system employed, and also upon the analytical technique used. Here the mass spectrometrist often has good opportunities to improve the performance of his instrument. The sensitivity of a mass spectrometer is a measure of the instrument's response to ions of a particular component at an arbitrary m/e value. Instrument response is measured, as mentioned already, either by the charge received at a measuring device or by the blackening of an ion-sensitive photographic plate. There are many ways to express sensitivity depending upon the problem involved and sometimes also upon convenience. For example, when a sample is evaporated into the ion source, instrument response per unit weight of material consumed per second may be used to express sensitivity. In the analysis of gas chromatography effluents, sample flux in g/sec may be employed. In residual gas analysis, sensitivity is often measured as amperes of ion current per torr pressure of the gas in the ion source. Still another method expresses sensitivity as detector response per unit charge of ions leaving the source.

It is stressed that sensitivity is always expressed for a particular peak and a particular sample. Various materials exhibit different efficiencies for ionization in the source, and there might be differences in the efficiency of the transmission of ions through the mass analyzer. Also, the detector may exhibit a higher or lower efficiency for a particular mass or type of ion. As a result, different output signals may be obtained for equal quantities of two substances, or the output signal of a given quantity of sample may be changed in the presence of another material (interference effect). Sensitivities are often expressed on a relative scale with respect to an arbitrarily selected sample. For example, residual gas analyzer sensitivities are usually expressed relative to that of nitrogen.

In trace analytical applications the *sensitivity limit* or *detection limit* of the instrument is of vital importance. Detection limits may be expressed in absolute or relative terms. The *absolute limit of detection* is defined as the minimum amount of element or compound, expressed in convenient units such as weight of material, or pressure in the ion source, that can be detected under given experimental conditions. *Concentrational detection limit* is the minimum concentration of an element or compound which can be detected in a matrix of another material. This type of sensitivity is normally expressed as parts by volume, weight, or number of atoms of the impurity per million (or billion) parts of the sample (ppm, ppb). Units must be carefully selected. In spark source mass spectrography, for example, results are usually expressed in atomic parts per million units. Often, however, values must be compared to results of other analytical techniques, expressed in weight parts per million, and conversion is necessary. (How does one convert from atomic ppm to weight ppm?)

There are two parameters of importance in connection with detection limits. The first is *instrument noise level*, i.e., the spurious instrument response not due to ions striking the ion collector or photoplate. The signal-to-noise ratio must be at least 2:1 in order to ascertain the presence of a signal above noise. The second parameter, *instrument background*, refers to the instrument response, at a given mass, due to influences extraneous to the sample being investigated. The *background spectrum* of an instrument is a mass spectrum obtained with operating conditions similar to those used for sample analysis, except that the sample is absent. The causes of instrument background, the influence of background on resolution requirements, the corrections for background, as well as the question of the signal-to-noise ratio are discussed in appropriate sections later in the text.

Sensitivity and resolving power are approximately inversely proportional to each other. This important fact can often be utilized for trade-off, and many modern instruments provide changeover from "high" to "low" sensitivity without extensive modifications. This changeover is usually accomplished by adjusting appropriate slit sizes and/or by employing different kinds of detectors. Of course, the price to pay for higher sensitivity is reduced resolution. It is emphasized once again, that corresponding resolution must always be considered when comparing or quoting sensitivity values.

Another performance characteristics of importance is the *mass range* of an instrument which, as the name implies, gives the lowest and highest masses that can be measured. The mass range of an instrument is not directly related to its resolution; thus an instrument may have a wide mass range, but poor resolution could severely reduce the useful portion. The lowest measurable mass is usually 1 or 2, while the highest measurable mass in modern instruments is as high as 3000.

The term "scanning" refers to the process whereby each successive ion beam is swept across the ion collector, in order of either increasing or decreasing mass. Scanning may be automatic, manual, or repetitive. The *scan rate* or scan speed is the rate at which the ion beams successively are focused onto the collector. Scan rates are normally expressed in seconds per mass decade, seconds per mass octave, seconds per stated mass range, or peaks per second. A mass decade means a factor of 10 in mass, e.g., from mass 1 to mass 10, or from mass 10 to mass 100, etc. There is, of course, no mass scanning when a photographic plate is employed for detection.

II. SCOPE OF APPLICATIONS

The development of mass spectrometry may be divided into three periods: (*1*) development of basic instrumentation and applications in nuclear physics; (*2*) development of commercial instruments and applications in chemical analysis; (*3*) high resolution mass spectrometry and applications in structural organic chemistry and biochemistry.

In 1898 Wien showed that positive rays ("Kanalstrahlen"), discovered by Goldstein only a few years before, could be deflected in magnetic and electric fields, and consist of positively charged particles. In 1910 J. J. Thomson designed the so-called parabola spectrograph which became the forerunner of all mass spectroscopes. This instrument was constructed for the study of positive rays as counterparts of electron rays for which the charge-to-mass ratio had just been determined. In 1912 Thomson reported the first direct experimental evidence that neon consisted of a mixture of two different isotopes (mass 20 and 22). The discovery of the existence of stable isotopes is generally considered to be the greatest achievement of mass spectrometry, and J. J. Thomson the "father" of mass spectrometry.

Thomson's work was continued by his student, F. W. Aston, who built the first true mass spectrograph in 1919 and confirmed Thomson's discovery of the isotopes of neon. The atomic weight of neon had been determined by vapor density measurements to be 20.2. The relative intensities of the mass spectrographic lines at m/e 20 and 22 were in the ratio of 9:1, yielding $[(9 \times 20) + (1 \times 22)]/10 = 20.2$ as the average atomic weight for neon, in agreement with vapor density results. The term "mass spectrum" was first used by Aston in 1920.

According to Aston's *whole-number rule*, atomic masses are, to a close approximation, integral multiples of the mass of hydrogen and any deviations from whole numbers (as determined by chemical methods) are due to isotopes, each of which has a whole-number atomic weight. The atomic weight of chlorine (35.45), for example, results from the two isotopes, ^{35}Cl and ^{37}Cl, occurring in a 3:1 ratio. This rule provided a confirmation of the famous Prout hypothesis (1815) that all atoms can be derived from hydrogen. Aston's greatest contribution was, however, the discovery that the integral nature of the isotopic weights is only approximately true. He found that the exact mass (nuclidic mass) is slightly different from the mass number, and called the difference the "mass defect." The mass defect results from energy release during the formation of the atom from its constituents. The actual mass loss corresponding to this energy release can be calculated from Einstein's formula, $\Delta m = \Delta E/c^2$, where c is the speed of light; conversely, the formula yields the energy corresponding to a given mass. Using ^{16}O as a standard, Aston systematically determined the whole-number divergences of many elements, and advanced the concept of "packing fractions." His packing fraction curve (Fig. 13-3) was intensively used in the development of modern nuclear physics.

During the next 20 years efforts were concentrated on the one hand on increasing instrument resolution and improving the precision of mass determinations, and on the other hand on developing reliable methods for relative abundance measurements. Instruments were designed for special purposes and various types of mass spectrometers soon became associated with the names of their originators, e.g., Dempster-type instrument, Nier geometry, etc. Such designations are commonly used even today, but, to avoid ambiguity, it is better to supply additional information in describing instrument types, e.g., "Nier-type, first-order direction focusing, 15 cm radius, 60° homogeneous magnetic sector field instrument." Those historical instruments which served as prototypes for present-day commercial instruments are discussed in Section 3-I.

Mass spectrometers are electromagnetic separators of ions and the importance of separating isotopes in "large" quantities was recognized in the early days of mass spectrometry. Efforts in this direction culminated in 1940 when Nier and his co-workers isolated and separated uranium-235 and uranium-238, thus paving the way to nuclear weapons as well as to nuclear power plants. With modern isotope separators, gram-size quantities of highly enriched stable isotopes are produced.

The appearance of commercial mass spectrometers in the early 1940's signaled a new era in mass spectrometry. Formerly mysterious monsters, which could be tamed only by their creators the university professors, mass spectrometers were to be installed in increasing number in industrial laboratories, particularly in the petroleum and petrochemical industries. Complex hydrocarbon mixtures could be quantitatively analyzed by trained technicians in a matter of hours, instead of the days necessary when fractional distillation was employed. The operational stability of commercial instruments made it possible to provide comparative data, and the mass spectrometer has become a routine analytical tool.

Today there is a new revolution in mass spectrometry. With the advent of modern electronics and ultrahigh vacuum technique, commercial instruments of extreme capability and versatility have become available. The current market for mass spectrometers is estimated by industrial sources as more than 10 million dollars yearly in the USA. Mass spectrometers are now employed in virtually every industry. They are used in research laboratories, industrial plants, and hospital rooms, and they are fired into outer space. The use of spark source mass spectrometers in the field of semiconductors is hardly 10 years old, but the technique is already considered as routine in the analysis and characterization of electronic materials.

The most active area in mass spectrometry today is the application of high-resolution techniques to organic structure elucidation. Hundreds of papers and almost a dozen books have been published during the last five years dealing with the structure of compounds ranging from simple hydrocarbons to complex natural products. High-resolution mass spectrometers are capable of routinely measuring ionic masses to an accuracy of within a few parts per million, making it possible to directly determine molecular constitution. The combination of mass spectrometers with gas chromatographs enables one to separate complex mixtures and than identify all components including those present only in minute quantities. The use of these techniques in biochemistry is still in a stage of infancy, but significant results have already been reported, e.g., the study of proteins (Section 13-III). The techniques of high-resolution mass spectrometry and combined gas chromatography–mass spectrometry usually provide a large amount of analytical data. Computer techniques are now being developed to handle the data efficiently and the day is not too far away when automatic interpretation of mass spectra will provide the organic chemist with a relatively narrow range of possible structures for an unknown organic substance.

In Part II mass spectrometric techniques are discussed and applications are illustrated in many branches of the physical sciences. An overall view of the scope of mass spectrometry may be presented on the basis of the kinds of information one can obtain from mass spectra.

Mass spectrometry, in a broad sense, is a branch of radiation chemistry since ions are normally obtained by the action of ionizing radiation. In other kinds of spectroscopy, such as infrared, ultraviolet, etc., the manner in which the sample affects the radiation is studied. In mass spectrometry the interaction between radiation and matter results in ion formation and the information obtained relates to the properties of isolated primary ions.

Ionic mass and abundance are the quantities directly measured by mass spectrometers. Such information coupled with the understanding of the fragmentation process makes mass spectrometry an invaluable tool in almost every area of physics and chemistry.

A. Determination of Mass

In the evaluation of any mass spectrum the first step is to determine the masses of the separated ions. There are several methods for mass determination depending on the accuracy needed, but all methods are based on calibration with known masses. In Figure 1-1, for example, the positions of the various masses are determined by introducing pure carbon dioxide into the mass spectrometer. Sometimes it is adequate to determine mass only to integral numbers, while in certain cases accuracy of a few parts per million is required.

Precise mass measurements can be used to investigate:

- Atomic masses of elements
- Nuclear binding energies
- Organic compounds:
    1) Molecular weight and empirical formula
    2) Structural analysis

B. Determination of Isotope Abundance

The unique ability of the mass spectrometer to separate isotopes and provide information as to their relative quantities can be applied in many fields:

- Isotopic constitution of the elements and variations in the natural abundances of the isotopes
- Geochronometry, geologic thermometry
- Stable isotope tracers:
    1) Analysis by isotope dilution (geochemical prospecting, inorganic trace analysis)
    2) Biochemical processes
    3) Reaction mechanisms (intermediates, reaction rates)
- Selective ion counting
    1) Trace impurities in solids
    2) Leak detection

C. Uses of Mass Spectral Patterns

Most mass spectrometers operate in such a manner that the energy available for ionization is significantly more than is needed for parent ion formation. For elements this results in the appearance in the spectra of doubly, triply, etc. charged ions in addition to the isotope peaks. In the case of chemical compounds, a variety of fragmentation processes may take place, and the spectrum of a large organic compound may exhibit hundreds of peaks. This feature is both a help and a hindrance. In qualitative analysis fragment peaks provide important information on the structure of the unknown. In the quantitative analysis of mixtures complex calculations become necessary to ascertain the contribution to the fragment peaks by the various components.

The array of peaks in the complete spectrum of a pure substance is referred to as a *cracking pattern*. Peak heights in a spectrum are usually

Table 1-1   Cracking Pattern of Carbon Dioxide[a]

m/e	% Abundance	m/e	% Abundance
12	5.40	29	0.07
13i	0.08	32	0.08
16	7.85	44p	*100.00*
22d	2.00	45i	1.14
28	6.60	46i	0.41

[a] p = parent peak, i = isotope peak, d = doubly charged ion.

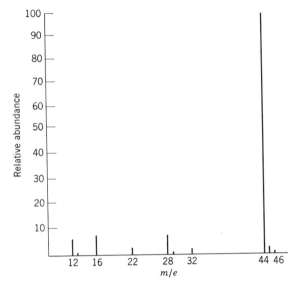

Figure 1-5. Line diagram of carbon dioxide spectrum. The abundance of the parent peak is taken as 100.

normalized by taking the largest peak in the spectrum (*base peak*) as 100. The base peak may or may not be the parent peak.* Standardized data may be presented in either tabular or line diagram form. Data from Figure 1-1 is shown in cracking pattern form in Table 1-1 and Figure 1-5. The tabular form provides numerical values for the abundances. This is the form used in actual calculations. Line diagrams, on the other hand, save space and show overall features of the spectrum at a glance. This form is becoming popular in publishing complex organic spectra.

Cracking patterns have been obtained for thousands of compounds and there are a number of compilations available. There are three important

* Normalization may be made with respect to *any* peak in the spectrum.

properties of cracking patterns: (*1*) Every chemical compound has its own distinctive cracking pattern ("fingerprint"); (*2*) cracking patterns are remarkably constant as long as experimental parameters are kept unchanged; and (*3*) when two or more components are present at the same time each will produce its own cracking pattern and the resultant spectrum is produced by linear addition of the components. These properties of mass spectral patterns can be utilized for a variety of purposes:

- Qualitative identification
  - *1*) Impurities in pure materials
  - *2*) Unknown gas chromatography peaks
  - *3*) Pyrolysis products of polymers
  - *4*) Residual gas analysis
- Quantitative analysis
  - *1*) Hydrocarbon mixtures
  - *2*) Trace constituents in gases, atmospheric pollutants
  - *3*) Gas content of metals, glasses, plastics, etc.
- High temperature chemistry
  - *1*) Thermodynamic properties
  - *2*) Reaction studies
- Investigation of ions from electric discharges, flames, shock tubes, etc.

D. Electron Impact Phenomena

The process and mechanism of ionization and dissociation of neutral particles by electron bombardment is the subject of studies in *electron impact phenomena*, a branch of physical chemistry.

When electrons with kinetic energy E_{kin} collide with neutral particles the collision may be elastic or inelastic. In elastic collisions no change occurs in the internal energy of the atom or molecule, the momentum and kinetic energy of translation are conserved, and, as may be easily shown, the electrons can lose no appreciable amount of energy. If the velocity of the impacting electrons is increased, the interaction will produce excited species, and when a certain energy is transferred to the neutral particle a valence electron will move from its ground state to an energy level above the lowest: the collision becomes inelastic. Depending on the complexity of the molecule, many quantum-mechanically permitted excited states exist and each is characterized by an excitation potential. However, atoms and molecules in such excited states are not charged and thus cannot be detected by mass spectrometry. The energy excess is eventually dissipated as radiation. Life of an excited state is $\sim 10^{-8}$ sec.

At even higher energies many configurational changes are possible (Fig. 1-6), the processes of ionization and dissociative ionization being the most important, as they lead to ion formation. At electron energies of 25–30 eV, stripping of two valence electrons becomes possible (producing doubly charged ions) and when the energy available becomes equal to the dissociation energy in one of the ion's degrees of freedom, fragment ions appear.

The *probability of ionization* is defined as the number of those collisions which result in ion production per centimeter of electron path through a

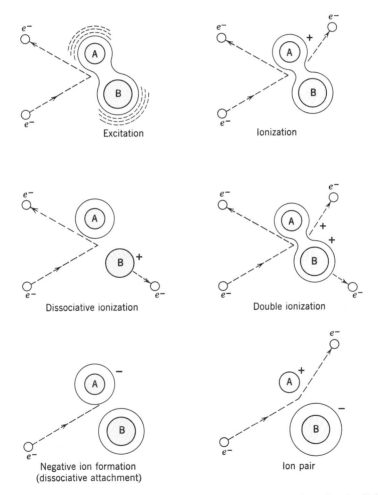

Figure 1-6. Types of collisions between an electron and a diatomic molecule. (After H. Massey, *Science & Technology*, July 1965, by permission of copyright owner, Conover-Mast Publications Inc.)

gas at a pressure of 1 torr at 0°C. This concept is needed because even when ionization is energetically possible under certain conditions, it may still be statistically improbable. The variation of ion intensity—in arbitrary units—as a function of electron energy (in electron volts) is depicted in the so-called *ionization efficiency curves.* Figure 1-7 shows ionization efficiency curves for a number of inorganic gases. It is seen that the maxima of the curves is located approximately at the same energy level (~70 eV) for most gases. These curves have many other features which will be discussed later.

The efficiency of ionization in the neighborhood of the ionization threshold is very sensitive to changes in electron energy (Fig. 1-8). The *minimum* energy that electrons must have to produce ions of a particular ionic species is called the *appearance potential* of the ion. Precise measurement of these values is difficult, demanding special instrumentation and technique (Section 12-I-C).

Ionization efficiencies and appearance potentials can be used in many ways to study electron impact phenomena:

- Mechanism of ionization and dissociation
- Calculation of chemical bond strengths
- Energy states of atoms, molecules, free radicals

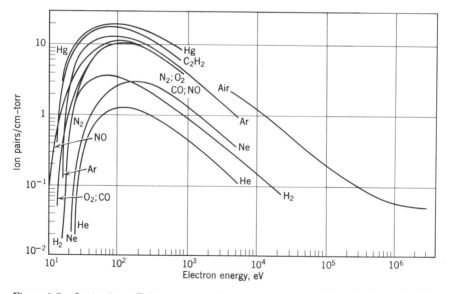

Figure 1-7. Ionization efficiency curves for inorganic gases. (After R. Jaeckel, "Allgemeine Vakuumphysik," in *Handbuch der Physik*, Vol. 12, S. Flügge, Ed., Springer-Verlag, Berlin 1958, p. 535.)

- Interactions of ions with matter
- Theory of mass spectra: statistical mechanics applied to calculate cracking patterns, prediction of fragmentation modes.

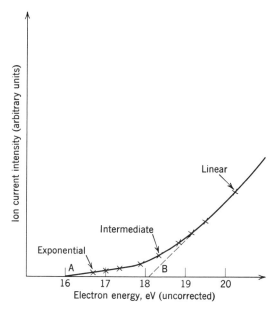

Figure 1-8.   Foot of ionization curve. Point *A* is the initial upward break, point *B* is extrapolated linear intercept.

Chapter Two

Mass Analyzers

I. INTRODUCTION

The primary purpose of a mass spectrometer is to measure the mass-to-charge ratios of ions, thus providing a means to identify them. While in optical spectroscopy frequency and intensity are the most important parameters, in mass spectrometry the number of parameters are four: mass, charge, velocity, and intensity. Mass and charge are closely related by the fact that fundamentally any mass spectrometer determines *only* the m/e ratio. If the actual magnitude of m is desired, the value of the charge must be determined by some other method.

Mass analyzers have two objectives: First, to *resolve* an ion beam of mass m from another beam of nearly the same mass $m + \Delta m$ and second, to *maximize* the resolved ion intensities. The former is a *dispersive* action, the latter a *focusing* one. Magnetic and electric fields thus assume the roles played by prisms and lenses in conventional optics. The action of magnetic and electric fields and their combinations on charged particles is studied in *ion optics* or ion ballistics. Since ions behave in electromagnetic fields the same way as electrons, but opposite in sign, ion optics can utilize directly the results of modern electron optics. Focusing systems in mass spectrometry are equivalent, most of the time, to cylindrical lenses.

To determine the trajectories of ions one must know their initial position and velocity as they enter the analyzers, and the spatial and/or temporal distribution of the field (fields) involved. Momentum, energy, and velocity are the most important properties of charged particles; to determine the m/e ratio only two of these quantities need be measured.

There are three basic principles upon which all analyzers are founded: (*a*) magnetic analysis, (*b*) electrostatic analysis, and (*c*) time-of-flight analysis. These methods are next discussed, starting with the electrostatic analysis.

II. ELECTROSTATIC ANALYSIS

A. Electric Field

An electric field is said to exist in a region in which an electrical force acts on an independent test particle (pole) brought into the region. The intensity, E, of the electric field at a certain point is defined as the force, F, exerted on a unit charge, e, at that point,

$$E = F/e \qquad (2\text{-}1)$$

No particular name is given to E. It is expressed in newtons per coulomb. The direction of the field is that in which a positive charge is urged. In practice the field is provided in the space between two metal plates, distance d apart, by applying a potential difference, V, between them. The strength of the electric field in units of volts per unit distance is given by

$$E = V/d \qquad (2\text{-}2)$$

B. Electrostatic Accelerators

In simple electrostatic accelerators electrons or ions move *parallel* to the direction of the field as they fall through a *known* potential difference. Such accelerators have two important applications in mass spectrometry. First, an electron beam of known and as nearly as possible homogeneous energy is needed in the *electron impact* ion sources where positive ions are produced from neutral gas molecules by electron bombardment (Section 4-II). As shown in Figure 2-1a, electrons which are initially at rest, or practically so (they are emitted from a hot filament), are made to fall through a potential difference, $V = Ed$. The kinetic energy, T_{kin}, acquired by the electrons is

$$T_{kin} = \tfrac{1}{2}m_e v^2 = eV \qquad (2\text{-}3)$$

where m_e is the mass of electrons and v is velocity. Here eV is equal to the work done, since the difference in potential is defined as energy acquired per unit charge, and under the action of the field all potential energy becomes kinetic energy. The energy content of electrons is usually expressed in units of *electron volts* (eV). One electron volt is defined as the kinetic energy of an electron accelerated from rest through a potential difference of 1 V. The work done by the electric field is 1 V times 1.6×10^{-19} coul or 1.6×10^{-19} joule. It must be kept in mind that an electron volt is a unit of energy, while a volt is a unit of potential difference. Conversion of electron volts to electrostatic units is sometimes necessary in mass spectrometry,

$$1 \text{ eV} = (4.8 \times 10^{-10} \text{ statcoul}) \times (1/300 \text{ statV}) \qquad (2\text{-}4)$$

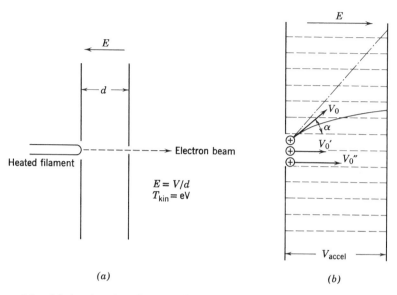

Figure 2-1. (a) Accelerating electrons through a known potential difference. (b) Accelerating ions through a known potential difference.

The problem of the various systems of units is discussed shortly. The velocity of an electron accelerated to 1 eV is, from equation 2-3, about 6×10^5 m/sec (mass of an electron is 9.11×10^{-31} kg) which is about 3 orders of magnitude higher than the molecular velocities of gases at room temperature.

The second application of electrostatic accelerators in mass spectrometry is to accelerate positive ions emerging from the ion source before they enter the analyzer section. The importance of this will become obvious in the discussion of the magnetic analyzers. The velocity, v_{ion} (m/sec), acquired by an ion falling through a potential difference of V volts is given by

$$v_{ion} = 1.39 \times 10^4 \sqrt{V} \sqrt{n/M} \qquad (2\text{-}5)$$

where n denotes the number of charges on the ion, M is the atomic mass (amu), and the constant enters to convert to convenient units. For example, the velocity of a H_2^+ ion ($n = 1$, $M = 2$) with kinetic energy of 1 eV is about 10^4 m/sec, a value only 6 times greater than the velocity of a neutral molecule due to thermal energy at room temperature.

When the ion entering the accelerator already has a velocity of v_0, the final velocity is given by

$$v_{ion} = \sqrt{v_0^2 + (2neV/m)} \qquad (2\text{-}6)$$

When two ions of the same mass, m, and same charge, ne, but of differing initial velocities (v_0' and v_0'') travel through a homogeneous electrostatic field of V volts (Fig. 2-1b), their final velocities, v_1 and v_2, are given by

$$v_1 = \sqrt{v_0'^2 + (2neV/m)} \quad \text{and} \quad v_2 = \sqrt{v_0''^2 + (2neV/m)} \quad (2\text{-}7)$$

Although the energy content of the ions increases during the process of acceleration, the absolute difference, ΔT, of their kinetic energies remains the same after acceleration as that before entering the accelerator, ΔT_0,

$$\Delta T = T_1 - T_2 = [(mv_0'^2/2) + neV] - [(mv_0''^2/2) + neV]$$
$$= [(mv_0'^2/2) - (mv_0''^2/2)]$$
$$\Delta T = \Delta T_0 \quad (2\text{-}8)$$

When, however, relative values, i.e., $\Delta T_0/T_0$ versus $\Delta T/T$, are evaluated, the change caused by the acceleration is significant. These relations must be considered carefully in ion source design. When ions enter the accelerator at an angle α to the field lines, the right-hand side of equation 2-8 must be multiplied by $\sin^2 \alpha$.

When two ions of different masses, m_1 and m_2, travel through a potential difference of V volts, their kinetic energies upon leaving the accelerator will be the same. The velocities, however, with which they travel through the field (and also their traveling times) will be different,

$$v_1/v_2 = \sqrt{m_1/m_2} \quad (2\text{-}9)$$

It is stressed that however complex an ion acceleration system may be, the final kinetic energy acquired is determined solely by the total potential difference through which the ions have moved.

C. Energy Filters

When positive ions are injected perpendicular to the lines of force into an electrostatic field between two parallel plates, they are accelerated toward the negative plate and describe parabolic trajectories. The deflection from normal at any time during the passage of a particle may be determined with elementary equations of the kinematics of translational motion.

More important is the behavior of ions in radial electrostatic fields (Fig. 2-2). The radial field is set up between two coaxial sector-shaped cylindrical electrodes. The outer plate is positive, the inner one negative. When positive ions of various masses and energies are injected midway between the plates and normal to the electric field, there will be ions which describe an exactly circular trajectory along the equipotential curve, PQ, the condition for this being that the ions must have such an energy content

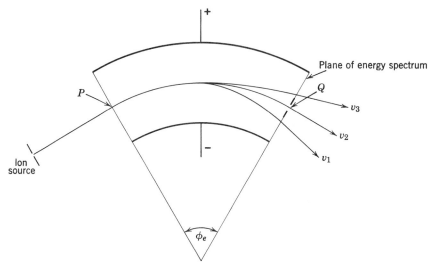

Figure 2-2. **Prism action of the electrostatic field on an ion beam of inhomogeneous energy.**

that the centrifugal force be exactly balanced by the acting electrostatic force,

$$mv^2/r_e = eE \quad \text{or} \quad mv^2/2 = r_e eE/2 \tag{2-10}$$

where v is the velocity of the injected ion of mass m, E is the field strength, and r_e is the radius. Replacing v by $v^2 = 2eV/m$ from equation 2-3 one obtains

$$m(2eV/m) = r_e eE \quad \text{or} \quad r_e = 2V/E \tag{2-11}$$

where V is potential difference. It is seen from this equation that such a system is an *energy filter*: only ions with a predetermined energy can describe a circular path of radius r_e, and by placing a small aperture at the exit plane an ion beam of nearly homogeneous energy may be obtained. Ions with other energies describe somewhat complicated trajectories and at the exit plane a kind of *energy spectrum* is formed.

Attention is called to the fact that m does not appear in the above equation, indicating that the electrostatic analyzer *does not* analyze mass. It does, however, analyze energy, since any variation in V results in a path of different radius. Also, ions of *any* mass will follow the same path for a given acceleration voltage and field strength. For example, singly charged ions accelerated to 2000 eV would require a field of about 160 V/cm to travel a radius of 25 cm.

The focusing action of electrostatic fields and the use of energy filters in double-focusing analyzers is discussed later in this chapter.

III. MAGNETIC ANALYSIS

A. Magnetic Field

A magnetic field is said to exist in a region in which a magnetic force acts on an independent test particle (pole) brought into the region. The path described by the motion of the pole is called a line of flux and is so drawn that a tangent to it at any point indicates the direction of the field. In the mks system the unit of magnetic flux is the weber which corresponds to 10^8 lines of flux (10^8 maxwells in the emu system). The number of flux lines per unit area permeating the magnetic field is the *magnetic flux density*, B. Units are the tesla and gauss in the mks and emu systems, respectively,

$$10^8 \text{ lines/m}^2 = 1 \text{ weber/m}^2 = 1 \text{ tesla (T)}$$
$$1 \text{ line/cm}^2 = 1 \text{ maxwell/cm}^2 = 1 \text{ gauss (G)}$$

Magnetic flux density is very frequently expressed in gauss units in mass spectrometry and conversion is needed when radii or other parameters are calculated in the mks system. To convert gauss into tesla divide by 10^4.

The oersted, which is the electromagnetic (and also Gaussian) unit of *magnetic field intensity*, H, i.e., the force exerted by the field on a unit north pole situated in the field, has the same dimension ($\text{cm}^{1/2} \text{ g}^{1/2} \text{ sec}^{-1}$) as the gauss in vacuum, and this may occasionally cause confusion. Conversion of oersteds into the mks system, where the unit is ampere-turn/meter, involves the permeability of free space which in the mks system is equal by definition to 4×10^7 henries/m; numerical value of the factor is 1.25×10^{-2}.

B. Force on Moving Charged Particles

When a particle with charge e and velocity v moves across a magnetic field of flux density B at right angles to the field, a force, F, is exerted upon it,

$$F_B = Bev \tag{2-12}$$

In mks units F_B is obtained in newtons when the velocity is expressed in meters per second and the charge in coulombs.

Force, flux density, and velocity are vector quantities, and in equation 2-12 they form a mutually perpendicular set. The direction of the force is thus at right angles to both the direction of the motion of the charge and the direction of the magnetic field. This relation for positive ions may easily be remembered by the *left-hand* rule (Fig. 2-3). In the general case, when there is an angle, ϕ, between B and v, one obtains

$$F_B = Bev \sin \phi \tag{2-13}$$

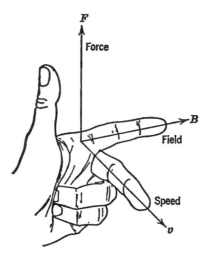

Figure 2-3.   The left-hand rule.

It is seen that the force is proportional to that component of the velocity perpendicular to the direction of the magnetic induction. When $\phi = 0$, no force is experienced by the moving charge, while with $\phi = 90°$ equation 2-12 results.

The orbit of a charged particle entering a (uniform) magnetic field in a direction at right angles to the field is described with reference to Figure 2-4. The magnitude of the velocity will not be affected by the force since they are perpendicular to each other. Since the magnitudes of B and e do not change as the particle moves, the magnitude of the force also remains unaltered. The particle is, therefore, under the influence of a force whose magnitude is constant but whose direction is always at right angles to the velocity of the particle (left-hand rule). The orbit of the particle thus

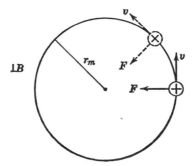

Figure 2-4.   Force on a positive ion moving normally to a uniform homogeneous magnetic field directed into the paper.

becomes a *circle*, and the force becomes equal to the centripetal force, F_C

$$F_B = Bev = F_C = mv^2/r_m \qquad (2\text{-}14)$$

where v is the tangential speed and r_m is the radius of the circle described.

If the direction of the initial velocity of the particle is not perpendicular to the magnetic field, the orbit is a helix, the cross section of which is a circle with a radius given by equation 2-14.

C. Momentum Filters

Rearranging equation 2-14 one obtains

$$mv = Ber_m \qquad (2\text{-}15)$$

It is seen that the magnetic field acts as a *momentum analyzer*. A beam of ions of a given charge and homogeneous in momentum can be obtained by positioning an exit slit at an appropriate point. A momentum distribution or momentum spectrum may be obtained by placing a photographic plate at the exit plane.

D. Analysis for Mass

A second, much more important, conclusion can be drawn from equation 2-14 by rearranging according to m/e

$$m/e = Br_m/v \qquad (2\text{-}16)$$

The momentum filter may thus become a mass filter when the conditions of equal ion energy and charge are introduced. Indeed, this property of the magnetic field is the basis of the majority of mass spectrometers. Although most ions are singly charged, multiply charged ions do occur and unless the charge is determined separately, only an m/e value is obtained. The constancy of ion energy is provided by making the ion beam pass an electrostatic analyzer after emerging from the ion source. The velocity of the ions, v, may now be expressed by equation 2-3 and equation 2-16 becomes

$$m/e = \sqrt{(m/2eV)} \times Br_m$$

or

$$m/e = B^2 r_m{}^2/2V \qquad (2\text{-}17)$$

Here m is the mass of the particle (kg), e is the charge (coul), B is magnetic flux density (T), V is acceleration voltage, and r_m is radius of circular trajectory (m).

At this point a digression into the various unit systems is in order. There is a considerable mixing of unit systems in everyday mass spectrometry and attention is needed to avoid confusion. Although the rationalized mks system of units is now generally accepted, the Gaussian system is still often used in mass spectrometry books and papers. In the Gaussian system electrical quantities such as charge, field strength, and current are expressed in electrostatic (esu) units, while all magnetic quantities are expressed in electromagnetic (emu) units. In the cgs electrostatic system, which is based on Coulomb's law, the unit of charge is the statcoulomb defined as the charge which repels an exactly similar charge, separated from it by 1 cm in a vacuum, with a force of 1 dyne. In the cgs electromagnetic system, which is based on the law of attraction between currents, charge is given in abcoulombs defined as the charge which passes a given surface in 1 second when a steady current of 1 abampere flows across the surface. Conversion from abcoulombs to statcoulombs involves division of the former by 2.998×10^{10}, the speed of light. Since both electrical and magnetic interactions occur in mass spectrometry, the use of the Gaussian system was justified, and the speed of light (c, cm/sec) frequently appeared in ion optical equations. In the rationalized mks system, electrical quantities such as voltage, current, and power are expressed in familiar practical units (volt, ampere, watt), and no conversion between magnetic and electric quantities need be made. In addition, since charge (Q, in coulombs) is a fourth selected physical dimension, no fractional exponents result in the expression of dimensions of various parameters. The numerical value of the charge of electrons is given in the three systems by

$$\text{electronic charge:} \quad \begin{aligned} &1.60186 \times 10^{-19} \text{ coul (mks)} \\ &4.80220 \times 10^{-10} \text{ statcoul (esu)} \\ &1.60186 \times 10^{-20} \text{ abcoul (emu)} \end{aligned} \tag{2-18}$$

The conversion units are: mks:esu:emu $1 : 3 \times 10^9 : 0.1$.

Returning now to equation 2-17, the units most frequently employed in mass spectrometry are atomic mass units for the mass, centimeter for the radius, gauss for the magnetic field, and electron volts for ion acceleration. Therefore, when considering that $1 \text{ kg} = 0.6023 \times 10^{27} u$, $1 \text{ T} = 10^4 \text{ G}$, and the electronic charge, $e = n e_e = n \times 1.602 \times 10^{-19}$, where n is the number of charges and e_e is the elementary electronic charge, we obtain

$$M/e = 4.83 \times 10^{-5} r_m{}^2 B^2 / V \tag{2-19a}$$

with r_m in cm, and

$$M/e = 3.12 \times 10^{-4} r_m{}^2 B^2 / V \tag{2-19b}$$

with r_m in inches. Once again, M is in amu using the ^{12}C scale, B is in gauss,

V is in electron volts, and e means the number of unit electronic charges, i.e., $e = 1$ for a singly charged ion. Equations 2-19 are often referred to as the basic mass spectrometer equations. Equation 2-19a may be rearranged to obtain the radius of curvature, r_m (cm)

$$r_m = 144(1/B) \sqrt{MV/e} \qquad (2\text{-}20)$$

This equation shows that the radius of curvature for particles of the same energy but of different masses varies according to the square root of the masses in a homogeneous magnetic field. For example, a $^{14}N^+$ ion accelerated to 2000 eV will be bent by a magnetic field of 1000 G into a curved path of 24 cm radius. For an $^{15}N^+$ ion the radius will, under the same conditions, be almost a full centimeter longer.

A mass spectrometer employing an electrostatic accelerator to obtain an ion beam homogeneous in energy, and a homogeneous magnetic field to effect mass dispersion, is shown in Figure 2-5. Here the ion collector is at a fixed position corresponding to a fixed value of r_m. The position of the collector is determined by the focusing properties of the magnetic field, to be discussed later in this chapter. The process whereby successive ions of mass M and charge e are brought onto the receiving collector is called *scanning* the spectrum. Scanning may be accomplished either by varying V, where $MV/e = $ const. for a fixed value of B, or by varying B, where $M/eB^2 = $ const. with V fixed. Increasing the magnetic field or decreasing the acceleration potential, while keeping the respective other variable constant, will focus heavier and heavier ions on the detector. The merits and shortcomings of magnetic field and voltage scanning are discussed under specific applications in Part II.

Table 2-1 gives the radii of the trajectories of a few typical ions accelerated to 500 and 3000 eV in magnetic fields of various sizes. Figure 2-6 gives the radii of curvature of 1000 eV ion trajectories as a function of

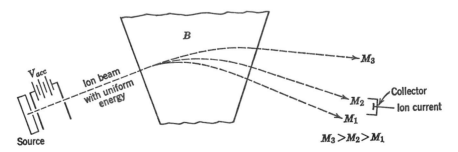

Figure 2-5. Prism action of homogeneous magnetic field on ions homogeneous in energy but differing in mass. Magnetic field is normal to the plane of the figure.

Table 2-1   Calculation of Radii Using the Basic Mass Spectrometer Equation: $r_m = 144\ (1/B)\sqrt{MV/e}$, with r_m in cm

Magnetic field:		50 G		300 G		1000 G		5000 G	
					Ion energy, eV				
Ion	Mass	500	3000	500	3000	500	3000	500	3000
H^+	1	64	158	10.7	26.3	3.2	7.9	0.64	1.6
CH_3^+	15	250	611	41.6	102	12.5	30.6	2.5	6.1
H_2O^+	18	273	669	45.6	112	13.7	33.5	2.7	6.7
N_2^+	28	341	835	56.8	139	17.0	41.8	3.4	8.3
CO_2^+	44	427	1046	71.2	175	21.4	52.4	4.3	10.5
Hg^{++}	100	644	1577	107	263	32.2	78.9	6.4	15.8
Xe^+	131	738	1805	123	302	36.9	90.4	7.4	18.1
$C_{18}H_{12}S^+$	260	1039	2543	173	425	52.0	127	10.4	25.5
$U_3N_4^+$	770	1787	4377	298	730	89.4	219	17.9	43.8

mass and magnetic field intensity. Figure 2-7 plots the radii of curvature of ion trajectories in 500 G magnetic field as a function of mass and energy. It is noted again that doubly charged ions behave as singly charged ions of half mass.

By logarithmic differentiation of equation 2-20, assuming the magnetic field strength to remain constant, one obtains

$$\Delta r_m/r_m = \tfrac{1}{2}(\Delta M/M) + \tfrac{1}{2}(\Delta V/V) \qquad (2\text{-}21)$$

This equation shows the *change* in the radius of curvature when the masses of the particles change from M to $M + \Delta M$, and when their energies change from eV to $e\,(V + \Delta V)$. It is concluded that in order to obtain the best separation for different masses (i.e., good prism action), one must provide an ion beam as homogeneous in energy as possible, so that ΔV becomes as small as possible. This problem is further discussed in connection with the resolving power.

Instead of the electrostatic accelerator one may employ a radial electrostatic field (Section 2-II-C) as the energy filter. This, as will be discussed in detail, is indeed done in double-focusing instruments. The third possibility is to use a velocity filter (see next section).

It is noted in passing that the radius, r (cm), for an electron in a homogeneous magnetic field is given by

$$r = 3.36\,\sqrt{V/B} \qquad (2\text{-}22)$$

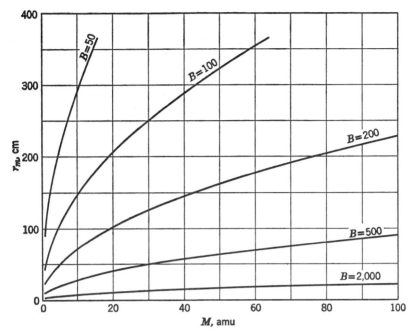

Figure 2-6. Radii of curvature of 1000-eV ion trajectories as a function of mass and magnetic field intensity (6).

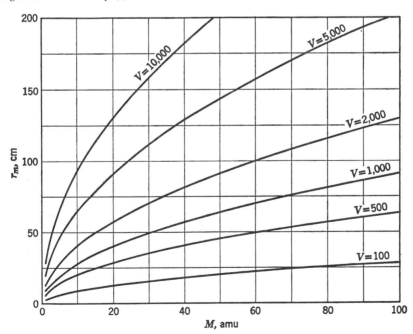

Figure 2-7. Radii of curvature of ion trajectories of 500-G magnetic field as a function of mass and energy (6).

E. Velocity Filters

Velocity filtering is obtained by the simultaneous use of electric and magnetic fields in such an arrangement that the force exerted by the magnetic field is cancelled by the force of the electric field. This can be realized by using magnetic and electric fields simultaneously, and adjusting the directions of the fields and the direction of ion beam projection to be mutually perpendicular (Fig. 2-8). The condition of force cancellation is

$$eE = Bev \qquad (2\text{-}23)$$

from which

$$v = E/B \qquad (2\text{-}24)$$

For given electric and magnetic field strengths only ions with velocity v can pass straight through the apparatus without deflection; particles with other velocities describe complicated trajectories.

Neither m nor e appears in equation 2-24, showing that the velocity is independent of these parameters. This was the basis of Thomson's and Wien's classical experiments in which the velocities of electrons and simple ions were first determined.

IV. TIME-OF-FLIGHT ANALYSIS

In a beam of ions of equal energy the heavy ions will travel more slowly than the light ones. A short pulse therefore will disperse as it moves in a long drift tube, into groups in which all ions have the same mass (Fig. 2-9a). This is the principle of the velocity spectrometer which can be used as a mass analyzer assuming a monoenergetic ion beam. There is no need for a magnetic field. The relationship between the time of flight and the m/e

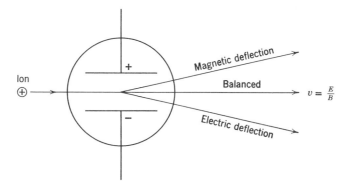

Figure 2-8. Velocity filter; magnetic field is directed into plane of figure.

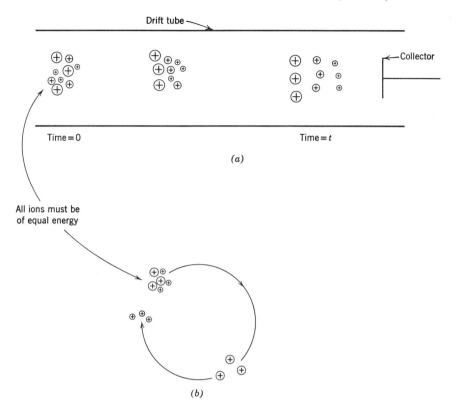

Figure 2-9. Time-of-flight principle: (*a*) linear path, no magnetic field; (*b*) circular ion path, requires magnetic field.

ratio of the ions is obtained by considering that the velocity of the ions results from letting them pass through an electrostatic analyzer. The acquired velocity is

$$v = \sqrt{2eV/m}$$

where V is the accelerating potential. The time of flight, t_f, is given by

$$t_f = d/v = d\sqrt{1/2V}\ \sqrt{m/e} \tag{2-25}$$

where d is the length of the linear path. In a given instrument where d is constant, and under given experimental conditions ($V =$ const.) the time of flight of the various ions is proportional to the square root of their respective m/e values. In practical units one obtains

$$M/e = 1.916 \times Vt^2/d^2 \tag{2-26}$$

where M/e is amu/electronic charge, V is voltage in volts, t is time in microseconds, and d is drift length in cm. For $^{15}N^+$ ions the time of flight in a 1 m drift tube is 6.1 μsec. For $^{14}N^+$ ions one obtains $6.1 \times (14/15)^{1/2}$ = 5.8 μsec; thus the time difference to be measured is 0.3 μsec.

The separation in time of arrival of adjacent masses decreases with increasing mass; at mass 200 the time difference to be measured is of the order of 50 nsec. Time-of-flight instruments of this kind, therefore, require sophisticated electronics, particularly when one considers that in addition to microsecond detecting, microsecond pulsing technique is needed for the production of the ion bunches. Several types of nonmagnetic time-of-flight mass spectrometers have been designed and constructed. The more important ones are discussed in Section 3-III.

Time-of-flight mass analysis may also be used in connection with magnetic field (Fig. 2-9b). The time interval, t (sec), for a charged particle to complete one full revolution in a magnetic field is

$$t = 2\pi r_m/v \tag{2-27}$$

Replacing r_m by mv/eB (eq. 2-15), one obtains

$$t = 2\pi(m/eB) \tag{2-28}$$

The frequency of rotation, f_c (rps), called the cyclotron frequency, is the number of revolutions per second,

$$f_c = 1/t = (1/2\pi)\,(eB/m)$$

When f_c is measured and e and B are known, the mass of the particle can be obtained from

$$f_c = 1.54 \times 10^3\, B(e/M) \tag{2-29}$$

where M is the atomic or molecular weight of the ion and e denotes the degree of ionization. For a $^{14}N^+$ ion the cyclotron frequency (rps) is

$$f_c = 1.1 \times 10^2\, B$$

For an electron the formula is

$$f_c = 2.8 \times 10^6\, B \tag{2-30}$$

These concepts are utilized in a number of rather ingenious mass analyzers; several are discussed in Chapter 3.

V. FOCUSING

A. Types of Focusing

Depending on the type of ion source employed, there is always a certain degree of inhomogeneity in both the initial direction and the energy of

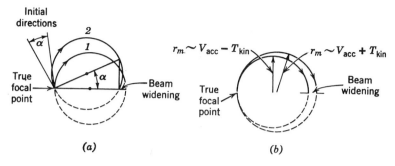

Figure 2-10. (*a*) Angular aberration; (*b*) energy aberration. V_{acc} is accelaration voltage, T_{kin} is kinetic energy, r_m is radius.

the ions as they enter the analyzer. Initial directions are not homogeneous (angular or spherical aberration) since the ion beam to be measurable must be relatively wide and not a line-like ribbon; the collimating slits, however narrow, must be of finite width (Fig. 2-10*a*). Energy aberration (chromatic aberration) results from the fact that ions leave the source at slightly differing velocities (Fig. 2-10*b*). The main causes of this effect are: (*1*) mole-cules before ionization possess a certain amount of translational velocity of thermal origin, (*2*) the energy attained during the ionization and/or fragmentation process is not equal, and (*3*) ion formation takes place throughout a region in which an electric field exists and ions pass through slightly different potentials before entering the main acceleration region. The degree of inhomogeneity depends on the construction of the ion source and the mode of ion formation (Table 4-1).

Focusing is thus required to eliminate overlapping and to maximize the resolved ion intensities. There are three types of focusing in magnetic mass spectrometers and their characteristics are summarized in Table 2-2. In *double focusing* ions homogeneous in mass but of various initial directions *and* velocities are focused to narrow images. Double-focusing systems always consist of two "components": one to achieve *direction focusing*, i.e., focusing

Table 2-2   Focusing Characteristics [a]

	Mass	Velocity	Direction
Direction	+	+	−
Velocity	+	−	+
Double	+	−	−

[a] + denotes homogeneous, − denotes heterogeneous.

of ions homogeneous as to mass and velocity but of different initial direction, and the other to provide *velocity focusing*, i.e., focusing of ions homogeneous as to mass and initial direction, but of different initial velocity. In addition to focusing, mass separation must also be provided by one of the components. In most *single-focusing* instruments only direction focusing is achieved.

The focusing or lens action of magnetic and electric fields is usually described in terms of optical *cylindrical* lenses to which their behavior is similar. Two-directional, i.e., point-to-point, focusing is equivalent to spherical lenses; their practical use is limited at this time. Focusing in the time-of-flight (TOF) mass spectrometers is briefly considered in the description of the various types of TOF instruments in Chapter 3. This section deals with the focusing properties of magnetic and electric fields and their combinations.

B. Focusing by Magnetic Field

1. Perfect Double Focusing

When an ion beam from a point source is injected into a magnetic field, all particles, regardless of their initial directions and velocities, will return to the point of origin after traveling a full circle (Fig. 2-10). Although this is perfect double focusing, it cannot be used directly for mass analysis because *all* masses are focused at the same point. Moreover, source and detector cannot be placed at the same point. The time interval, however, that is required for the completion of one or n full revolution(s) does depend on the mass (eq. 2-28), and there are instruments based on this principle. Perfect two-dimensional velocity and direction focusing and spacial mass dispersion are achieved simultaneously when ions are moving in crossed homogeneous magnetic and electrostatic fields. These so-called trochoidal (cycloidal) instruments are discussed in Section 3-II-B-1.

2. First-Order Direction Focusing

In direction focusing it is assumed that the ions entering the magnetic field are homogeneous as to energy. Methods for obtaining such beams have already been discussed in Section II of this chapter. As shown in Figure 2-10, an approximate direction focusing takes place in a homogeneous magnetic field after the beam has traveled through 180°. Since at the same time the mass dispersion property of the magnetic field is still in effect, the 180° magnetic field, combined with a proper source and detector, can be used as a mass spectrometer. Such an arrangement is shown in Figure 2-11.

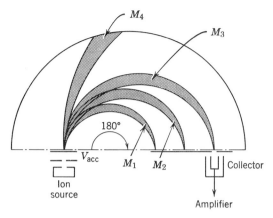

Figure 2-11. Mass dispersion and focusing in 180° magnetic analyzer.

The focusing is only approximate after a 180° deflection. The magnitude of *angular aberration* may be evaluated with reference to Figure 2-12 where two trajectories are shown, both having the same radius, r, but diverging by an angle α. The central ray which intersects line AC at a right angle is represented by ABC. The center of this trajectory is at point *1*. The center of a divergent ray, ADE, making an angle α with the central ray, is at point *2*. The error in focusing is EC. Since AF is the diameter of the circle of which the divergent ray (ADE) is a part, one obtains

$$EC = AC - AE$$

and

$$AC = AF = 2r$$

Since

$$\angle\ EAF = \alpha$$

and

$$AE = 2r \cos \alpha$$

hence

$$EC = 2r(1 - \cos \alpha)$$

When α is small, it is justified to employ the first two terms of the series

$$\cos \alpha = 1 - \alpha^2/2! + \alpha^4/4! - \alpha^6/6! + \cdots$$

and so the error in focusing is given by

$$EC \approx r\alpha^2 \tag{2-31}$$

where α is the half-angle of angular divergence. In perfect focusing the image of the entrance slit would be exactly reproduced on the line AEC.

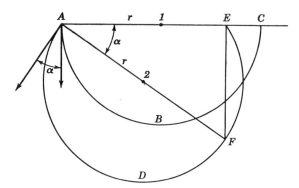

Figure 2-12.   Angular (spherical) aberration.

Due to the aberration the image is *broadened* by an amount given by equation 2-31. For example, assuming $\alpha = 3°$ (0.05 radian), the line broadening due to angular aberration in an analyzer of 10 cm radius will be 0.25 mm, a distance comparable with the width of the collector slit. Additional image broadening arises from chromatic aberration which is caused by energy inhomogeneity, from the nonhomogeneity of the magnetic field, and possibly other reasons. These are discussed in connection with resolution.

3. General Sector Field Theory

Refocusing at a deflection of 180° is, as proven by Barber (1) and Stephens (2), only a special case of the focusing action of any wedge-shaped magnetic field. The general theory of focusing was developed by Herzog (3). In the most general case, a beam of ions which is homogeneous in energy but contains various masses is injected from a point source with small initial direction divergence into a magnetic field. As shown in Figure 2-13, the distance of the source from the prism may be selected arbitrarily, and also the angle, ϵ, between the direction of the entering ray and normal to the field edge can be of any (reasonably small) value. In practice, *normal entry*, i.e., $\epsilon = 0$, is employed most frequently. The magnetic field acts both as a prism and as a lens and the ions separated into groups according to their mass are focused at points along an *image curve* as shown in Figure 2-13. The trajectories of the ions in the field are described by their radii which can be calculated from the basic mass spectrometer equation (eq. 2-20). The trajectory of the central ray, corresponding to the mean mass and velocity, is called the optic axis (AO in Fig. 2-13). The angle α of a neighboring ray is considered positive if the outward-drawn normals from the field boundaries lie in the opposite side of the optic axis from the center of curvature of the orbit in the field.

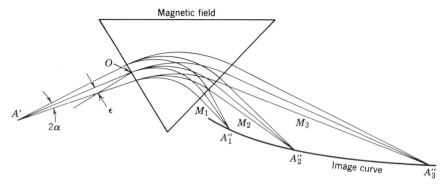

Figure 2-13. Direction focusing and mass separation in general magnetic sector field. (After H. Ewald, *Massenspektroskopische Apparate,* in *Handbuch der Physik,* Vol. 33, S. Flügge, Ed., Springer-Verlag, Berlin, 1956, p. 552.)

The positions of these foci on the image curve can be simply determined by a *graphical* method developed by Cartan (4). The technique is described with reference to Figure 2-14, where only a single ray is shown for simplicity. Ions from the source enter the field at point A, describe a circular path whose center is C, and leave the magnet at B along line BC. The problem is to find the image point I, which is also the first-order focusing point for rays making a small angle $\pm\alpha$ with the central ray. Draw lines AD and BE perpendicular to the magnet faces, and draw a perpendicular to AS which crosses AD at point S'. Next, connect S' to C and continue the line until it intercepts BE at point I'. Construct a perpendicular from I', and its interception with line BC at point I will be the image point of the source slit. It is repeated that I is also the first-order focusing point for rays making a small angle $\pm\alpha$ with the central ray.

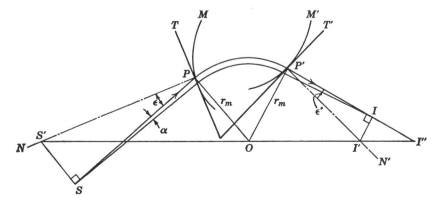

Figure 2-14. Determination of image point by Cartan's method.

It is seen from the figure that if ϵ' and ϵ'' are zero, i.e., if the beam enters and leaves the magnetic field at a right angle to the magnet face, points S and S' as well as I and I' coincide, and the source, the center of curvature of the central ion beam, and the focal point all lie in a straight line (Fig. 2-15). This is known as *Barber's rule* for the normal entry case and it is valid for *any* angle of deflection in the magnetic field.

The relationship between the object and image distances and the magnetic field parameters is expressed by Herzog (3) in terms of the "focal length," in analogy to geometrical optics (thick lenses). The equation of first-order direction focus for an ion beam of one particular mass, whose central beam enters and leaves the magnetic field at right angles (or nearly right angles), is given by

$$f_m^2 = (l' - g_m)(l'' - g_m) \tag{2-32}$$

where l' and l'' are object (source) and image (detector) distances measured to the *effective* face of the magnetic field, and f_m and g_m are defined by

$$f_m = r_m \csc \phi \tag{2-33}$$

and

$$g_m = r_m \cot \phi \tag{2-34}$$

where f_m is the focal length, r_m is the radius of curvature of the ions in the field, and ϕ is the angle through which the beam is deflected. The most commonly employed deflection angles are 180°, 90°, and 60°. The object and image distances are often kept equal, $l' = l''$ (*symmetrical* analyzer).

When both the source and detector slits are placed in the planes of the magnet entrance and exit faces, respectively, one obtains from equation 2-32

$$g_m = \pm f_m \quad \text{or} \quad \phi = n\pi \tag{2-35}$$

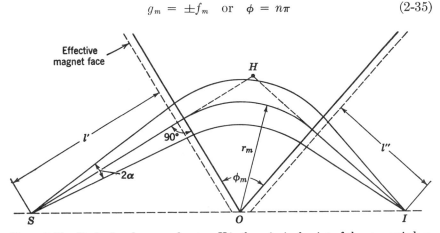

Figure 2-15. Barber's rule, normal entry. H is the principal point of the magnetic lens.

where n is an integer. This is, of course, the case of the 180° analyzer (Fig. 2-11), used by Dempster in one of the very first mass spectrometers (Section 3-I-C). The condition for focusing when the angle of deflection is 90° is also simple,

$$l' = l'' = r_m(\csc 90° + \cot 90°) = r_m(1 + 0) = r_m \qquad (2\text{-}36)$$

When $\phi = 60°$, simple subsitition into the above equation yields

$$l' = l'' = r_m(\csc 60° + \cot 60°) = 1.732\ r_m \qquad (2\text{-}37)$$

Thus the object and image distances are 1.732 r_m units away from the faces of the magnet. The 60° symmetrical arrangement has been used very frequently and is commonly known as "Nier-type" geometry (Fig. 2-16). The resolving power, as discussed later, *does not* depend on the angle of deflection. The merits and faults of the various deflection angles are discussed in Chapter 3.

4. Ideal Focusing Field

The complex problems of calculating the focusing properties of various field shapes and of evaluating the means to improve focusing beyond first

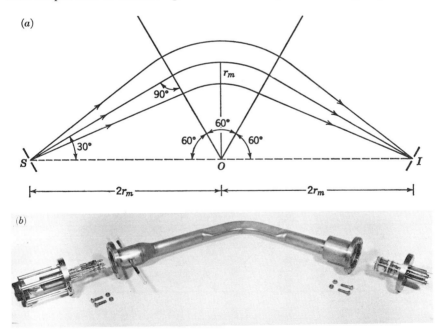

Figure 2-16. (a) "Nier-type" geometry, 60° sector magnet, normal entry, symmetrical; (b) A 6 in. radius 60° sector mass spectrometer tube. The portion of the envelope which is located within the magnet gap is flattened. The ion source is on the left side, the collector on the right side. (Courtesy of Nuclide Corp.)

order, have been approached by Kerwin (5,6), who advanced the concept of an *ideal* focusing field. The boundary, M, of a magnetic field which gives theoretically perfect focusing for a widely divergent beam is given by

$$y = x(a - x)/(r^2 - x^2)^{1/2}$$

where x and y are the coordinates of the point where ray 1 enters the field and a is one-half the straight line distance between the object and image points (symmetrical case). The equation is derived on the basis of simple trigonometric relations (Fig. 2-17) by considering that $x = r \sin \theta$ and $y = (a - x) \tan \theta$, where θ is the angle between beam 1 and SI. The shape of the field depends on the selection of parameters r and a. In practice, of course, one is limited by problems in machining, field inhomogeneities, and fringing-field effects. The amount of defocusing at I is determined by how closely the actual field approximates the ideal field M at any single point. For example, a straight-line approximation, N, yields the field described by Barber and Stephens, i.e., when the entering beam is perpendicular at P_n to the face of the magnet. The aberration in this so-called *normal* case is calculated (for details see ref. 6) to be $r\alpha^2$ in agreement with equation 2-31. Such straight-line approximation may be applied anywhere on the ideal curve and produces first-order approximations. It is emphasized that the aberration in this case is independent of the angle of deflection, i.e., the aberration in 180°, 90°, and 60° instruments is the same when conditions are comparable.

A second type of linear approximation is the *inflection* case where the beam enters the field at the inflection point, P_i. This point may be found

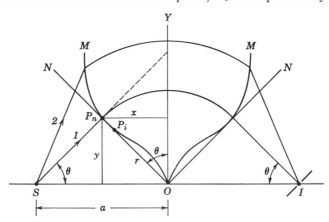

Figure 2-17. An ideal magnetic field is approximated by a straight line, N. Beam 1 enters normally at point P_n. Another type of linear approximation is the case when the beam enters at the inflection point, P_i (6).

by equating the second derivative of the equation of the ideal focusing field to zero (6). The focusing here is of second order and the α^2 aberration is eliminated. The required magnet is smaller than for the normal case, but the mass dispersion is also smaller because less magnetic flux is traversed. For details on circular approximations (e.g., normal-circle case) and asymmetric cases the reader is referred to reference 6.

C. Focusing by Electrostatic Field

The directional focusing action of the radial electrostatic field on ions emerging from a source with a small angular divergence in the radical direction is shown in Figure 2-18. The similarity with the magnetic field focusing is evident from the general equations describing the relation between object and image distances (l_e' and l_e'') and the focal length, f_e,

$$f_e{}^2 = (l_e' - g_e)\,(l_e'' - g_e) \tag{2-38}$$

$$f_e = r_e/(\sqrt{2}\,\sin\,\sqrt{2}\,\phi_e)$$

$$g_e = (r_e\,\cot\,\sqrt{2}\,\phi_e)/\sqrt{2}$$

where r_e is the radius of curvature of the ion beam in the electrostatic field and ϕ_e is the angle of deflection.

Symmetrical focal points for the 90° deflection case are obtained by

$$l_e' = l_e'' = r_e(\cot\,127.28° + 1/\sin\,127.28)\,\sqrt{2} = 0.35\,r_e$$

This equation shows that the object and image distances must be 0.35 r_e units away from the effective ends of the electrostatic field. The value 127.28° is the result of multiplying 90° by $\sqrt{2}$.

When the focusing angle of the electrostatic field is selected to be 127°17′, the object and image distances become zero. Thus, as shown in Figure 2-19a, source and detector are placed at the edges of the field. In the case of

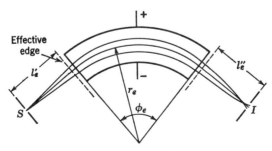

Figure 2-18. Electrostatic focusing action. Aberration is the result of the fact that outer rays are bent more than needed for perfect focusing.

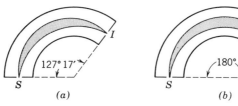

Figure 2-19. Direction focusing in a 127° electrostatic condenser (a) corresponds to direction focusing through 180° in a magnetic field (b).

magnetic focusing, a 180° deflection is required to achieve the same action (Fig. 2-19b). Thus, for equal "size," the focal point produced by the electrostatic field is closer to the field boundary than that formed by the magnetic field. When the focusing angle of the electrostatic field is $\pi/4 \sqrt{2} = 31°50'$, solution of equation 2-38 shows that if the object distance is selected as $l_e' = r_e/\sqrt{2} = 0.707 \; r_e$, then the image is formed at infinity, i.e., the ion beam leaving the electrostatic analyzer is essentially parallel.

It has been mentioned several times that mass analysis cannot be made by electrostatic fields since the mass dispersion for all ions of constant energy is zero. The focusing cases just described have important applications, however, in linear combinations with magnetic fields to form double-focusing systems.

Radial electrostatic fields have focusing properties only in the central plane of deflection and not perpendicular to it (i.e., not in the z direction). Both radial and axial focusing properties are found in the *toroidal* condensers. There are two sets of lens equations, corresponding to radial and axial focusing. The radial field is actually a special case of the toroidal, a case when $r_z = \infty$. The case where $r_e = r_z$ is commonly called the *spherical* condenser. Detailed discussion of toroidal fields is beyond the scope of this book (7).

VI. RESOLUTION

A. Dispersion of Magnetic Field

When two monoenergetic ion beams composed of mass M and ΔM are analyzed in a homogeneous magnetic sector field, they will proceed along (mean) radii r_m and $r_m + \Delta r_m$, and will be focused at points I_1 and I_2 on the image curve (Fig. 2-20). Both ion beams have a finite width, W_b, as they reach the collector. This width should be equal to the source slit or beam defining slit width (0.001–0.5 mm) as long as the *magnification* of the magnetic lens is unity (symmetrical arrangement). In analogy with light optics, magnification is defined as the ratio of the image size to the object size. In practice W_b is, of course, wider than the source slit width, S_1, due to aberrations to be discussed soon.

Instead of the distance $I_1 I_2$, the distance $I_1 Q_2$ is employed in practice as a measure of the mass separating ability of the analyzer. This so-called *dispersion* is measured in the plane S_2–S_2, perpendicular to the direction of the beam in focus. The cross section of the beam at point Q_2 is essentially the same as at the true focal point I_2, and it is practical to place the collector slit at the S_2–S_2 plane (Fig. 2-21). As mentioned above, the beams have a finite width even at the true focal points due to the width of the entrance slit and the change in cross sections along the stretch $Q_2 I_2$ is negligible compared to the entrance slit width plus aberration distance.

In case of the 180° field the dispersion is simply the difference between the diameters of the trajectories (see Fig. 2-11),

$$D_m = 2\Delta r_m \tag{2-39}$$

Using equation 2-21 for equal energies one obtains

$$D_m = r_m(\Delta M/M) \tag{2-40}$$

The dimension of D_m is, of course, the same as that of r_m.

This equation can be shown to be true for all types of symmetrical $(l_m' = l_m'')$ magnetic sector fields where the beams are directed *normal* to the magnetic field. Attention is called to the fact that the angle of deviation does not appear in the equation, the dispersion *does not* depend on the angle of deviation. For example, a mass spectrometer of 15 cm radius produces a mass dispersion of 1.5 mm for a mass difference of 1%. If the radius is

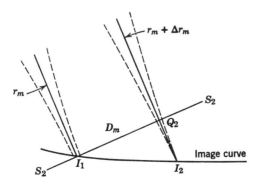

Figure 2-20. Mass dispersion in magnetic field.

increased to 50 cm, dispersion increases to 10 mm. These results are the same at any angle of deflection. One must not forget, however, that the "size" of the magnet which involves such parameters as total described path length and required pole face area, does depend on the magnitude of ϕ. It is repeated that dispersion refers to the actual physical distance between the centers of two neighboring ion beams. When the beams are recorded on a properly positioned photoplate, dispersion may be measured directly.

The general dispersion equation which is valid for nonsymmetrical arrangements is given by

$$D_m = \frac{\Delta r_m}{r_m} \left\{ r_m (1 - \cos \phi) + l_m'' \left[\sin \phi + (1 - \cos \phi) \tan \epsilon_m'' \right] \right\}$$

$$= \frac{\Delta r_m}{r_m} K_m'' \tag{2-41}$$

where ϵ_m'' is the angle made by the mean emergent ion beam and the normal to the field boundary (for normal exit $\tan \epsilon_m'' = 0$); K_m'' is called the coefficient of dispersion.

B. Resolution: Theoretical

The inherent resolving power of a mass spectrometer depends on two factors: mass dispersion and beam width at the focal plane. As long as the dispersion is greater than the beam width there is a separation between the two beams, so they can be resolved. The dispersion of an analyzer system is determined by its optics, and once selected cannot be changed. The total image width, however, can be improved upon by reducing entrance slit width and by reducing or eliminating aberrations. An important contributor to peak broadening is the energy inhomogeneity of the beam, $\Delta V / V$, which can be reduced by using high accelerating voltages, or practically eliminated by employing an energy filter. When W_b (Fig. 2-21) becomes equal to the dispersion, the two beams begin to touch and the limit of resolution of the instrument is reached. This condition is attained when

$$d = D_m - W_b > S_2 \tag{2-42}$$

where d is the distance between the outer edges of the ion beams and S_2 is the width of the collector slit.

The theoretical resolution or resolving power, R, of a mass spectrometer is defined as the reciprocal of the minimum mass difference $(\Delta M / M_{\min})$, which can just be separated. In the general case it is given by

$$R = \left(\frac{M}{\Delta M} \right)_{\max} = \frac{1}{2(|S_1 G| + |S_2|)/|K_m''| + |\Delta V/V|} \tag{2-43}$$

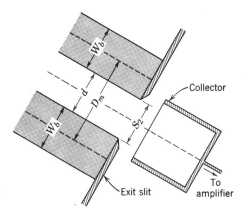

Figure 2-21. Correlation between ion beam width, W_b, dispersion, D_m, and size of exit slit, S_2. (After C. Brunnee and H. Voshage, *Massenspektrometrie*, Verlag Karl Thiemig, Munich, 1964, p. 21.)

where S_1 and S_2 denote the source and exit slit widths, respectively, G is the magnification of the ion optical system, and K_m'' is the velocity dispersion coefficient. Both G and K_m'' can be calculated by ion-optical methods using such geometrical parameters as radius and angle of deflection.

For symmetrical analyzers with normal beam entrance, $K_m'' = 2r_m$ and $G = -1$, thus

$$R = \frac{M}{\Delta M} = \frac{1}{(S_1 + S_2)/r_m + \Delta V/V} \qquad (2\text{-}44)$$

This equation applies when ions are detected by electrical means, i.e., when there is a collector slit of width S_2 with a plane located perpendicular to the direction of the incoming beam. For photographic detection $S_2 = 0$. Single-focusing instruments usually employ electrical detection. An instrument with $r_m = 10$ cm, $S_1 = S_2 = 0.2$ mm, and $\Delta V/V = 0.001$ has a resolution of 200.

Two characteristics of equation 2-44 should be noted. First, resolution is independent of mass. This is true for all magnetic sector field instruments, double-focusing instruments, and cycloidal analyzers. In time-of-flight and radiofrequency instruments, however, resolution is mass dependent and thus will have to be evaluated as a function of mass. Also the practical methods of resolution determination have to be altered for this condition. The second interesting feature of equation 2-44 is that it does not contain ϕ_m, i.e., the resolution does not depend on the angle of deflection.

Equation 2-44 shows the dependence of resolution on slit widths and on the energy homogeneity of the ion beam. The total image width, W_b, has a number of additional components which tend to increase it. The various aberrations are usually summed up and an additional term, β, is added to the denominator of the resolution equation,

$$R = \frac{M}{\Delta M} = \frac{r_m}{S_1 + S_2 + \beta r_m} \qquad (2\text{-}45)$$

The aberration caused by imperfect focusing has already been discussed. It is called α-aberration, and it may be of the second or higher order. Assuming $\alpha = \pm 2°$, the resolution of the instrument mentioned above would be reduced by about 20%.

Other possible aberrations include those caused by fringing field effects, field boundaries with curved contours, space charge, and surface charge. Fringing field aberrations are caused by the fact that magnetic and electric fields do not possess well-defined boundaries. The first-order effect of stray magnetic fields is to displace the image position without altering its quality. Several methods have been proposed to calculate *fringing field* effects, and some of them are rather involved. An approximate correction for the displacement of the focal point is obtained by taking the virtual boundary to be *one pole gap width* beyond the actual boundary. When the contours of equal field intensity do not remain parallel to the boundary of the field-forming structures, both the location and quality of the image will be adversely influenced.

Space charge, which may be present both in the source and analyzer, alters the potential and gradient distribution. In the source, space charge results in what is known as interference, i.e., the presence of one mass changes the sensitivity of the source to another mass. This is a problem in trace analysis where linearity may be destroyed by the large quantity of the matrix material (Section 11-II). Space charge in the analyzer broadens the ion beam resulting in loss in resolution. *Surface charge* is the result of ion bombardment on high-resistance surfaces. Surface charge, which can seriously reduce resolution, can be avoided by careful construction of the instruments.

A second-order fringing field effect which contributes to the total image width is that ions with paths not in the medial plane suffer an additional deflection from the component of the fringing field. The amount of line broadening is given by Berry (8) as $A_z = z^2/r_m$, where z is the half-gap width and r_m is the radius. In an instrument of 10 cm radius and 1 cm magnet gap width, the fringing field aberration is 0.25 mm, resulting in a resolution loss of about 10%. This aberration is absent in the important Mattauch-Herzog double-focusing geometry.

A possible means to improve focusing is the use of *asymmetrical* arrangements. Since any field may be considered as being composed of two halves, one converting a divergent beam into a parallel beam and the other reversing the process, the aberration produced by the entrance boundary may be cancelled at the exit. Instruments have been constructed with asymmetric fields but they have not come into general use.

Many methods have been suggested to eliminate or reduce various other effects that contribute to the beam width. The selection of optimum geometrical conditions for the best resolving power is discussed by Kerwin (5,6). Sometimes one aberration may be cancelled out by another; at other times required beam intensity considerations prevent the use of certain techniques to eliminate aberrations. Such theoretical calculations, at any rate, are only used in instrument design. The actual resolution of a given instrument, which may be a far cry from the expected one, must be determined from the recorded peaks. Methods for this are considered in the next section.

C. Resolution: Practical

The basis of practical resolution determinations is the fact that certain distances on a recorded peak are proportional to dimensions in the analyzer. As shown in Figure 2-22 (see also Fig. 2-21), when the ion beam moves onto the collector slit during the scanning of the spectrum, the current

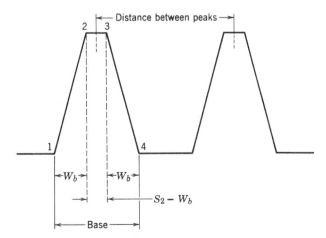

Figure 2-22. Illustration of what the recorded peak represents in terms of analyzer dimensions.

measured on the collector starts to increase at point *1*. The current increases until the entire beam is on the slit (point *2*), and if the width of the collector slit (S_2) is larger than the beam width (W_b) the signal stays at a maximum until point *3* where the beam starts to move off the slit. At point *4* the beam has left and the ion current returns to zero. When $W_b = S_2$ the maximum is instantaneous. The region *2–3* is called *flat-top*.

It is seen that the base of the peak is proportional to $S_2 + W_b$, while the distance between the maxima of the two peaks is proportional to the dispersion. In the resolution equation (eq. 2-44) S_1 plus the sum of all aberrations can be designated as W_b, i.e., the actual beam width, resulting in

$$R = M/\Delta M = r_m/(W_b + S_2) \qquad (2\text{-}46)$$

Now, since the base of the peak is proportional to $W_b + S_2$, and the distance between peaks is proportional to the dispersion, i.e., to $r_m \Delta M/M$, one obtains

$$R \propto \frac{\text{Distance between peaks}}{\text{Base of peak}} \times \frac{M}{\Delta M} \qquad (2\text{-}47)$$

Here M is the nominal mass of the peak whose base is measured, and ΔM is the mass difference between the two peaks. When the distance between the two peaks equals the base of the peaks, $R = M/\Delta M$, corresponding to a zero valley definition.

The shape of most recorded peaks is not trapezoidal. Due to aberrations, in all but the "perfect" double-focusing analyzers (see later), peaks have *tails*, i.e., peaks are rounded at the top and flare at the baseline, somewhat in the shape of a bell as shown in Figure 1-3. Accordingly, all definitions of resolution must consider the shape of the peaks in addition to the relative mass difference of the two ion masses that are just separable.

The experimentally determined resolution is understood to be $M/\Delta M$, where M is the mass of the observed peak and ΔM is the amount by which the mass of another peak of equal intensity must be smaller or larger than the observed peak in order for the two peaks to be considered separated. The degree of peak separation needed varies according to the problem in question, and various ways have been suggested to obtain numerical values for resolution.

Most popular is, perhaps, the *% valley* definition (Fig. 1-3c) where the intensity between two peaks of equal height is expressed as a fraction of the maximum intensity, taking the top of peak equal to 100%. The emphasis here is on the ability to distingush two distinct peaks (qualitative analysis). The fraction, x, at which the base is measured is arbitrary, 10% and 2%

being used most frequently. It is important to note that when the base is measured at 5% peak height, the actual valley will be 10% since the superimposed parts are additive. Similarly, a valley of 2% means that the base is measured at 1% peak height. Conversion from resolution determined at some % valley to the equivalent at some other valley is almost impossible (except for triangular peaks) because of the complex relationship between peak base and height when the top of the peak has a finite width.

In the *peak width* method, the amplitude (breadth) of the peak is measured at a level of a certain percent of the maximum intensity (Fig. 1-3d). The value of y is selected arbitrarily, 5% and 50% being used most frequently. It is recalled from Section 1-I-D that the 5% peak width definition is technically equivalent to the 10% valley definition provided the peak is symmetrical and the system linear in the range between the 5% and 10% levels of the peak. Resolution expressed on the 50% peak width basis is about double that based upon the 10% valley definition. As mentioned in the introductory section, the main advantage of the peak width definition is that resolution may be determined on isolated peaks in the mass spectrum.

When the peaks have long "tails," extending much beyond the nominal mass number, there may be considerable addition to the top of one peak from the tail of another (Fig. 2-23a). This can be a particularly difficult problem in trace analysis where small peaks must be ascertained in the presence of large ones. Another case where tail contribution is a problem concerns the determination of low abundance isotopes adjacent to masses of high abundance. The ability of mass spectrometers to handle such problems is measured by the *abundance sensitivity*, a performance measure discussed later in the text.

For quantitative analysis, the determined peak intensities must be accurate measures of the intensity of the ion beam producing the peak. In other words, contributions from adjacent peaks must be minimal. Using the *peak height contribution* or *cross-talk* definition of resolution, the intensity with which one peak contributes to the intensity of the other at its *highest* value is given as a fraction of the peak height (Fig. 2-23b). Unit resolution is thus defined as extending to the point where one peak will add less than z% to the height of an adjacent peak of equal height. The value of z is usually selected as 1%. If the overlap of the two peaks were 50%, which is equivalent to 0% cross-talk, the distance between the peaks, D, would equal one-half of the measured base of one of the peaks. In the case of a 1% cross-talk, the dispersion is equal to one-half the base of the peak, measured at 1% of peak height. Thus, resolution is measured the same way

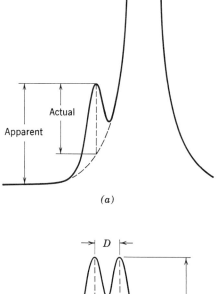

Figure 2-23. (*a*) Peak tailing. (*b*) Cross-talk definition.

as before, but in equation 2-47 the base of peak (measured at 1% peak height) is divided by 2,

$$R = 2\,\frac{\text{Distance between peaks}}{\text{Base of peak}} \times \frac{M}{\Delta M} \qquad (2\text{-}48)$$

Depending upon the peak shape, a 1% cross-talk definition may correspond to peaks which are separated, have an inverted valley, or appear as one composite peak (Fig. 1-3*a*). For close to triangular peaks, however, a 5% cross-talk definition is about twice a 10% valley definition. It is concluded that for quantitative analysis a rigorous 2% valley definition is advisable to minimize cross-talk.

It would be desirable to achieve standardization on the expressions for resolution. In the meantime, care must be exercised when comparing reported resolution values. Resolution requirements for various problems in applied mass spectrometry are discussed in connection with instrument selection in Section 8-II.

D. Transmission

Another important performance factor is transmission which can be defined as I/I_0, where I_0 is the ion intensity of mass M behind the entrance slit and I is the intensity behind the exit slit (i.e., separated ions). Transmission is normally expressed as a per cent value. Transmission is greatly influenced by such factors as charge and energy distributions and instrumental geometry. In spark-source mass spectrometry, for example, the energy pass band of the double-focusing geometry is quite small compared to the total energy distribution resulting in intensity problems. Both resolution and transmission must be considered when evaluating the quality of an instrument for analytical purposes. A figure-of-merit value is the product of resolution and transmission; the problem is discussed in connection with high resolution instruments by Berry (9). Research work in mass spectrometer design has always been directed to improve the resolution–transmission product.

E. Resolution of Single-Focusing Instruments

According to equations 2-44 and 2-45, the parameters which determine the resolution of a single-focusing magnetic mass spectrometer are the radius of curvature, the entrance and exist slit sizes, the homogeneity of the energy of the entering ion beam, and the variety of aberrations in the ion optical system. Resolution can be increased by increasing the sector radius; however, the size of the instrument cannot be increased indefinitely. In addition, some of the aberrations also increase with the increase of the radius. Resolving power can also be increased by narrowing slits. The price for increased resolution is a decrease in sensitivity. Since the energy spread, ΔV, is relatively constant for a given ion source arrangement, the energy spread aberration may be somewhat reduced by employing the highest possible accelerating voltage.

Attainable resolving power is limited by aberrations. Assuming a radius of 20–30 cm, source and collector slits of the order of 10^{-3} cm, $e\Delta V$ of the order of 1 eV, and acceleration voltage in the 1–3 kV range, the maximum attainable resolution of single-focusing magnetic deflection instruments is of the order of 3000. This resolution has been more than adequate for many problems; until recently, in fact, the resolution of the majority of instruments was below 1000. Achievement of high resolution in magnetic instruments is based on application of the double-focusing principle which is discussed after a short digression into the dispersion and energy resolution of electrostatic fields.

F. Dispersion and Energy Resolution of Electrostatic Sector Field

By logarithmic differentiation of equation 2-11 one obtains

$$\Delta r_e / r_e = \Delta V / V \tag{2-49}$$

This equation shows that while the electrostatic sector field cannot be used for mass analysis, it can be used as both an energy filter and an energy spectrometer. For the dispersion one may derive equations very similar to equations 2-39 and 2-41 for the symmetrical and general cases, respectively,

$$D_e = \Delta r_e \qquad (l_e' = l_e'') \tag{2-50}$$

and

$$D_e = \frac{\Delta r_e}{2} \left[(1 - \cos \sqrt{2}\,\phi_e) + \frac{l_e'}{r_e} \sqrt{2} \sin \sqrt{2}\,\phi_e \right] = \frac{\Delta r_e}{2} K_e'' \tag{2-51}$$

where K_e'' is the coefficient of velocity dispersion in a radial electrostatic field.

The energy resolution, R_e, is derived as

$$R_e = \left(\frac{V}{\Delta V} \right)_{\max} = \frac{r_e}{S_{e1} + S_{e2}} \tag{2-52}$$

This is also similar to the magnet cases, S_{e1} and S_{e2} being the proper slit widths.

The radial electrostatic analyzer has been widely used as an energy spectrometer to determine the energies of particles in transmutation reactions. In mass spectrometry the main use of such fields is in linear combination with magnetic fields in double-focusing analyzers where they serve as energy selectors.

VII. DOUBLE-FOCUSING PRINCIPLE

When an ion traverses a radial electric field, the field acts as an energy separator and one could filter out a particular energy with the aid of a suitably located slit. The portion of the beam that passes this slit could next be introduced into a sector magnetic field which would thus receive a monoenergetic beam. Very sharp lines (i.e., high resolution) would result after mass dispersion. The only problem is that of a sharp loss in intensity, since only a very small portion of the original beam would reach the magnetic field. This intensity loss can be reduced by "properly dimension-ing" the fields, i.e., by combining the focusing properties of the fields. The double-focusing principle will be developed by first considering velocity focusing and dispersion focusing in tandem electric and magnetic fields separately, and then combining the two.

In *velocity focusing* it is assumed that the beam is very narrow, i.e., that angular divergence is negligible. This was achieved by Aston in his first mass spectrograph (Section 3-I-B) by passing the ion beam through two narrow slits before entrance into the electrostatic field (S_1 and S_2 in Fig. 3-3). Restricting ourselves to a beam containing only one specific mass, let us consider a beam heterogeneous in energy (Fig. 2-24). The electrostatic field will produce an energy spectrum in the plane B, at which the beam appears as if it originated from point E; the energy dispersion is given by equation 2-51. Passing a slit placed at B the beam enters the magnetic sector field where ions of different velocities again undergo different de-flections. Ions of similar mass but differing in energy recombine at a single point, the *point of velocity focusing*. If the original ion beam contained ions of different specific charge, the magnetic field will also act as a prism, and a different velocity-focusing point will be formed for each mass. The locus of the velocity-focusing points is called the velocity-focusing curve (line g, Fig. 2-24).

In *direction focusing* it is assumed that the ion beam is monoenergetic. Considering only one mass, the ion beam which enters the electrostatic field with a small angular divergence is brought to a point focus by the direction-focusing properties of the field, according to equation 2-32. From this point the beam diverges again but after going through the magnetic field it is brought once more to a direction focus. For ions of different mass the magnetic field will, of course, also act as a prism, resulting in as many direction-focusing points as there are mass components. The locus of the direction-focusing points is called the direction-focusing curve.

The conditions for *double focusing*, i.e., simultaneous correction for both direction and velocity inhomogeneities, can be derived with reference to Figure 2-25. Here the beam coming from the source has a small angular divergence and also a small energy spread. After emerging from the electrostatic field, ions of different velocities are brought to direction focusing in the plane B. Considering particles of energy $eV(1 - \beta)$ and $eV(1 + \beta)$, where β is a small number, energy focusing occurs at points A_{e1} and A_{e2}, each at a distance l_e behind the electrostatic field. The slit at B (energy slit) is wide enough to permit all ions within this energy range to continue into the magnetic field, but cuts out ions with larger energy values. The *central* rays corresponding to different velocities are brought

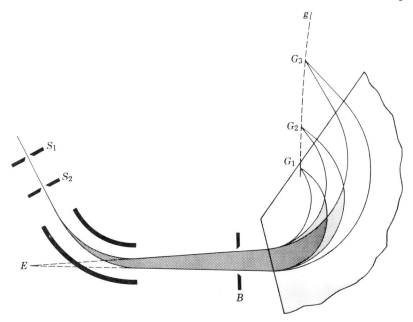

Figure 2-24. Velocity-focusing principle. (After H. Ewald and H. Hintenberger, *Methoden und Anwendungen der Massenspektroskopie*, Verlag Chemie, Weinheim, 1952, p. 73.)

to focus on the velocity focusing curve by the magnetic field which is selected such as to compensate for the velocity spread in the electrostatic field. For one particular mass the focusing point is M_2, while for others the points lie on the velocity-focusing curve (line g). Those ions of the different velocity groups, however, which enter the magnetic field with an angular divergence, undergo direction focusing, and are focused on the cure of direction focusing (line r). The point corresponding to M_2 is at A_2''.

By the proper selection of the fields, the velocity- and direction-focusing curves may be made to intersect, and even to overlap for extended regions. At the point of intersection (A_1'') double focusing is achieved, while at other points focusing is not sharp and line broadening occurs. At point M_2 the broadening is caused by the directional spread of the ions of different velocities, while at A_2'' the broadening is caused by velocity spread.

When the direction-focusing electric field is in tandem with the direction-focusing magnetic field, the condition for velocity focusing is given by Mattauch and Herzog as

$$K_e'' = \pm K_m'' \tag{2-53}$$

or

$$r_e(1 - \cos \sqrt{2}\ \phi_e) + l_e''\ \sqrt{2} \sin \sqrt{2}\ \phi_e$$
$$= \pm r_m(1 - \cos \phi_m) + l_m' \sin \phi_m \tag{2-54}$$

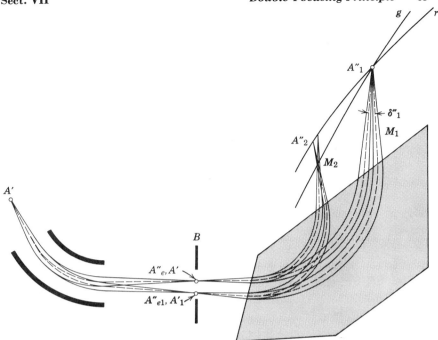

Figure 2-25. Double-focusing principle. (After H. Ewald, *Massenspektroskopische Apparate*, in *Handbuch der Physik*, Vol. 33, S. Flügge, Ed., Springer-Verlag, Berlin, 1956, p. 581.)

The angle between the ion beam incident to the magnetic field and the normal to the field (ϵ_m') is taken as zero.* The positive sign is valid when the ion deflections are the same in both fields, while the negative signs apply for opposite deflections. The condition for double focusing is that the velocity dispersion suffered by the ions in the electrostatic field must be exactly compensated by the magnetic field maintaining, at the same time, the direction-focusing conditions of both fields. This means that equations 2-32, 2-38, and 2-54 must be satisfied simultaneously. In addition, mass dispersion must, of course, also be obtained; this is determined solely by the magnetic field.

In general, a point of double focusing is achievable for only one radius of curvature in the magnetic field. Therefore, an electrical detector must be placed at the point of double focusing, and the strength of the magnetic field must be changed to bring ions of various masses into focus. In one particular arrangement (Section 3-II-B-3), simultaneous double focusing for all masses can be achieved in the same plane, permitting the use of a photographic plate.

* When $\epsilon' \neq 0$, the second term in the right-hand side of equation 2-54 is replaced by l_m' [sin ϕ_m + (1 − cos ϕ_m) tan ϵ_m'].

The major objective of the double-focusing arrangement is to eliminate the $\Delta V/V$ term in equation 2-43 resulting in

$$M/\Delta M = K_m''/2S_m'G_m \qquad (2\text{-}55)$$

where S_m' is to be put equal to the width of the image of the entrance slit produced by the radial electric field. The resolving power of a double-focusing apparatus is derived, utilizing relations which can be obtained from basic ion optical equations (7,10), as

$$\frac{M}{\Delta M} = \frac{r_e}{2S_e'}\left(1 + \frac{l_e'' - g_e}{f_e}\right) \qquad (2\text{-}56)$$

where S_e' is the entrance slit width and r_e is the radius in the electrostatic analyzer. It is seen that the *resolution* of double-focusing mass spectrometers depends *only* on the constants of the *electrostatic* analyzer and on the entrance slit width. Resolution is directly proportional to r_e. *Mass dispersion*, on the other hand, is determined *solely* by the constants of the *magnetic field*.

The numerical value of the expression in the parenthesis in equation 2-56 lies usually between 1 and 2, thus resolution is given by

$$M/\Delta M \approx r_e/S_e' \qquad (2\text{-}57)$$

For example, if $r_e = 25$ cm and $S_e' = 0.005$ mm, the resolving power is of the order of 50,000.

Various double-focusing arrangements are described in Section 3-II-B.

References

1. Barber, N. F., *Proc. Leeds Phil. Lit. Soc. Sci. Sec.*, **2**, 427 (1933).
2. Stephens, W. E., *Phys. Rev.*, **45**, 513 (1934).
3. Herzog, R., *Z. Physik*, **89**, 447 (1934).
4. Cartan, L., *J. Phys. Radium*, **8**, 453 (1937).
5. Kerwin, J. L., *Rev. Sci. Instr.*, **20**, 36 (1949); **21**, 96 (1950); also *Can. J. Phys.*, **30**, 503 (1952).
*6. Kerwin, J. L., in *Mass Spectrometry*, C. A. McDowell, Ed., McGraw-Hill, New York, 1963, Chapter 5.
*7. Duckworth, H. E., and S. N. Ghoshal, in *Mass Spectrometry*, C. A. McDowell, Ed., McGraw-Hill, New York, 1963, Chapter 7.
8. Berry, C. E., *Rev. Sci. Instr.*, **27**, 849 (1956).
9. Berry, C. E., *11th Ann. Conf. Mass Spectrometry*, ASTM Committee E-14, *Proc*, p. 199, 1963.
10. Mattauch, J., and R. Herzog, *Z. Physik*, **89**, 786 (1934).

Chapter Three

Types of Mass Spectrometers

I. HISTORICAL INSTRUMENTS

A. Parabola Mass Spectrograph

The first systematic investigation of positive rays was undertaken during the years 1910–1920 by Thomson (1) with his parabola mass spectrograph. This instrument should really be considered as the first mass spectrograph, although a number of years passed before Aston coined the expression. Thomson's objective was to determine the specific charge of positive rays in a manner similar to that used earlier for the study of electrons.

A schematic drawing of a parabola spectrograph is shown in Figure 3-1. Ions are obtained in a high-voltage discharge tube, and the particles have initial velocities ranging from zero to that corresponding to full discharge voltage (50,000 V). The ion beam is collimated either by passing it through

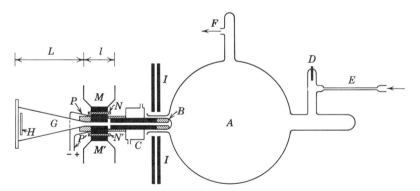

Figure 3-1. Thomson's positive ray parabola apparatus (2). *A*, discharge tube; *B*, cathode; *C*, water jacket for cooling cathode; *D*, anode; *E*, gas inlet; *F*, pump lead; *G*, photographic plate; *I*, magnetic shield; *MM'*, magnetic poles; *NN'*, mica for electrical insulation of *PP'*, which are pieces of soft iron to serve both as condensor plates and to define the magnetic field.

a long, narrow tube (a hypodermic needle was used in the first experiments) or a pair of defining apertures. The beam then enters a short region of parallel and concurrent electrostatic and magnetic fields with strengths E and B, respectively. The length of the field is l. The pole pieces of the electromagnet are placed outside the discharge tube, just behind the cathode, and are insulated from the rest of the electromagnet (by thin mica sheets) so that they can be used as the plates of a capacitor by being connected to a battery. In this field combination the ions are deflected simultaneously by the two fields and as a result, as shown below, particles can be analyzed according to their m/e ratios. Leaving the field region the particles continue on a straight line and strike, at a distance $L \gg l$, either a fluorescent screen where they produce a spot of light, or a photographic plate where they are permanently recorded. The detector is placed perpendicular to the ion beam.

Assume a positive particle moving perpendicular to the plane of the paper in Figure 3-2. Without any deflecting fields the ion would describe a straight path from source to detector and would strike at point O. With the application of a uniform electrostatic field, E, the particle can be deflected to the right so as to strike the paper at point x along the X axis. The displacement Ox is given by

$$x = k_1(eE/mv^2) \tag{3-1}$$

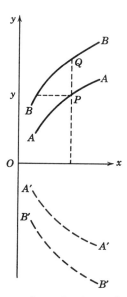

Figure 3.2. Spectrum formation in parabola spectrograph.

where v is the velocity of the particle, m and e are its mass and charge, respectively, and k_1 is a constant the magnitude of which depends on the dimensions of the instrument. Employment of a magnetic field, B, on the other hand, results in vertical displacement along the Y axis, with a displacement given by

$$y = k_2(eB/mv) \tag{3-2}$$

where k_2 is another constant depending on instrument geometry. It is noticed that the deflection produced by the magnetic field varies inversely as the first power of the velocity, while the deflection produced by the electric field varies inversely as the second power of the velocity.

When both fields are applied simultaneously the particle will be deflected to point P whose coordinates are x and y. Eliminating v from equations 3-1 and 3-2 one obtains

$$v^2 = k_1(eE/mx) \qquad \text{and} \qquad v = k_2(eB/my)$$

or

$$k_1(eE/mx) = k_2{}^2(e^2B^2/m^2y^2)$$

from which

$$y^2/x = K(e/m)(B^2/E) \tag{3-3}$$

where $K = k_2{}^2/k_1$.

Considering now an ion beam which consists of ions of the same m/e, and keeping E and B at constant value, equation 3-3 reduces to

$$y^2/x = \text{constant} \tag{3-4}$$

Equation 3-4 describes a parabola. It is thus seen that all ions of the same m/e will fall on a parabola whose coordinates are given by the above equation. Every point on curve AA (Fig. 2-2) represents a particle of the same mass with a particular velocity. Fast particles are deflected only slightly, while slow ones are deflected to a considerable extent. The minimum displacement, which corresponds to maximum kinetic energy, is set by the full potential drop in the discharge tube. All parabolas begin from points nearly equidistant from point O. The extension of the parabolas should, in principle, continue to infinity (kinetic energy of zero). In practice the intensities of low velocity ions decrease rapidly. The intensity distribution along a parabola represents the energy distribution of the ions of a particular mass. The continuity of the observed parabola AA indicates the presence of ions with all possible velocities within a limited range.

Another ion beam of different m/e describes another parabola. The greater the charge-to-mass ratio, the greater y^2/x, the higher the position on the curve. Curve BB thus corresponds to ions having a higher e/m value than those giving curve AA. The coordinates for the various parabolas are determined by the knowledge of the origin and the x axis. The point of origin is easily recognized from the spot produced by the undeflected beam, while the position of the x axis is fixed by reversing the direction of the magnetic field (electric field kept constant) which results in parabolas $A'A'$ and $B'B'$. Comparison of the various parabolas is done by considering two points on the ordinate at constant x. For example, at points P and Q

$$(y_A/y_B)^2 = [(e/m)_A/(e/m)_B] \tag{3-5}$$

Assuming the charges to be equal,

$$(y_A/y_B)^2 = m_B/m_A \tag{3-6}$$

For mass or specific charge determination the instrument must be calibrated, i.e., the ordinates of known substances must be determined. There is usually sufficient oxygen and carbon monoxide in every discharge to provide calibration lines, but traces of any desired substance may be introduced into the discharge tube to establish reference lines.

The first element to be analyzed by the parabola method was neon of atomic weight 20.2. The spectrum of neon (Plate 1) revealed the presence of a faint line at mass 22. At first it was believed to be a doubly charged carbon dioxide line; however, its intensity did not change when all carbon dioxide was removed by cooling with liquid air. Next, the line was considered to be NeH_2, but this was in disagreement with the chemical properties of neon. It remained for Aston to prove, some 10 years later, that the line at mass 22 was due to an isotope of neon.

A tremendous wealth of information can be obtained from the developed photoplates of a parabola mass spectrograph: (*1*) Every parabola corresponds to a different e/m value. (*2*) Every point on a parabola corresponds to a different ion velocity. (*3*) Those points of the parabolas which are on a line parallel to the magnetic axis of deflection (i.e., same electrical deflection) represent particles of the same kinetic energy. (*4*) Particles of the same momentum are on a parallel to the axis of electric deflection. (*5*) Particles of the same velocity but of different mass lie on straight lines going through the origin.

While there is no other instrument that provides information on so many properties of an ion at the same time, there are several drawbacks of parabola spectrographs. Most importantly, there is *no focusing* action in parabola spectrographs, and the only way to increase resolving power is to drastically increase the collimation of the incoming ion beam which, in turn, results in a severe loss in ion intensity. This is a serious limitation,

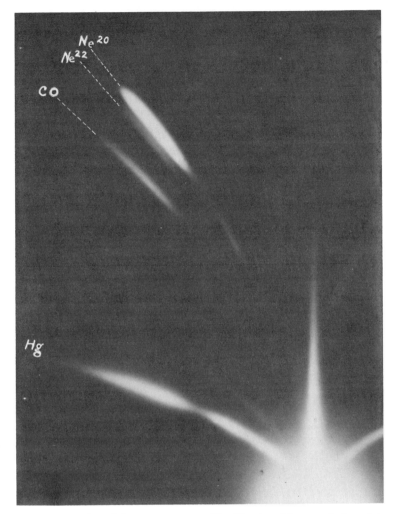

Plate 1.   The parabolas of the neon isotopes (2).   Reprinted by permission of Arnold & Co.

and today the parabola spectrograph is largely superseded by instruments with focusing action. It is used only occasionally to study ion production processes in discharges.

B. Aston's Mass Spectrograph

Aston's first attempts in 1913 to concentrate by fractional distillation the substance of relative mass 22 which had been found by Thomson using the parabola spectrograph were unsuccessful. Next, he tried fractional diffusion at low pressures through claypipe stems and succeeded in separating two fractions with molecular weights of 20.28 and 20.15 as determined by density methods. The two fractions, however, exhibited different spectra and no definite conclusions could be made. Finally, with an improved parabola spectrograph Aston obtained two positive ray parabolas for neon at relative masses 20 and 22 with an intensity ratio of 10:1; this was in excellent agreement with the known molecular weight of neon, 20.18. To further study the nature of isotopes, Aston developed an instrument with the main objective of improved resolution without detrimental intensity loss. Using an optical analog he called his instrument a "mass spectrograph." Aston's first mass spectrograph, built in the Cavendish Laboratory at Cambridge in 1919, is shown in Plate 2. Compared to the streamlined instruments of today, it appears indeed amateurish and primitive. Yet this and a handful of other home-built mass spectroscopes were among the most significant experimental tools in the establishment of modern nuclear physics.

A schematic drawing of Aston's mass spectrograph is shown in Figure 3-3. The main improvement on Thomson's method is that all positive ions of the same mass fall on a single line on the photographic plate instead of being spread out into a parabola. This is achieved by *velocity focusing*. The positive ions formed in the discharge tube leave through the hole in the cathode and, after passing through two defining slits (S_1 and S_2) enter

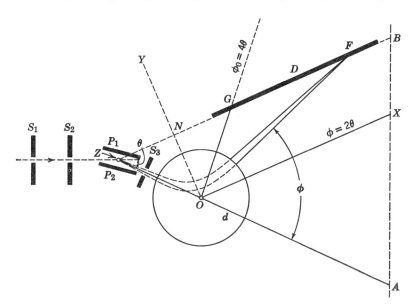

Figure 3-3.   Aston's velocity-focusing mass spectrograph (2).

Plate 2.   Photograph of Aston's original mass spectrograph set up in the Cavendish
aboratory in Cambridge in 1919 (2). *A*, anode connected to high potential terminal of
induction coil below table; *B*, discharge tube; *C*, reservoir containing gas to be analyzed;
G, Gaede-type rotating mercury pump connected to the camera and the discharge tube
by glass tubes and stopcocks; *H*, magnet circuit ammeter; I_1, I_2, charcoal–liquid air
tubes exhausting slit system and camera; *L*, leads from high tension battery to electric
plates; *M*, Du Bois-type electromagnet; *O*, magnet circuit control resistances; *S*, soft
iron plates to shield discharge from stray magnetic field; *T*, pea lamp for photographing
fiducial spot; *V*, vacuum-tight and light-tight control for moving photographic plate;
W, camera, showing light-tight cap on the left. Reprinted by permission of Arnold & Co.

into the electric field established between capacitor plates P_1 and P_2. The deflections of the ions depend upon their energies and those with kinetic energy $\frac{1}{2}mv^2$ are deflected by an amount

$$d_1 = \frac{1}{2}(eE/mv^2)L^2 \tag{3-7}$$

where L is the length of the path between the plates. After emerging from the electric field, the beam may be taken (first-order approximation) as radiating from a virtual source, Z, half-way through the field on the line S_1S_2. Behind the plates there is a diaphragm, S_3, which permits only those ions deflected through d_1 to pass and enter the magnetic analyzer at right angles to the plane of the paper. The deflection angle, θ, is taken to be small. Next, the beam is deflected in the opposite direction, through more than twice this angle, by a uniform circular magnetic field centered at O (large circle in Fig. 3-3). The amount of deflection, d_2, is given by

$$d_2 = \frac{1}{2}(eB/mv)b^2 \tag{3-8}$$

where b is the length of path in the magnetic field. The objective is to bring ions of the same e/m but of slightly different initial energies to a single focus point on photoplate F. This is accomplished for ions of one particular mass in the following way: In the electrostatic field (eq. 3-7) ions possessing smaller energies are deflected downward more than those ions with larger energies (deflection is inversely proportional to the square of velocity). In the magnetic field, on the other hand, particles are deflected upwards (deflection is inversely proportional to the first power of velocity, eq. 3-8) so that there will be a point, F, where the paths of the slow-moving ions will intersect those of the faster ions. This point, whose exact location depends on the geometry of the instrument, is a focal point for ions of a particular mass. For ions having a different mass but the same range of initial energies there is another point of focus on the photoplate.

It can be shown that the foci for different masses lie along line ZB which is drawn parallel to OX. Z is the virtual source of the radiation (at the center of the capacitor plates). Point G on the photoplate corresponds to the image of Z in OY. At this point $\phi = 4\theta$ (ϕ = angle of deflection in the magnetic field), and the mass scale on the photographic scale is very nearly linear. In practice, masses were determined empirically by comparing distances on the photoplates using known lines as reference points. Aston developed several techniques for mass determination, including the "coincidence" and "bracketing" methods (2). Present-day methods for mass determination are discussed in Section 10-III.

Aston's first mass spectrograph had a resolution of 130 with a mean dispersion of 1.1 mm for 1% mass difference; the resolution of the parabola spectrograph was about 12. His second mass spectrograph (1925) possessed a resolution of 1 part in 600 (dispersion of 2.2 mm for 1% mass difference), and its accuracy in mass determination was as high as 1 part in 10^4, a remarkable achievement. The velocity focusing in Aston's mass spectrograph was a considerable improvement over the parabola instrument. However, direction focusing was still lacking and the broadening of the image by the angular spread was controlled only by the collimating system.

C. Dempster's Mass Spectrometer

Working concurrently with Aston, Dempster followed a different approach in constructing a mass analyzer. His instrument is based on the principle (first advanced by Classen in 1907) that charged particles of a given mass and energy diverging from a slit in a magnetic field will be approximately focused after deflection through 180° (Figs. 2-9, 2-10, and 2-11). This is called *direction focusing*. In Aston's instrument the fields acted as prisms; here the magnetic field acts as a lens. Dempster's mass spectrometer is governed by the basic mass spectrometer equation (eq. 2-20).

Dempster's mass spectrometer is shown schematically in Figure 3-4 (3). Positive ions were obtained either by directly heating a salt on a platinum filament (thermal source) or by bombarding the sample with electrons from a heated filament. The positive ions pass through a small hole in iron plate P and enter an electrostatic accelerator with a potential of 500–2000 V between P and S_1, where they acquire nearly the same energy as given by $\frac{1}{2}mv^2 = eV$ (eq. 2-3). Next, the accelerated ions enter (through slit S_1) a semicircular magnetic field; the magnetic field B (approx. 3000 G) is perpendicular to the paper. The ion beam is focused on slit S_2 (radius given by eq. 2-20) behind which a quadrant electrometer detector is placed. This was the first true mass spectrometer employing an electrical means of detection.

Dempster held the magnetic field at a constant value and scanned the spectrum by varying the acceleration potential. On the recorded spectrum relative ion intensities were plotted against the acceleration potential and calibration was experimentally performed using known mass peaks. The resolution of this instrument was 1 part in 100.

With the introduction of direction focusing, using a monoenergetic ion source and employing electrical detection, Dempster opened the way for

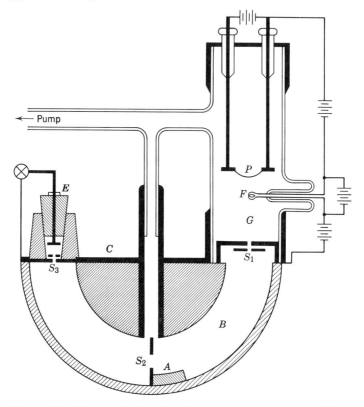

Figure 3-4.   Dempster's direction-focusing mass spectrometer (2).

accurate abundance determinations. He discovered many isotopes and determined their abundances (4).

D. Bainbridge's Mass Spectrograph

Bainbridge (5) removed the monoenergetic source of Dempster and replaced it with a velocity filter (Section 2-III-E). His instrument is shown schematically in Figure 3-5. Ions from a discharge tube enter the velocity filter through slit S_2. The crossed electric and magnetic fields filter out all ions except those with $v = E/B$ (eq. 2-24) which enter the magnetic analyzer through S_2. The 180° magnetic deflection of Dempster's instrument was retained, but the electrical detector was replaced by a photoplate. It is seen from equation 2-16 that at constant ion velocities and constant magnetic field strength the radius is proportional to mass, resulting in a convenient linear mass scale. Bainbridge's work on the reaction

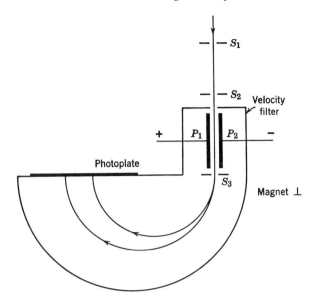

Figure 3-5. Bainbridge's mass spectrograph with velocity selector (2).

$^1H + {}^7Li \rightarrow 2\,{}^4He$ provided the first experimental proof of the Einstein mass–energy relationship (6).

II. MAGNETIC DEFLECTION INSTRUMENTS

A. Single Focusing

1. Sector Instruments

Although Barber and Stephens showed in the early 1930's (refs. 1 and 2 in Chapter 2) that refocusing at 180° is only a special case of the focusing action of any wedge-shaped magnetic field, almost all mass spectrometric work prior to 1940 which involved relative isotope abundance measurements was performed with 180° magnetic analyzers combined with nearly monokinetic ion sources and electrometer-type detectors. Indeed, 180° instruments are still very much in use, and there are several commercial models available (Section 8-III).

A true milestone in mass spectrometer instrumentation was reached by Nier in 1940 with the introduction of a sector-type analyzer (7). This instrument and an improved version described in 1947 (8) have served as the models for many subsequent designs. Nier's instrument, designed for gas and isotope analysis, was a 60° angular deflection instrument with a 15-cm radius of curvature, and a resolution of 1 part in 100. The "Nier geometry" is shown in Figure 2-16. The ion source was a relatively simple electron bombardment source, which later became known as the "Nier-type" source (Section 4-II). The detector design, utilizing a double collector arrangement with a null method of determining isotopic abundance ratios, has also become famous (Section 5-I).

As shown in Figure 2-16, the ion beam enters and leaves the wedge-shaped magnetic field at right angles to the boundaries. Source and detector are equidistant from the magnet (symmetrical case), and the object and image points and the apex of the magnetic field all lie on a straight line, as required by Barber's rule (Section 2-V-B). The relationship between radius and m/e ratio is given, in terms of the acceleration voltage and magnetic field strength, by the mass spectrometer equation (eq. 2-20). Object and image distances can be calculated using equation 2-37. The spectrum may be scanned either by varying the acceleration voltage or the magnetic field, keeping the other at a constant value. In either case, the position of the collector is fixed and the various masses are brought onto the collector by successively changing their radii in the field. It is repeated that only direction focusing is present and a nearly mono-kinetic ion source is required.

As the sector angle is reduced, the area of required magnetic field for a given ion radius of curvature is also reduced. Therefore, a much smaller magnet is required for equal performance when a 60° deflection is employed compared to the 180° type. At the same time, required magnet gap width is also reduced since both source and detector are quite far away from the magnetic field. This, of course, is also an advantage in the design of these structures. Also, mass-discriminating effects (particularly important in absolute isotope ratio measurements) are reduced. It is noted, however, that the ion source section is usually not completely free of magnetic field because a weak auxiliary field (about 1/10 of the main field) is needed for electron collimation (Section 4-II-B). On the debit side for the sector fields is the fact that ions must cross the magnetic field boundaries twice during their flight from source to detector. These boundaries are not sharp and well-defined, and a fringing field exists which is difficult to calculate. In practice, provision is normally made to mechanically adjust the magnet with respect to the spectrometer tube for best empirical focusing.

Several general-purpose and special-purpose sector instruments have been described (9,10). An instrument with a resolution of 1000 was constructed by Inghram et al. (11) in 1953 for the analysis of uranium and other heavy elements. Although the 60° deflection angle has been the most popular, several instruments with 90° and 120° deflection angles have also been used.

2. Multistage Analyzers

In the discussion of resolution (Section 2-VI-B) the importance of the energy spread of the beam and of the various aberrations has been included. A certain amount of image diffuseness may be caused by *small-angle elastic scattering collisions* of the ions with gas molecules along the ion flight path in the analyzer and also by charge transfer collisions in the ion acceleration region. This results in "peak tails," the presence of which reduce the accuracy with which relative isotopic abundances may be determined. The magnitude of the *abundance sensitivity* of an instrument is a measure of its ability to measure faint isotopes adjacent to abundant ones. Abundance sensitivity is defined as the ratio of ion beam current of mass M to the "background current" at adjacent mass spectral positions $M \pm 1$,

$$\text{Abundance sensitivity} = \frac{\text{Peak ion current at mass } M}{\text{Ion current at } M \pm 1} \qquad (3\text{-}9)$$

Abundance sensitivity is normally mass dependent, decreasing with increasing mass. The best first-order direction focusing instruments can detect a faint isotope next to an abundant one only if the former is present to at least 1 part in 10^5 of the abundant one (at around mass 100).

Inghram and co-workers (11) eliminated, or at least considerably reduced, the abundance sensitivity limitation caused by scattering in a design employing two first-order single-focusing analyzers in series. As shown in Figure 3-6, the first analyzer acts as an isotope separator. The ions passing a *discriminating slit* placed at the position of focus of the first analyzer are further accelerated to 10,000 V energy (initial acceleration is to about 1500 V) and then pass through the second analyzer before final collection. Those ions from the strong ion beam which are scattered in the first analyzer and would cause peak "tails" are resolved into separate peaks in the second analyzer. Also, space-charge scattering is reduced

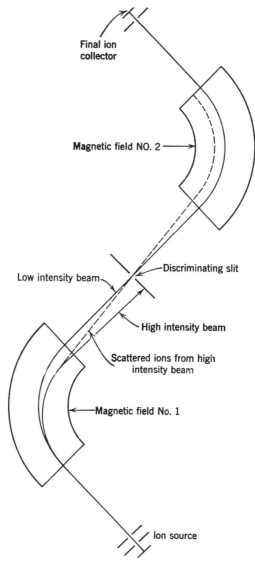

Figure 3-6.   Two-stage analyzer.

since the ion current of the major peak is very much less intense in the second analyzer during the time when the minor peak is being collected. The end result is that the tails of the major peaks are suppressed by a factor of 1000. (Explain why the tails are somewhat more completely suppressed on the high mass side than on the low mass side of the large peak!) The configuration shown in Figure 3-6 is an "S-type" design. The two analyzers may also be put into a "C-type" configuration. The dispersion of the S-type is twice that of the C-type. Both types appear to provide the same abundance sensitivity.

The abundance sensitivity achievable with two-stage analyzers is about 10^7. White, Collins, and Rourke (12) examined many elements for the presence of naturally occurring isotopes of very low abundances. The ^{180}Ta isotope, for example, is present only at the 0.0123% level next to the large ^{181}Ta isotope (Appendix I). A three-stage analyzer was designed by White, Rourke, and Sheffield (13), and a four-stage analyzer has recently been described by White and Forman (14). The latter, which has an abundance sensitivity of 10^8, is described in connection with the Bainbridge-Jordan geometry in Section II-B-2 of this chapter.

A 15-in. two-stage tandem mass spectrometer has been equipped with a retardation lens by Freeman and Daly (15). Energy selection is provided by connecting the gridded retardation lens, located after the collector slit of the second magnetic stage, through a variable bias voltage (± 150 V) directly to the acceleration voltage. Abundance sensitivity measured on the low mass side of a Cs^+ peak increased from 1.6×10^6 to 4×10^9. On the high mass side the abundance sensitivity (10^9) is not affected by the technique because high energy ions are present in this region.

Multistage mass spectrometers should not be confused with double-focusing instruments. The former are designed to achieve high abundance sensitivity so that low intensity peaks adjacent to large intensity peaks can be measured. The abundance sensitivity of double-focusing instruments is normally not significantly higher than that of single-focusing instruments, although their resolution, of course, is much higher.

It is noted in passing that mass scanning in multistage instruments is usually accomplished by variation of the ion acceleration voltage rather than by scanning the magnetic fields.

3. Isotope Separators

It has been stated at the very beginning of this text that mass spectrometers are essentially electromagnetic separators of isotopes. The purpose of the separation in mass spectrometers is to measure the relative abundances of the separated isotopes or to precisely measure the masses. Another obvious application is to obtain "large" quantities of the separated isotopes.

From equation 2-21, considering both the accelerating voltage and the magnetic field to remain constant, we obtain

$$dr/r = dM/2M \qquad (3\text{-}10)$$

In a 180° analyzer, as shown in Figure 3-7, the physical separation, ds, between two particles whose radii differ by dr will be just $2dr$. Therefore,

$$ds = r(dM/M) \qquad (3\text{-}11)$$

This means that if the radius of the instrument is 1 m, ions differing in mass by 1% will arrive at the 180° focal points 1 cm apart. If "collection pockets" are placed at the proper points, separated isotopes can be collected and later removed by scraping or by appropriate chemical dissolution.

Large-scale electromagnetic separation of isotopes started in the early 1940's in connection with the United States atomic energy development project. Separators are often called *calutrons* after the first instruments built at the University of California. The large calutrons at the Oak Ridge National Laboratory in Tennessee and at the Atomic Energy Research Establishment at Harwell, England are 180° deflection instruments with a radius of 24 in. Figure 3-8 shows the ion beam separation (in./mass unit) as a function of mass number at the focal point of ion beams in a 180° separator of 24 in. radius. It is seen that separation becomes more and more difficult at higher masses and the only practical solution is to increase the instrument radius.

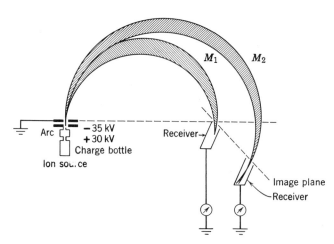

Figure 3-7. Schematic diagram of the ion paths and principal component parts of a 180° electromagnetic separator. (After A. E. Cameron, "Electromagnetic Separations," in *Physical Methods in Chemical Analysis*, Vol. IV, W. E. Berl, Ed., Academic Press, New York, 1961, p. 122.)

Some of the advantages of sector instruments mentioned previously also apply to separators; however, most sector-type separators have been designed for lower current operation for experimental or research purposes. The methods of collecting the separated isotopes and production rates are discussed in Section 13-I-D.

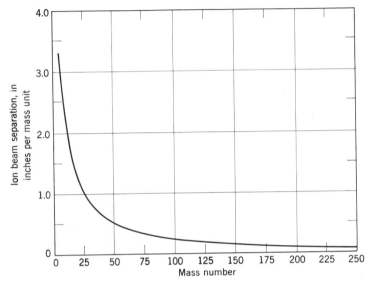

Figure 3-8. Separation in inches per mass unit as a function of mass number at the focal point of ion beams in an electromagnetic separator of 24-in. radius. (After A. E. Cameron, "Electromagnetic Separations," in *Physical Methods in Chemical Analysis*, Vol. IV, W. E. Berl, Ed., Academic Press, New York, 1961, p. 121.)

B. Double Focusing

1. Trochoidal Instruments

It was mentioned in Section 2-V-B that the 360° circular deflection in a homogeneous magnetic field cannot be used for mass analysis, although perfect double focusing is produced, because the focal point is not mass dependent. Addition of an electrostatic field perpendicular to the magnetic field, however, provides a linear transverse movement superimposed on the circular one, and the ion trajectories in such *crossed* fields become trochoidal, resulting in both spatial mass dispersion and perfect double focusing in the plane normal to the magnetic field.

As shown in Figure 3-9, a *trochoid* is the locus of a point, $P(x,y)$ at a distance b from the center of a circle of radius a as the circle rolls along with an angular velocity ω on a fixed straight line. The equations of a

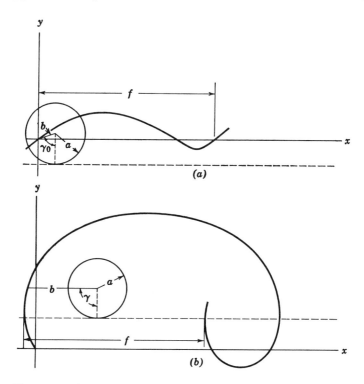

Figure 3-9. Trochoid curves: (a) curtate cycloid; (b) prolate cycloid.

trochoid are

$$x = a\gamma - b \sin \gamma \tag{3-12}$$

$$y = a - b \cos \gamma \tag{3-13}$$

where γ is the angle subtended at the center of the circle. If $b < a$ the curve is called a *curtate* cycloid, if $b > a$ it is a *prolate* cycloid, and if $b = a$ the path is called simply a cycloid.

The equation for motion for ions of velocity v is

$$m(dV/dt) = e(vB - E) \tag{3-14}$$

where v, B, and E are vector quantities. When ions enter the field at $x = 0$, $y = 0$ the focal distance (cycloidal pitch), f, is obtained as

$$f = 2\pi n(E/B^2)(m/e) \tag{3-15}$$

The distance f is measured from the point of origin along the x axis, and at the focal point $y = 0$. The factor n is a positive integer and $n = 1$ corresponds to one revolution. The time, t, required to reach the focal point is

$$t = 2\pi n/\omega = 2\pi n m/eB \tag{3-16}$$

Equation 3-16 does not contain terms connected to either initial velocity or direction; thus perfect double focusing is achieved. Since f is directly proportional to m/e the mass scale is linear (when used as a spectrograph), as compared to magnetic deflection instruments where the specific mass is a quadratic function of the radius. The theoretical resolving power of cycloidal instruments is thus twice that of magnetic instruments of comparable size. The difference is actually larger due to the perfect focusing, particularly for small instruments. Here, due to the fact that small magnetic deflection instruments must use small ion energies, variations in the energy spread (e.g., thermal energy) become relatively more important.

Among the three possible cycloidal paths the prolate cycloid was found to give the best performance. Here ions enter almost normal to the entrance slit. The common cycloidal paths, as well as curtate ones are baffled out by instrument design to prevent space charge defocusing. An isometric view of a cycloidal focusing analyzer with prolate cycloid paths is shown in Figure 3-10. Ions are usually collected on an electrical detector as shown, although a photoplate could also be used in an appropriate design. The magnetic field is provided by a permanent magnet (B = constant) while mass scanning is performed by varying the electric field between two parallel plates. A stack of guard rings, fed from a voltage divider, is used to homogenize the field. By making the acceleration voltage proportional to E, all ions follow the same trajectory as the mass scale is scanned. Only two fixed slits are needed (source and detector), and they are coplanar in an equipotential plane.

After the original design of Bleakney and Hipple in 1938 (16), a number of experimental versions were built. In 1956, Robinson and Hall (17) described a small instrument with f = 2.7 cm and a resolving power of about 150. Two versions of this design are now commercially available (Section 8-III).

Mass peaks in cycloidal instruments have trapezoidal shapes and flat tops, and there are no tails. Due to the absence of cross contributions, unit resolution is often given as "less than 1% contribution to the top of one peak by an adjacent peak of equal magnitude."

2. Dempster and Bainbridge-Jordan Machines

In 1935 Dempster reported (18) the construction of a double-focusing instrument where the positive ions coming through slit S_2 (Fig. 3-11) pass through a radial electric field between two cylindrical capacitor plates (C_1 and C_2) and then enter the magnetic field perpendicular to the plane

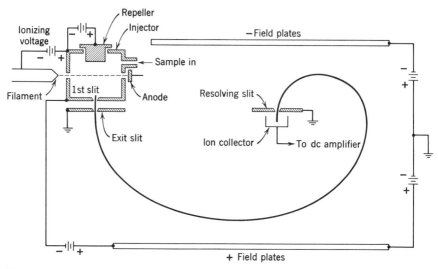

Figure 3-10.   Cycloidal mass spectrometer. (Courtesy of Consolidated Electrodynamics Corp.)

of the diagram; ions are recorded on the photographic plate. This instrument is double-focusing for one mass only.

The ion beam enters and exits both electrical and magnetic fields normally and image formation is symmetrical. The image of the entrance slit formed by the electrostatic field is located at the entrance boundary of the magnetic field which provides 180° deflection so that the final image is formed on the photographic plate placed at the boundary of the magnetic field. This exit boundary constitutes the direction-focusing curve which is intersected by the velocity-focusing curve at an angle of 41°. This is the point of double focusing. Relatively sharp lines can be obtained over a mass range of about 10% from this point.

Dempster's original instrument employed 8.5 and 9.8 cm radii for the electrostatic and magnetic fields, respectively. These dimensions resulted in a resolving power of about 3000 with an entrance slit width of 0.0025 cm (dispersion 0.098 cm per 1% mass difference). More recently, several instruments of this type, but with considerably larger radii (by a factor of about 30), have been constructed. Resolving power as high as 100,000 has been achieved and the peak-matching technique (Section 10-III) has been utilized for the accurate determination of atomic mass differences (19).

In the Bainbridge-Jordan mass spectrograph (20) the ions are first bent through an angle of $\phi_e = \pi/\sqrt{2}$ radians $= 127°17'$ in an electrostatic field, then travel a field-free length of $l_m' = \sqrt{3}\ r_m$, after which they are bent

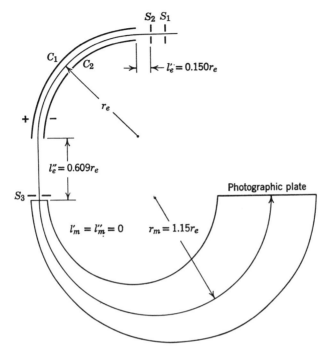

Figure 3-11. Dempster's double-focusing mass spectrograph (18).

through an angle of $\phi_m = \pi/3$ radians $= 60°$ in a sector magnetic field followed by another straight travel through a field-free region of $l_m'' = \sqrt{3}\ r_m$ at the end of which the photographic plate is located. Ion optics shows that $r_m = r_e$ for the velocity-focusing condition when symmetrical fields are used. Double-focusing in this instrument is again achieved at only one point. The photographic plate is placed at an angle of approximately 30° to the direction of the mean ion beam. The mass scale in this design is approximately linear.

In the first instrument, built by Bainbridge and Jordan, $r_m = r_e = 25$ cm, resulting in a resolving power of about 10,000 with a principal entrance slit of 0.0025 cm employed (dispersion 0.50 cm per 1% mass difference). Several later instruments of this type have been constructed both in Japan and in Russia and resolution as high as 70,000 can be achieved.

A four-stage mass spectrometer based on the Bainbridge-Jordan design has recently been built by White and Forman (14). As shown in Figure 3-12, the first half of the instrument is a complete double-focusing system with both electrostatic and magnetic fields providing 90° deflection. The mean radius of curvature is 50.8 cm for each element. At point S_3 there is first-order energy *and* momentum focusing. The beam emanating from S_3 enters the second double-focusing system and is finally collected on an electron multiplier placed at the image plane at S_5. The system is a "C-type" multistage spectrometer. Isotopic abundance sensitivity of 10^8 is achievable with the instrument, and it can be programmed for many kinds of investigations.

3. Mattauch-Herzog Spectrograph

In 1936 Mattauch described (21) a mass spectrograph which was, to the first order, double-focusing at all masses. The design was based on earlier theoretical considerations of Mattauch and Herzog (22). In recent years, the *Mattauch-Herzog geometry* has become extremely popular, and there are several commercial instruments employing this design (Section 8-V). The main advantage of the Mattauch-Herzog design is the possibility to

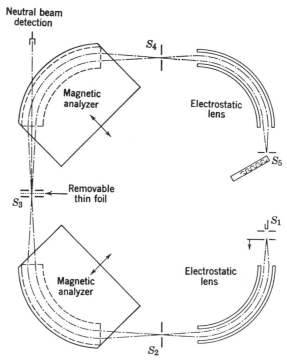

Figure 3-12. Four-stage double-focusing mass analyzer (14). Magnets can be moved in direction indicated, and all slits are variable in aperture. Thin foils or other targets may be introduced at S_3 for investigating charge exchange phenomena, or for ion range measurements.

detect the entire mass spectrum simultaneously with all masses focused sharply on the photographic plate.

The velocity-focusing equation (eq. 2-54) defines the shape of the exit-pole boundary of the magnetic field. It is recalled that in equation 2-54 the positive sign refers to deflection in the same sense in the two fields, and the negative sign denotes opposite deflections. To achieve double focusing for all masses, the exit-pole boundary curve, which normally has a rather complicated shape (23), must be made a straight line so that a photographic plate can be employed for detection. The cause of the complicated shape is the presence in equation 2-54 of r_m, the radius of curvature in the magnetic field, which introduces mass dependency. It has been shown (22) that all terms including r_m can be eliminated when the image distance in the electric field, l_e'', is made to approach infinity. Under this condition one obtains

$$\sin \phi_m = \pm \sqrt{2} \sin \sqrt{2} \, \phi \qquad (3\text{-}17)$$

Here the plus sign refers to fields causing deflections in the opposite sense, and the minus sign refers the deflections in the same sense. This equation is independent of r_m.*

The ion optical considerations concerning double focusing for all masses may be summarized as follows: (*1*) the entrance slit of the electrostatic analyzer should be located at the principal focus of that field; (*2*) ions should emerge from the electrostatic field as a parallel beam and should enter the magnetic field as a parallel beam; (*3*) the detector should be placed at the principal focus of the magnetic field; (*4*) a linear plane photographic plate placed at the exit-pole boundary of the magnetic field permits the detection of all ions under double-focusing conditions.

A schematic diagram of the Mattauch-Herzog geometry is shown in Figure 3-13. In this design the entry into the magnetic field is normal ($\epsilon_m' = 0$). To eliminate the need for the ion beam to cross stray magnetic fields, the photographic plate is placed at the exit boundary of the magnetic analyzer, so that $\phi_m = \pi/2 = 90°$. This choice automatically determines the value of ϕ_e. From equation 3-17 we obtain $\pi/4 \sqrt{2}$ radians $= 31.8°$ for the deflection angle within the electrostatic field. The photographic plate, as indicated, is placed in coincidence with the exit field boundary, which makes a 45° angle to the plane perpendicular to the beam entry point.

The resolving power, as shown in equation 2-57, depends upon the radius of the electrostatic analyzer, r_e, and the size of the entrance slit, S_e' but

---

* When $\epsilon_m' = 0$ the term $(1 - \cos \phi_m) \tan \epsilon_m'$ must be added to the left side of equation 3-17.

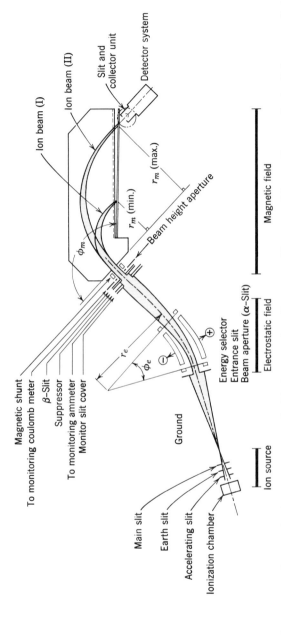

Figure 3-13. Mattauch-Herzog geometry; double-focusing for all masses. (Courtesy of Japan Electron Optics Laboratory Co.)

does not depend on the radius of the magnetic analyzer. Mass dispersion, on the other hand, is given by $0.00707r_m$ for 1% mass difference and depends entirely on the radius of the magnetic analyzer.

In the first instrument the radius of the electrostatic analyzer was selected to be 28 cm, resulting in a resolution of about 5600 when an entrance slit of 0.0025 cm was employed. The radius of the magnetic sector was 24 cm, yielding a maximum mass dispersion of 0.17 cm for 1% mass difference. Dispersion varies along the photoplate since the value of r_m is different for different masses.

The *mass scale* in the Mattauch-Herzog design is *quadratic* (24). The radius is proportional to the square root of the mass as long as the ion energy remains the same. The actual determination of the mass on the photoplates is accomplished by calibration with known mass lines (Section 10-III).

Several Mattauch-Herzog geometry mass spectrographs have been built for research purposes. Some of these were exact copies of the original design using careful adjustments to improve resolution. Others employed much larger radii, and resolution as high as 100,000 have been achieved. Still other designs modified the basic geometry by using an oblique ion entrance angle, by employing toroidal condensers, etc. For the description of these instruments, and also for discussion of the various aberration effects, the reader is referred to more advanced texts and journal references (24). Commercially available Mattauch-Herzog instruments are described in Section 8-III-C.

4. Nier-Johnson Spectrometer

The Nier-Johnson geometry (Fig. 3-14) employs a 90° electrostatic analyzer using symmetrical object–image arrangement, followed by a 60° magnetic analyzer used asymmetrically. The result is (25) direction focusing to the second order plus double focusing to the first order. This means the achievement of three ion optical conditions: (*1*) first-order angular focusing, i.e., coincidence of the final image in the plane of the collector slit; (*2*) second-order angular focusing, i.e., the elimination by cancellation of the $\alpha^2 r$ image broadening (Section 2-VI-B) of the individual analyzers; and (*3*) first-order energy focusing, i.e., zero lateral displacement of the ion beam at the collector for a small change in ion accelerating voltage.

The instrument is a mass spectrometer, i.e., all masses are recorded at a fixed r_m so that simultaneous focusing over a wide mass range is not needed. As shown in Figure 3-14, the intermediate image formed by the electrostatic field is the same distance from the field as the source. A slit

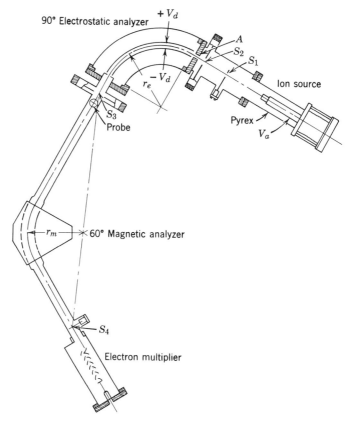

Figure 3-14. Nier-Johnson double-focusing mass spectrometer; direction focusing of the second order (27).

is placed at this point (S_3). From ion optical considerations one obtains the condition that velocity focusing requires the ratio r_m/r_e to be 0.81.

The first instrument built by Nier and Roberts (26) was relatively small (r_e = 18.8 cm, r_m = 15.2 cm) and provided a resolution of about 14,000 with $S_1 = S_4 = 0.00125$ cm. In a much larger instrument (27), employing r_e = 50.3 cm and r_m = 40.6 cm, resolution as high as 75,000 (half-width) could be achieved (slits were 0.0001 cm). The Nier-Johnson geometry is now employed in commercial instruments (Section 8-V). It is noted here that precise mass measurement with this geometry is possible by the peak-matching technique (Section 10-III).

III. DYNAMIC MASS SPECTROMETERS

A. General

In a recent text (28), Blauth defines dynamic mass spectrometers as "those in which the time-dependence of one or more parameters of the systems, e.g., electrical field strength, magnetic field strength, or ion movement, is fundamental to the mass analysis." The basic principle of mass separation is thus seen to involve the time dispersion in the ion motion. A seemingly heterogeneous group of mass spectrometers belong in this group, and concise classification is difficult. Blauth's classification scheme includes four broad classes based upon the principle by which the time dispersion of the ion motion is used for mass analysis: (*1*) energy balance mass spectrometers, (*2*) time-of-flight mass spectrometers, (*3*) path stability mass spectrometers, and (*4*) characteristic frequency generator spectrometers.

A further narrowing of this classification is achieved by considering in each class three basic types of ion motion: (*1*) linear direct motion, (*2*) linear periodic motion, and (*3*) circular periodic motion. Table 3-1 summarizes those basic designs which have found practical applications. These are discussed in the text that follows. A few other designs of lesser present-day importance are also shown with references. For example, the feasibility of the characteristic frequency generators, which rely on self-excited oscil-

Table 3-1   **Dynamic Mass Spectrometers** [a,b]

Energy balance	Time-of-flight	Path stability	Characteristic frequency generator
Linear direct			
Bennett Redhead Topatron	Ion Velocitron Bendix	Smythe-Mattauch Monopole	Klystron*
Linear periodic			
Palletron* Double-well*	Tempitron*	Farvitron Quadrupole Reflektron* Ion cage*	Barkhausen-Kurz* Oscillotron*
Circular periodic			
Omegatron	Chronotron Trochoid TOF* Synchrometer Hipple-Thomas*	Falk-Schwering*	Magnetron*

[a] Table based upon classification by Blauth (28).

[b] Instruments marked by * are not discussed in this text. The reader is referred to reference 28. Several other designs described by Blauth have not yet been built.

lations of the ions in an appropriate field configuration to give mass analysis, has been demonstrated repeatedly, but no practical instruments have been reported. For a detailed discussion of the many versions of dynamic mass spectrometers developed in recent years, the reader is referred to Blauth's text (28).

B. Energy Balance Spectrometers

In the energy balance spectrometers energy is supplied (acceleration spectrometers) or withdrawn (deceleration spectrometers) to the ions in an interaction with a radiofrequency field. Maximum exchange is achieved in the *resonance condition*, and mass scanning is facilitated by varying the mass-dependent resonance conditions. The motion of the ions may be linear direct, linear periodic, or circular periodic.

1. Bennett Radiofrequency Spectrometer

The Bennett rf spectrometer, of which several versions have been described (29), operates similarly to a linear accelerator. Ions are formed from neutral gas molecules by conventional electron bombardment (Section 4-II). The ions formed are accelerated to a known voltage and are drawn into the first stage of the spectrometer (Fig. 3-15). The first

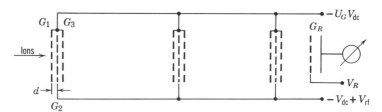

Figure 3-15.   Bennett-type rf spectrometer.

stage consists of three grids, all at the same dc potential. When an rf voltage is applied to the grid in the middle, equal but opposite radiofrequency fields are established between grids G_1-G_2 and G_2-G_3. Some of the ions entering this region will be accelerated, while others will be decelerated. The parameters to be considered are: the velocity of the ions, v, the distance between the grids, d, the frequency of the rf voltage, f, and the amplitude of the rf voltage in the instant the ions pass through the grids (phase angle). At the resonance condition the ions take up maximum energy from the field. This condition is given by

$$d/v = 0.74(1/2f) \tag{3-18}$$

Expressed in words, at the resonance condition the time of flight of the ions between the grids equals the duration of a half-wave of the rf oscillation. The factor 0.74 results from exact calculation of the optimum phase angle. In order for ions to be accelerated between both G_1-G_2 and G_2-G_3 the polarity of the rf field must be changed at the exact moment when the ions are at the center grid. However, the ion source provides a continuous stream of ions (continuous "duty cycle"); therefore an optimum phase angle is always present.

The mass of the ions enters into the equation through the ion velocity. Replacing v in equation 3-18 with the expression given in equation 2-3 one obtains the mass (amu/electronic charge)

$$M/e = 0.27 \times 10^{12}(V/f^2d^2) \tag{3-19}$$

where V is acceleration potential in volts, d is in cm, and f is in cps.

Three such acceleration stages are shown in Figure 3-15. When the distance between these stages is properly selected, i.e., when the time of flight of the ions in the field-free space is equal (or an integral multiple of) the period of oscillation $(1/f)$, mass separation will be improved as the ions are progressing toward successive stages. At the end of the tube there is a retarding grid, G_R, at a positive potential, V_R, which permits the passage of only those ions which obtained maximum acceleration during their passage. Mass scanning is accomplished by varying the rf field or by varying the acceleration voltage. Resolution is changed by adjusting the retarding voltage. When no retarding voltage is employed, the instrument becomes a total current detector.

The resolution of the Bennett-type rf mass spectrometers is relatively low (about 15–30), although improved versions have exhibited resolution up to 250. In Redhead's version (29) some 20 equidistant grids are used and the transmission factor is as high as 95%; several similar designs have also been reported (28). The main advantages of the Bennett radiofrequency spectrometer are the absence of the magnet and the general simplicity. These immediately suggest applications in upper atmosphere research (Section 13-II-B). Other applications include residual gas analysis and analysis of ions in flames.

2. Topatron

The topatron, developed by Váradi, Sebestyén, and Rieger (30) in 1958, is an rf spectrometer in which ions are accelerated throughout their linear motion. The instrument is primarily intended for both total and partial pressure measurement in high-vacuum systems (*to-pa*-tron) in the 10^{-3} to

10^{-7} torr range. As shown in Figure 3-16, total pressure is measured on grid G_1, whose negative potential attracts a portion of the ion current leaving the source. The fraction measured is directly proportional to the total ion current and hence to the total pressure. The rest of the ions pass grid G_1 and pass through the spectrometer to be collected on T; the spectrometer is essentially the same type as that described in the previous section. Resolution ($M/\Delta M$ = const), referred to 20% of the profile height, is 40–50, mass range is 2–100. Commercial topatron tubes are about 20 cm long and 4 cm in diameter, are made of glass, and can be connected directly to vacuum systems.

3. Omegatron

The omegatron is an energy balance spectrometer with circular periodic acceleration. Mass analysis is accomplished by the use of crossed homogeneous magnetic and rf alternating electric fields. The principle involved is similar to that used in cyclotrons, where charged particles move in circular paths and are accelerated at the edges of the "dees" twice per revolution. First developed, and named, by Sommer et al. (31) for measuring mass constants, omegatrons became widely used as partial pressure analyzers after Alpert and Buritz (32) demonstrated their advantages.

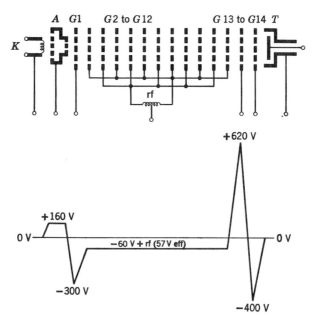

Figure 3-16. Topatron (30). (Manufactured by Leybold Co., Cologne.)

The basic arrangement is shown in Figure 3-17. Electrons for ionization are obtained from a heated filament and, after acceleration to about 90 V, travel across a small cubic box (about 2 cm edge), to be collected by an electron trap which is biased positively to suppress secondary emission. A parallel magnetic field of 3000–5000 G exerts a strong collimating action on the electron beam; the electron emission current, I^-, is a variable operating parameter (see later). As the electrons pass through the chamber they ionize molecules of the gas present in the ionization chamber. When the omegatron is used as a partial pressure analyzer, molecules enter from the vacuum system of which the omegatron becomes an integral part.

The motion of ions in an omegatron has been treated intensively (28,33). The orbital angular velocity, ω_c (rad/sec), of an ion with a particular m/e moving in a plane perpendicular to a homogeneous magnetic field is given by

$$\omega_c = eB/m \qquad (3\text{-}20)$$

Since ω_c is mass dependent (ω = omega, hence the name omegatron) determination of ω_c may be used for mass analysis. This is accomplished in the following manner: A sinusoidally alternating electric field, $E = E_0$ sin ωt, is provided perpendicular to the magnetic field by the rf plates at the top and the bottom of the box. Those ions for which $\omega_c = \omega$, where ω is the angular frequency of the rf field, are accelerated by the combined action of the fields. The trajectory is an Archimedes-type spiral with ever-increasing radius; the ions finally impinge on a suitably located collector plate. The m/e of the ions is determined from the resonance frequency, while the ion current, measured by a vibrating reed electrometer (Section

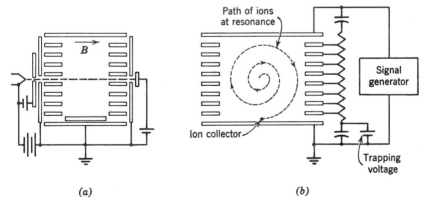

Figure 3-17. Omegatron (31). Magnetic lines of force are (a) horizontal and (b) perpendicular to the drawing plane.

5-I-B), is proportional to the abundance of ions. Mass scanning is accomplished by sweeping either the rf frequency or the magnetic field strength (the former is more practical). Unlike the cyclotrons, no harmonic peaks are formed in the omegatron because ion acceleration is continuous.

Those ions for which $\epsilon = |\omega - \omega_c| \ll \omega_c$ describe an approximately spiral trajectory with an angular frequency of $(\omega + \omega_c)/2$ and a radius of

$$r = (E_0/\epsilon B) \sin (\epsilon t/2) \tag{3-21}$$

It is seen from this equation that the radii will pass through successive minima and maxima for all values of ϵ as the rf field changes. The only exception is the case for which $\epsilon = 0$, the resonance condition, at which the radius linearly increases to infinity,

$$r_t = E_0 t/2B \tag{3-22}$$

where r_t is the radius of the resonant ion orbit at time t. If the ion collector is located at a distance r_0 from the central axis of the electron beam, then there is a critical value of ϵ for which ions will reach the collector. This critical value, ϵ', is given by

$$\epsilon' = E_0/r_0 B \tag{3-23}$$

Accordingly, an ion with a cyclotron frequency of ω_c will be collected if

$$\omega - \omega_c \lessgtr E_0/r_0 B \tag{3-24}$$

For a given value of ω, i.e., a given rf frequency, the range of collection, $\Delta\omega_c$ is given by

$$\Delta\omega_c = 2E_0/r_0 B \tag{3-25}$$

The resolution, $M/\Delta M$, of the omegatron is obtained as (eqs. 3-20 and 3-23),

$$M/\Delta M = \omega_c/2\epsilon' = eB/m \times r_0 B/2E_0 = er_0 B^2/2mE_0 \tag{3-26}$$

or, in practical units,

$$M/\Delta M = 4.8 \times 10^{-5}(r_0 B^2/E_0 M) \tag{3-27}$$

where M is in atomic mass units, r_0 is in cm, B is in G, and E_0 is in V/cm. The resolution may be measured experimentally because the $2\epsilon'$ term in equation 3-26 is the width of the resonant ion peak measured at the base of the peak. Resolution is thus inversely proportional to mass and diminishes rapidly toward high masses. It could be increased by employing a stronger magnetic field or a smaller rf amplitude. The former is limited by physical size, the latter by sensitivity and reproducibility requirements.

The length of an Archimedes's spiral is approximately equal to the number of revolutions times the average orbit circumferences. The path length, L (cm), of the spiraling ions is given by

$$L = n\pi r_0 = 2r_0(M/\Delta M) \qquad (3\text{-}28)$$

Here n is the number of revolutions an ion makes in reaching the collector. The right-hand side of the equation is obtained by considering that (eq. 3-20) $\omega_c = eB/m = 2\pi f_c = 2\pi n/t_0$, where f_c is the cyclotron frequency and t_0 is the time it takes the resonant ion to reach a radius r_0.

From equation 3-28 it is easy to calculate that at a resolution of 44 (carbon dioxide) the total path length for a 1 cm collector distance is nearly 1 meter. This is a severe limitation in omegatrons: to prevent scattering effects, the instrument must be operated at pressures below 10^{-5} torr. If one attempts to increase resolution by decreasing E_0 below about 1 V/cm, the path length will increase and maximum operating pressure must further be reduced. Similarly, space charge effects limit the use of high ionizing electron current. Operation at 10 μA is possible only at pressures below 10^{-7} torr.

Omegatrons thus operate at a resolution around 44. This is achieved using a collector distance of 1 cm, magnetic field of 3000–5000 G, and an rf field of $E_0 = 1$ V/cm. The frequency range to be covered from hydrogen to argon is about 3 Mc/sec to 150 kc/sec. To prevent the drifting of ions, due to their thermal velocities, in the axial direction of the magnetic field, a dc trapping voltage of about 1 V is applied to the shield in such a manner as to be positive with respect to the rf plates.

The sensitivity, S, of omegatrons is given by (34)

$$S = I^+/I^- P \qquad (3\text{-}29)$$

where I^+ is the measured positive ion current, I^- is the electron beam current, and P is the pressure existing within the omegatron. The value of I^+ is proportional to the cross section of ionization at a given electron energy, the gas density, the effective electron path length, and the electron beam current. Typical sensitivity is 10^{-14} A/10^{-9} torr at 1 μA emission current.

Omegatrons are small, compact instruments with glass envelopes which can be directly sealed to electronic devices, high vacuum systems, etc; when the magnet is removed, they may be thoroughly baked. There are several commercial versions available. They have wide-scale application in residual gas analysis. For many references on both theory and applications the reader is referred to reference 28.

4. Syrotron

The operation of the Syrotron (trademark of Varian Associates, after the Greek word for "sweep") is based, like that of the omegatron, on the principle of ion cyclotron resonance. All basic equations of the previous section apply.

A schematic drawing of the Syrotron is shown in Figure 3-18 (35). Here ion source and analyzer sections are separated, and the ions produced in the source are moved into the region of the analyzing fields by the action of a small static electrical field at a right angle to the magnetic field. In the analyzer section there are three mutually perpendicular fields: a magnetic field, an oscillating electric field, and a static electric field. With the static electric field within the analyzer region (trapping fields) one may change from positive to negative ion analysis. The importance of the separate ion production is that resolution- and sensitivity-limiting space charge effects are reduced. In omegatrons the resonance conditions are disturbed by the presence of both electrons (at energies just above the ionization potential) and the ions themselves (at higher electron energies) within the analyzing area.

The ions slowly ($\sim$10 msec) move from the source into the analyzer region, and in the absence of the rf field the total ion current may be measured on a suitable electrode (Fig. 3-18). When the rf field is "on," total ion current less resonant current is measured. The mass spectrum is scanned either by varying the frequency of the rf field or by sweeping the magnetic field at fixed frequency. The measurement of the resonating ions is not based upon the collection of the ions on a collector electrode, as in the case of the omegatron, but upon the absorption of the rf energy implied by the expanding motion of the ions. When the natural cyclotron

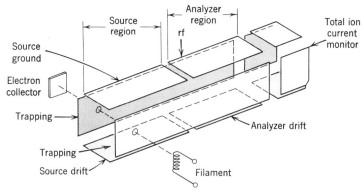

Figure 3-18. Syrotron. (Courtesy of Varian Associates.)

frequency of an ion becomes equal to the rf exciting frequency, the ion will absorb energy from the rf field and will start to move on an Archimedes spiral. It is stated (35) than an individual ion of mass 100, at a magnetic field of 100 kG, absorbs about 10^{-17} joules of rf energy in reaching a radius of 1 cm. The energy absorbed by the ions is measured by making the analyzer a part of a tuned oscillator circuit of high Q. At the ion resonance point there is an increase in the power lost by the circuit which leads to a decrease in the value of Q, which in turn decreases the level of oscillation. This decrease is plotted as a function of the magnetic field. In order for ac amplifiers to be employed, either the ion beam or the magnetic field must be modulated. The ion beam may be modulated by turning the ion beam, the drift voltages, or the trapping voltages on and off. The resulting ion bunches will modulate the amplitude of the rf oscillator, providing the means to present the mass spectra in a conventional graphic form after ac amplification and demodulation. When the magnetic field is modulated by a superimposed sine wave, the spectra are presented in a derivative fashion.

In evaluating performance, one must consider, as with all other mass spectrometers, the conditions of operation and the objectives of the particular application. A resolution of 1000 (50% width) is readily achievable in the Syrotron at mass 200. The magnetic field dispersion (G/amu) is from 50 to 500. At maximum magnetic field (14,000 G) the mass range may be extended to 280. Since the Syrotron mass spectrometer has been available for a relatively short time, little is known about practical applications. It appears, however, that the instrument will find considerable interest, particularly in ion–molecule reactions where the relatively long (10 msec) lifetime of the ions, combined with almost 100% detection efficiency, can be utilized advantageously.

C. Time-of-Flight Spectrometers

In time-of-flight mass spectrometers the time required for an ion to travel a predetermined distance is measured. Starting with a monoenergetic ion beam, the time of flight is proportional to the square root of m/e (eq. 2-25). The ion motion may be any of the three types mentioned, and instruments for each type have been described and constructed (Table 3-1).

1. Ion Velocitron

The "pulsed" mass spectrometer of Stephens and the ion "velocitron" of Cameron and Eggers (36) are direct linear time-of-flight instruments consisting of a source which provides short pulses of ions homogeneous in energy or momentum, a long (about 3 m) evacuated drift tube, and a detector connected to a pulse amplifier with oscilloscope readout; no magnet is needed.

The ion source releases microseconds long ion bursts at milliseconds intervals. Ions of a given energy traverse the drift tube with a velocity given by equation 2-6, while the time of flight is given by equation 2-25. The time of flight for singly charged ions of constant momentum, p, is dm/p, where d is the length of the tube and m is the mass. The difference in the transit time for two ions of different masses is proportional to the product of d and the difference of the square roots of the two masses. When a particular ion packet reaches the collector, the latter is sensitized for a period Δt and the resolution becomes proportional to $t/\Delta t$ for constant-energy ions. For ions having constant momentum the resolution is double that of the constant-energy type. However, the influence of the thermal energies of the ions is the true limiting factor in resolution. The amplified signal is displayed on the vertical deflection plates of the oscilloscope. The resolution of the early instruments was very poor. These instruments are now completely superseded by more complex ones offering higher resolution.

2. Bendix TOF Mass Spectrometer

In 1955 Wiley and McLaren (37) described an improved version of the Cameron-Eggers spectrometer; their instrument later became commercially available from Bendix Corporation. In what follows the basic operation of the Bendix TOF instrument is described (Fig. 3-19).

Ionizing electrons, obtained from a heated tungsten or rhenium wire, enter the ionizing region through a control grid and a collimator slit. The control grid is negatively biased and electrons can only pass through when a positive pulse of 0.1–1.5 μsec duration is applied at the beginning of a cycle. The collimator has two roles: First, it collimates the electron beam. Second, the potential difference between the collimator slit and the filament determines the energy of the electron beam. The collimator is normally grounded; thus, electron energy is varied (0–100 V) by adjusting the dc potential on the filament. The electrons travel through the ionization region toward the electron trap, and ionization and dissociation occurs in the same way as in conventional impact sources (Section 4-II). A weak magnetic field assists in electron collimation.

The second step in the cycle, after the interruption of the electron beam, is the pulsing of the ion *focus grid* to about −250 V. This pulse

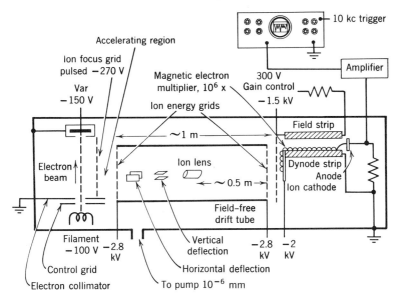

Figure 3-19. Bendix time-of-flight mass spectrometer. (After R. W. Kiser, *Introduction to Mass Spectrometry and Its Applications*, Prentice-Hall, Englewood Cliffs, N. J., 1965, p. 68.)

lasts for 1.5–5 μsec and during this time positive ions are removed from the ionization region and are drawn into an acceleration zone. The acceleration grid, located about 1.6 cm away from the electron beam, has a potential of −2.8 kV with respect to ground. The accelerated ion beam is next projected into the field-free drift tube (1–2 m long) and this is the place where mass separation takes place. Ions of different mass are separated because of the mass dependency of their velocities, as shown in equation 2-24. Light ions arrive at the collector first, followed by the heavier ones. Since the distance between arriving peaks is of the order of nanoseconds, an amplifier system with a bandwidth of about 100 Mc/sec is required. Description of the Goodrich-Wiley magnetic electron multiplier used in the Bendix instruments is given in Section 5-I-C. The mass spectrum is obtained in the form of ion current versus arriving time. It is displayed as a voltage waveform on an oscilloscope, the horizontal deflection of which is synchronized with and triggered by the ion drawout pulse. Since this is about 10 kc/sec, one obtains 10,000 complete mass spectra per second. Any portion of the spectrum may be viewed in detail by properly adjusting the horizontal sweep speed controls on the cathode ray oscilloscope and the continuously variable delay on the spectrometer between the ion acceleration pulse and the scope trigger. The spectra may also be displayed on a multiple channel analog scanning readout (Section 5-I-C).

The focusing action in the time-of-flight spectrometer is complex (28,37). Focusing is required, first, because of the initial space distribution of the ions (space resolution) and second, because of the initial kinetic energy spread (energy resolution). Those ions farther away from the -270 V grid fall through a larger potential during the process of ion ejection than those nearer the grid. Thus the trailing ions have a chance to eventually overtake those in front due to the greater velocity they acquire during acceleration. This "crossover point" may be made to coincide with the collector (i.e., focus) by the proper adjustment of the fields in the ionizing and accelerating regions. Velocity focusing is improved by making the final velocities of the ions large compared to their initial velocities and by introducing a time lag between ion formation and acceleration.

Many modifications have been made on the Bendix instrument during the last few years (38). Focusing conditions, and consequently resolution, can be improved markedly by operating the source in a continuous mode. This modification requires the addition of another grid between the drawout and acceleration grids, together with changes in the physical spacings between backing plate, electron beam axis, and the first grid. The energy focusing may be modified this way so that less energetic ions are now trapped in the potential well of the electron beam, while the more energetic ones are drawn to the backing plate where they are neutralized. When the drawout pulse is applied to the first ion grid, the electron beam is deflected away from the ionization region, and the ions are drawn out. Since an ionization current as high as 100 μA can be used in the continuous mode, as compared to 1 μA in the pulsed mode, significant increase in sensitivity results—together with an increase in resolution due to the better focusing.

Other modifications were directed to improve accuracy by increasing the stability of the (rather complex) electronic circuits, particularly the electron multiplier. Repetition frequency can now be extended to 100 kc/sec when needed, and much development work has been done to improve the recording of single-cycle spectra at high repetition rates.

Time-of-flight mass spectrometers are mechanically simple and quite rugged, and no magnet is required. Resolution depends on temporal rather than geometric factors, and improvements are made with practically every new discovery in pulsed electronic circuitry. With present commercial instruments resolution of 300 is readily attainable (1% overlap of adjacent peaks definition), and the mass range is up to 1500 amu. Sensitivity is better than 10^{-9} torr and may be as high as 10^{-12} in certain applications.

With recent improvements and with the many available accessories, TOF mass spectrometers are now widely used, particularly in chemical physics. Their speed is well utilized in the study of fast gas-phase reactions, explosions, flames, etc. Many uses are described in Section 12-III.

3. Chronotron and Mass Synchrometer

The orbital period, t_c, of an ion circling in a uniform magnetic field was given by equation 2-28 as

$$t_c = 656(M/B) \text{ } \mu\text{sec} \tag{3-30}$$

where M refers to the weight of the ions in amu and B is magnetic field strength in gauss. It is seen that while t_c is proportional to the mass, it is independent of velocity and initial direction. This relation makes it possible to separate and determine masses by measuring orbital periods in a given magnetic field. According to the equation, the time-of-flight difference between two masses 1 amu apart is 6.56 μsec, regardless of the actual masses. The time of flight of an ion of mass 150 in a magnetic field of 100 G is about 1000 μsec. Since t_c increases with the size of the ions in a given magnetic field, the technique appears useful for heavy atoms. The mass scale is linear.

The first instrument of this type was proposed by Goudsmit in 1948; later it was named "chronotron" (39). This is a time-of-flight instrument where the ion motion is circular periodic in a magnetic field. Ions from the source are drawn out by a very short (½ μsec) voltage pulse, and start to move in helical paths in a magnetic field of large volume and low intensity. The collector, or a secondary multiplier, is located directly beneath the source and the amplified current is displayed on the vertical plates of an oscilloscope. The measurement consists of determining the time difference between, say, the second and eleventh revolutions. The accuracy of time measurement is about 0.01 μsec, making it possible to determine masses with an accuracy of 10^{-3} amu. The instrument was used to measure the masses of heavy elements, e.g., ^{208}Pb and ^{209}Bi.

The *mass synchrometer* of Smith (40) is a circular periodic time-of-flight mass spectrometer with a phase relationship introduced. The principle involved is simple: at a certain stage along the orbit the ion beam is velocity modulated at rf frequency, and the orbital period is measured in terms of the modulation frequency. The synchrometer (so named because of its resemblance to the synchrotron accelerator) differs from the chronotron in that the ions move in a plane perpendicular to the magnetic field and also in the phase relationship mentioned.

As shown in Figure 3-20, a pulse of ions (of limited mass range) is in-

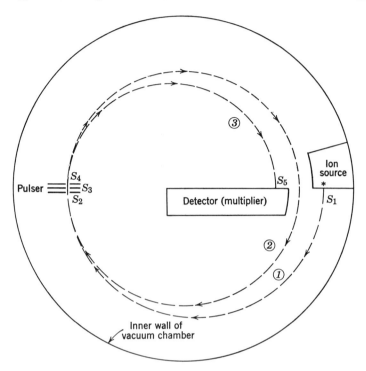

Figure 3-20. Mass synchrometer. (After L. G. Smith in *Mass Spectrometry in Physics Research*, Natl. Bur. Std. Circ. 522, Washington, D.C., 1953, p. 117.)

jected into a uniform magnetic field at right angles to the direction of the field. After describing one-half an orbit, the ion beam is led through a "pulser" consisting of three slits; the middle slit is connected to a pulse generator, the other two are grounded. The first pulse on S_3 is such that the ions are forced to move on path *2* thus missing the source upon completion of the first revolution; this is achieved by negative rectangular pulses of 1 μsec duration. After n orbits, i.e., after an interval of nt_c, a second pulse on S_3 decelerates the ions still in orbit, forcing them to describe path *3* onto the detector through slit S_5. Mass is determined from equation 3-30.

Smith and Damm (41) further improved the synchrometer by introducing additional slits and using rf modulation to increase beam intensity. They also developed the peak-matching technique (Section 10-III) for precise mass measurement. Resolving power as high as 25,000 (half-width) has been reported with the mass synchrometer for all mass numbers below 250. Accuracy of mass measurement is 1 part in 10^7.

D. Path-Stability Spectrometers

Path-stability mass spectrometers depend on an rf field to modify the motion of the ions so that only those ions having a certain narrow range of velocities, phases, or masses can remain within (or leave) a prescribed region of space. The paths described by the ions may again be linear direct, linear periodic, or circular periodic.

An example of the velocity-stability, linear direct path mass spectrometer is the design proposed by Smythe and Mattauch some 35 years ago (42). In this instrument a sequence of coplanar rf fields were applied at right angles to the direction of the motion of the positive ion beam in such a manner that only ions of a selected velocity could pass through the filter. The beam emerging from the filter was further analyzed by a radial electrostatic field. The importance of this instrument is historic: it was the first design employing rf fields and introducing time dependence.

1. Farvitron

A purely electrical ion resonance spectrometer is the Farvitron, designed by Tretner (43) in 1959 for the analysis of residual gases in ultrahigh vacuum systems. This instrument operates in the following manner: Ions formed in an electron impact source are injected by a suitable voltage into a dc potential well where the frequency of oscillation of the ions within the parabolic field is mass dependent. When a radiofrequency is superimposed on the dc field, ions of a particular m/e value corresponding to the given frequency will gain in the amplitude of their motion and escape the potential well to reach a cup-shaped collector electrode. The rf current to the collector is amplified, then rectified, and the dc voltage obtained is displayed on the vertical deflection electrodes of an oscilloscope. The radiofrequency is varied periodically by a 50 cps wobbler signal over the range 0.1–1.8 Mc/sec. Mass is proportional to the square of the frequency. Mass range can be varied from 2 to 250. Resolution is about 20. Sensitivity is about 5 cm peak height on the screen per 10^{-7} torr pressure. The Farvitron, which is commercially available, can only be used to obtain a general idea of the partial pressures in a vacuum system, or to follow qualitatively fast processes in a vacuum system; however, the instrument is simple and inexpensive. Several improvements have been suggested by Blauth (28).

2. Quadrupole Analyzer

In the quadrupole analyzer (mass filter), developed by Paul and coworkers in 1953 (44), mass separation is achieved solely with electric fields.

It is a path-stability spectrometer where the "quality" determining the stability of the ion paths is the specific charge. An ideal quadrupole analyzer consists of four long hyperbolic cyclinders in a square array with the inside radius of the array equal to the smallest radius of curvature of the hyperboles. In practice, as shown in Figure 3-21, this is approximated by four parallel cylindrical rods mounted precisely at the corners of a square at a distance $2r_0$ from each other, with opposite rods electrically connected. A two-dimensional quadrupole field with a potential

$$\varphi(x,y,z,t) \;=\; (U \,+\, V \cos \omega t)(x^2 \,-\, y^2)/r_0{}^2 \tag{3-31}$$

is produced along the axis of this system by superimposing an rf potential, $V \cos \omega t$, and a dc potential, U, on the rods. The potentials applied to the two pair of rods are equal in magnitude, but the dc potentials are opposite in sign and the ac potentials are shifted in phase by 180°. The result is a set of negative dc rods lying in the yz plane and a set of positive dc rods lying in the xy plane. The potential equation represents a constant dc term plus a time-varying ac term, both multiplied by a term expressing the off-axis position of the point whose potential is being determined. When x and y are equal to zero in equation 3-31, the potential is zero along the

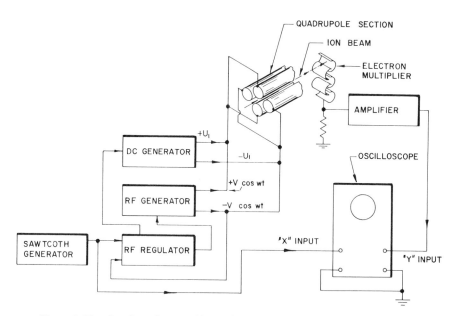

Figure 3-21.   Quadrupole mass filter. (Courtesy of Electronic Associates, Inc.)

z axis. The potential is also zero when $x = y$ (two planes intersecting at the z axis and at $45°$ to the xz and yz planes).

The source of ions is placed on one end of the quadrupole array on the z axis. When an ion beam accelerated to V_{acc} is injected through a diaphragm along the axis of the field and the voltage amplitudes are adjusted to have the relationship $U/V = 0.17$ (see later), there is a frequency at which only those ions having a specific m/e can move along the z axis to be collected at the other end of the instrument. The relationship between mass and the frequency of the rf field ($f = \omega/2\pi$) is given in practical units for singly charged ions by

$$M = 0.136V/r_0^2 f^2 \tag{3-32}$$

where M is in amu, V in volts, r_0 in cm, and f in megacycles. Ions of different masses undergo increasing oscillations and eventually lose their charges on hitting the rod electrodes. Mass scanning is accomplished either by varying the rf or dc voltage, keeping their ratio and also f constant or, less frequently, by varying the rf frequency at fixed voltages.

The equations describing the ion trajectories* in the quadrupole field can be transposed into the so-called Mathieu equations (treated in many texts on advanced mathematics)

$$(d^2f/d\phi^2) + (\alpha + 2q\cos 2\phi)f = 0 \tag{3-33}$$

by introducing the following dimensionless parameters,

$$\alpha = 8eU/m\omega^2 r_0^2; \quad q = 4eV/m\omega^2 r_0^2; \quad \phi = \omega t/2 \tag{3-34}$$

One thus obtains

$$(d^2x/d\phi^2) + (\alpha + 2q\cos 2\phi)x = 0 \tag{3-35}$$

$$(d^2y/d\phi^2) - (\alpha + 2q\cos 2\phi)y = 0 \tag{3-36}$$

The axial motion of the ions is determined by integrating $m\ddot{z} = 0$. Since $m\dot{z} = $ constant, it is concluded that the axial velocity of all ions is constant and is not affected by the voltages applied to the rods. The magnitude of the axial velocity is determined by V_{acc}, the ion acceleration voltage.

As the ions proceed with constant velocity in the z direction, they also perform oscillatory motion in the x and y directions. The amplitude of these oscillations is determined by the stable solutions of equations 3-35

---

* $m\ddot{x} + 2e(U + V\cos \omega t)(x/r_0^2) = 0$

$m\ddot{y} - 2e(U + V\cos \omega t)(y/r_0^2) = 0$

$m\ddot{z} = 0$

and 3-36 for certain values of a and q. A stability diagram (Fig. 3-22) shows that for certain values of a and q the oscillations are unstable, i.e., the amplitude of the oscillations increases with time until the ions finally hit the rods and lose their charges. The values of a and q for which the solutions of both Mathieu equations are simultaneously stable define a region of stable oscillations. The conditions for this are independent of both the initial velocity and position of ions and the phase of the rf at the instant of their injection into the quadrupole field.

An operating line is selected by establishing a constant a/q ratio. At $a = 0.237$ and $q = 0.706$ one obtains $V = 7.219Mf^2r_0^2$ and $U = 1.212$-$Mf^2r_0^2$ from which

$$U/V = 0.1678 \qquad (3\text{-}37)$$

and the working line touches the apex of the stable region. At this point the resolution is infinity but the intensity is zero. Thus, the operating line must be selected such that the U/V ratio is only in the vicinity of 0.17 (variable from 0 to 0.2) and the stable region is cut near the top. The mass range that passes through can be determined from the two points where the line cuts the boundaries. In the region of high resolution, peaks are triangular, whereas at low resolution (smaller a/q) peaks have trapezoidal shapes.

In order for mass separation to occur at all, a minimum number of oscillations is required. In other words, the ions must spend a sufficiently long period of time inside the quadrupole field to achieve the removal of

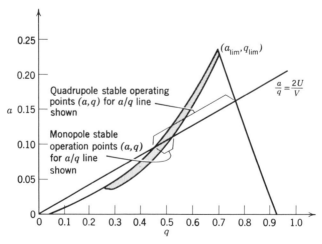

Figure 3-22. Stability diagram for quadrupole and monopole spectrometers (49).

the unstable particles. This requirement sets an upper limit to the ion acceleration voltage, V_{acc} (volts),

$$V_{acc} \leq 4.2 \times 10^2 \, L^2 f^2 \Delta M \qquad (3\text{-}38)$$

where L (meters) is the length of the quadrupole lens and ΔM the separable mass difference. Apart from this requirement, the energy range of the ions is not critical and simple, highly efficient ion sources may be employed. The angular divergence of the injected ions, for which there is also a maximum permissible value, is set by the injection-aperture diameter, a,

$$a \approx (\Delta M / M) r_0 \qquad (3\text{-}39)$$

At the collector end no defining slit is necessary.

There are strict mechanical and electrical requirements for good resolution. The electrode rods (stainless steel) must have extremely uniform cross section and must be positioned parallel to each other with a tolerance of only a few microns. Typical mechanical dimensions in a quadrupole analyzer designed for residual gas analysis are: rod diameter 8 mm, field diameter 7 mm, rod length 200 mm. A second problem is to provide the required potentials with good accuracy and stability and with fine adjustments. Typical electrical data are: rf amplitude 0–800 V, dc voltage 0–140 V, rf frequency 3–4 Mc/sec, maximum rf output 30 W. Stabilization of 1 part in 10^5 for the amplitudes and 1 part in 5×10^3 for frequency are minimum requirements. Electron multipliers are normally employed for high sensitivity and fast scanning. Spectra may be displayed on an oscilloscope, an X–Y plotter, an oscillograph, or a fast strip chart recorder (Section 5-I-C). A quadrupole analyzer is shown in Figure 3-23.

As mentioned already, transmission and resolution can be varied by changing the U/V ratio of the dc and rf voltages. In a typical instrument for residual gas analysis, a maximum resolution of 100 is achieved at a 10% transmission level with a sensitivity of 10^{-5} A/torr using a dc amplifier. For 100% transmission, the resolution must be lowered to about 35. The partial pressure detection limit may be lowered into the 10^{-12} torr range with electron multipliers.

The main advantages of quadrupole analyzers are: (*1*) independence (below a limiting value) from the energy distribution of the ion beam, (*2*) high transmission rate (particularly at low resolution) because of the continuous focusing action along the field, resulting in high sensitivity, and (*3*) opportunity for fast scanning.

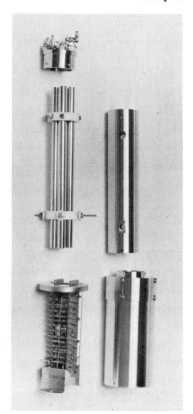

Figure 3-23. Picture of quadrupole analyzer, 5-in. mass filter. *A*, ionizer (large axial beam); *B*, quadrupole mass filter in stainless steel can (rods supported by ceramic spiders); *C*, electron multiplier (14-stage CuBe); *D*, mounting flange with feedthroughs; *E*, electrical connectors; *F*, ceramic spacers for leads; *G*, leads for ionizer voltages; *H*, leads for rf and dc voltages. (Courtesy of Electronic Associates, Inc.)

Paul and co-workers reported resolutions as high as 1500 in a new instrument (44); they also constructed a large instrument for precision mass measurement (45). Brubaker and Tuul (46) investigated the use of quadrupole instruments for satellite use and made performance studies. There are now several commercial versions available (Section 8-III) featuring resolution up to 500. Quadrupole instruments are becoming increasingly popular, and considerable evaluation and development work is in progress in many laboratories.

3. Monopole Analyzer

In 1963 von Zahn (47) designed a mass spectrometer which is of the quadrupole type but consists of only two electrodes, one of a circular cross section and the other of right angle. As shown in Figure 3-24, one-quarter of a quadrupole field is set up ("monopole") by superimposing dc and rf voltages. Although the instrument is a development of the quadrupole mass filter, its properties have been shown to be considerably different. Ion trajectories in the monopole spectrometer have been calculated by Lever (48). The basic equations for the potential and the motion of the ions are similar to those of the quadrupole.

Only those ions with a and q values placing them within a narrow band on the right side of the y-stability curve (Fig. 3-22) can reach the collector.

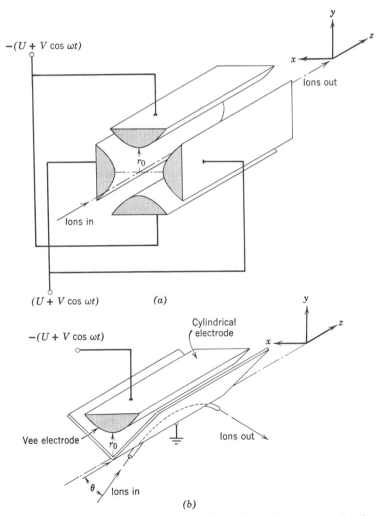

Figure 3-24. Quadrupole and monopole configurations compared (48).

For an ion to pass through, y must be larger than $|x|$, and the field length must be less than one-half the length of the total beat. About 50% of the injected ions hit the right-angle electrode in one or two rf periods and become lost.

The U/V ratio does not necessarily have to be kept constant, and operation is not restricted to the apex region of the stability curve. Relatively good resolution is achieved over a wide range of a and q. Mass scanning is accomplished by varying the rf frequency while keeping U, V, and the injector voltage (V_{acc}) constant. These features are believed to result in a reduction of electronics requirements as compared to the quadrupole mass spectrometer.

The monopole mass spectrometer is a relatively new instrument and there is considerable controversy at the present time concerning its usefulness in comparison with the quadrupole instrument. A commercial version is available (Section 8-III). Performance characteristics have been discussed by Hudson and co-workers (49).

References

1. Thomson, J. J., *Rays of Positive Electricity and Their Application to Chemical Analyses*, Green & Co., London, 1913.
2. Aston, F. W., *Mass Spectra and Isotopes*, Arnold, London, 1942.
*3. Dempster, A. J., *Phys. Rev.*, **11**, 316 (1918).
4. Dempster, A. J., *Phys. Rev.*, **18**, 415 (1921); **20**, 631 (1922).
5. Bainbridge, K. T., *Phys. Rev.*, **43**, 103 and 1056 (1933); *J. Franklin Inst.*, **215**, 509 (1933).
*6. Bainbridge, K. T., *Phys. Rev.*, **44**, 123 (1933).
7. Nier, A. O., *Rev. Sci. Instr.*, **11**, 212 (1940).
*8. Nier, A. O., *Rev. Sci. Instr.*, **18**, 398 (1947).
9. Graham, R. L., A. L. Harkness, and H. G. Thode, *J. Sci. Instr.*, **24**, 119 (1947).
10. Hagstrum, H. D., *Rev. Sci. Instr.*, **24**, 1122 (1953).
*11. Inghram, M. G., and R. J. Hayden, *A Handbook on Mass Spectroscopy*, Nuclear Science Series, Rept. No. 14, National Research Council, Washington, D.C., 1954.
12. White, F. A., and T. L. Collins, *Appl. Spectry.*, **8**, 169 (1954); White, F. A., T. L. Collins, and F. N. Rourke, *Phys. Rev.*, **101**, 1786 (1956).
13. White, F. A., F. N. Rourke, and I. C. Sheffield, *Appl. Spectry.*, **12**, 46 (1958).
14. White, F. A., and L. Forman, *Rev. Sci. Instr.*, **38**, 355 (1967).
15. Freeman, N. J., and N. R. Daly, *J. Sci. Instr.*, **44**, 956 (1967).
16. Bleakney, W., and J. A. Hipple, *Phys. Rev.*, **53**, 521 (1938).
*17. Robinson, C. F., and L. G. Hall, *Rev. Sci. Instr.*, **27**, 504 (1956); also Voorhies, H. G., C. F. Robinson, L. G. Hall, W. M. Brubaker, and C. E. Berry, in

Advances in Mass Spectrometry, J. D. Waldron, Ed., Pergamon, New York, 1959, p. 44.

18. Dempster, A. J., *Proc. Am. Phil. Soc.*, **75**, 755 (1935); Duckworth, H. E., *Rev. Sci. Instr.*, **4**, 532 (1950).

19. Isenor, N. R., R. C. Barber, and H. E. Duckworth, *Can. J. Phys.*, **38**, 819 (1960); R. C. Barber, R. L. Bishop, L. A. Cambey, W. McLatchie, and H. E. Duckworth, *Can. J. Phys.*, **40**, 1496 (1962).

20. Bainbridge, K. T., and E. B. Jordan, *Phys. Rev.*, **50**, 282 (1936).

*21. Mattauch, J., *Phys. Rev.*, **50**, 617, 1089 (1936).

22. Mattauch, J., and R. Herzog, *Z. Physik*, **89**, 786 (1934).

23. Herzog, R., and V. Hauk, *Z. Physik*, **108**, 609 (1938); *Ann. Phys.*, **33**, 89 (1938).

*24. Duckworth, H. E., and S. N. Ghoshal, in *Mass Spectrometry*, C. A. McDowell, Ed., McGraw-Hill, New York, 1963, Chapter 7; also I. Takeshita, *Z. Naturforsch.*, **20a**, 624 (1965); **21a**, 9, 14 (1966); *Rev. Sci. Instr.*, **38**, 1361 (1967).

*25. Johnson, E. G., and A. O. Nier, *Phys. Rev.*, **91**, 10 (1953).

26. Nier, A. O., and T. R. Roberts, *Phys. Rev.*, **81**, 507 (1951).

*27. Quisenberry, K. S., T. T. Scolam, and A. O. Nier, *Phys. Rev.*, **102**, 1071 (1956).

*28. Blauth, E. W., *Dynamic Mass Spectrometers*, Elsevier, Amsterdam, 1966.

29. Bennett, W. H., *J. Appl. Phys.*, **21**, 143 (1950); also in "Mass Spectroscopy in Physics Research," *Natl. Bur. Std. Circ.* **522**, 111 (1953); Redhead, P. A., *Can. J. Phys.*, **30**, 1 (1952); Readhead, P. A., and C. R. Crowell, *J. Appl. Phys.*, **24**, 331 (1953).

30. Váradi, P. F., L. G. Sebestyén, and E. Rieger, *Vakuum Technik*, **7**, 13 (Part I), 46 (Part II) (1958).

*31. Sommer, H., H. A. Thomas, and J. A. Hipple, *Phys. Rev.*, **82**, 697 (1951).

32. Alpert, D., and R. S. Buritz, *J. Appl. Phys.*, **25**, 202 (1954).

33. Berry, C. E., *J. Appl. Phys.*, **25**, 28 (1954); McNarry, L. R., *Can. J. Phys.*, **36**, 1710 (1958); Schuchhardt, G., *Vacuum*, **10**, 373 (1960); Brodie, I., *Rev. Sci. Instr.*, **34**, 1271 (1963); Petley, B. W., and K. Morris, *J. Sci. Instr.*, **42**, 492 (1965).

34. Lichtman, D., *J. Appl. Phys.*, **31**, 1213 (1960); Marklund, I., and H. Danielsson, *Rev. Sci. Instr.*, **37**, 319 (1966).

35. Note in *Chem. Eng. News*, July 21, 1865, p. 55; *Technical Information Bulletin*, Varian Associates, Palo Alto, Calif., Fall 1965.

36. Stephens, W. E., *Phys. Rev.*, **69**, 691 (1946); Cameron, A. E., and D. F. Eggers, *Rev. Sci. Instr.*, **19**, 605 (1948); also Katzenstein, H. S., and S. S. Friedland, *Rev. Sci. Instr.*, **26**, 324 (1955).

*37. Wiley, W. C., and J. H. McLaren, *Rev. Sci. Instr.*, **26**, 1150 (1955); Wiley, W. C., *Science*, **124**, 817 (1956); Harrington, D. B., in *Encyclopedia of Spectroscopy*, C. F. Clark, Ed., Reinhold, New York, 1960, p. 628.

*38. Damoth, D. C., in *Advances in Analytical Chemistry and Instrumentation*,

Vol. 4, C. N. Reilley, Ed., Wiley, New York, 1965, p. 371; also in *Mass Spectrometry*, R. I. Reed, Ed., Academic Press, New York, 1965, p. 61.

39. Goudsmit, S. A., *Phys. Rev.*, **74**, 622 (L) (1948); Hays, E. E., P. I. Richards, and S. A. Goudsmit, *Phys. Rev.*, **84**, 824 (1951).

40. Smith, L. G., *Rev. Sci. Instr.*, **22**, 115 (L) (1951); *Phys. Rev.*, **81**, 295 (1951); **85**, 767 (1952).

41. Smith, L. G., and C. C. Damm, *Phys. Rev.*, **90**, 324 (L) (1953); **91**, 481 (1953); *Rev. Sci. Instr.*, **27**, 638 (1956).

42. Smythe, W. R., and J. Mattauch, *Phys. Rev.*, **40**, 429 (1932); Smythe, W. R., *Phys. Rev.*, **45**, 299(1934); Hintenberger, H., and J. Mattauch, *Z. Physik*, **106**, 279 (1937).

43. Tretner, W., *Z. Angew. Phys.*, **11**, 395 (1959); **14**, 23 (1962).

*44. Paul, W., and H. Steinwedel, *Z. Naturforsch.*, **8a**, 448 (1953); Paul, W., and H. Raether, *Z. Physik*, **140**, 262 (1955); Paul, W., H. P. Reinhard, and U. von Zahn, *Z. Physik*, **152**, 143 (1958).

45. Paul, W., and U. von Zahn, *Phys. Verhandl.*, **12**, 222 (1961).

46. Brubaker, W. M., and J. Tuul, *Rev. Sci. Instr.*, **35**, 1007 (1964).

*47. Zahn, U. von, *Rev. Sci. Instr.*, **34**, 1 (1963).

48. Lever, R. F., *IBM J. Res. Develop.*, **10**, 26 (1966).

49. Hudson, J. B., and R. L. Watters, *IEEE Trans. Instr. Meas.*, **IM-15**, 94 (1966); Grande, R. E., R. L. Watters, and J. B. Hudson, *J. Vac. Sci. Tech.*, **3**, 329 (1966).

Chapter Four

Ion Sources

I. GENERAL CONSIDERATIONS

There are numerous methods for ion production and their relative importance is continuously changing as new methods are developed and old ones revitalized. Mass spectrometric techniques are often described with reference to the source technique employed, e.g., spark source technique for solids analysis. The choice of the appropriate source is dictated by the nature of the sample to be investigated (solid, gas, quantity available, etc.) and the kind of information sought (coverage, sensitivity required, bulk versus surface analysis, etc.).

From an instrumental point of view, the functions of an ion source are to produce as many ions as possible from the neutral particles present and to form, shape, and eject an ion beam that is suitable for entrance into the analyzer. From an analytical point of view, the main objective of a source is to produce an ion beam the composition of which accurately represents that of the sample. Ion sources have the following important characteristics: (*1*) energy spread, (*2*) sensitivity, (*3*) ionic species produced, (*4*) background and memory, (*5*) mass discrimination, and (*6*) ion current stability and noise.

Perhaps the most important characteristic of an ion source is the *energy spread*, in eV, of the ion beam it produces. This determines whether a single- or double-focusing analyzer is required for mass analysis. For an energy spread less than 10 eV a simple magnetic filter is adequate, but a larger spread necessitates a double-focusing analyzer. The lower limit in energy spread is set by the Boltzmann distribution of thermal motion which is about 0.1 eV at a temperature of 1000°K.

Source sensitivity or *yield* may be defined as the ratio of ions produced per neutral sample atom or molecule introduced. When the sample is a gas or a vapor, the number of neutral particles present is characterized by the sample pressure in the source. In the analysis of solids, as discussed later in this chapter, it may be difficult to estimate the number of neutral particles available for ionization, although weight loss may be used as a starting point. In certain cases, such as in electron impact and thermal ionization sources, the total number of ions formed may be *predicted* by utilizing such data as ionization potentials, ionization cross sections, and such experimental parameters as absolute temperature of sample, ionizing electron current, and ionizing path of electrons (eqs. 4-2 and 4-3). In any event, the total number of ions leaving the source may be determined by *measuring* the positive ion current at the object slit. Many instruments have provisions for such measurements. The *overall yield* in a given instrument, i.e., the yield involving the final available ion current at the collector, is naturally much smaller than the source yield, because of the transmission factor of the analyzer and the response factor of the detector.

Source sensitivity must carefully be considered when selecting a source for a particular problem. When *individual* elements are analyzed, the main objective is to select a source which provides high ionization efficiency for the element in question. For example, thermal ionization is extremely efficient for certain elements, and quite inefficient for others (Section 4-III). In the analysis of mixtures, the composition of the ion beam arriving at the collector may not be representative of the original sample when the efficiency of ionization is greatly different for different components in the mixture. When yields for different components (e.g., a trace element and the matrix) are compared, *relative sensitivity coefficients* are obtained. Such coefficients are usually determined in calibration runs. Ion sources are sometimes characterized by the term *selectivity*, making a qualitative statement about relative sensitivities. When all sensitivities lie within a factor of 3–5, the source is said to have a wide coverage.

The knowledge of the *ionic species produced* is important in the evaluation of the mass spectra. In trace analysis, for example, a spectrum consisting solely of singly charged species of the parent molecules would be desirable for maximum sensitivity and minimum interference. In organic structure determinations, on the other hand, the presence of fragment ions is often essential.

The term *source background* refers to all ions produced which do not originate from the sample. Residual gases in the vacuum system, gases desorbed from construction materials (e.g., ions from a hot sample container), etc. may significantly alter the composition of the ion beam or they may set a poor sensitivity limit. Low background is mostly a matter of good vacuum practice and proper programming of the sequence of analyses. The same considerations apply to the so-called *memory effects*, i.e., cross contamination between successive samples.

Once ions are formed, the source should not exhibit *mass discrimination*. Mass discrimination results from the fact that the various electrical and magnetic fields present in the source do not act in identical fashion on ions of different masses. The electrode systems of ion sources are usually designed in an empirical way with emphasis on highest sensitivity. Relatively little work has been reported on the ion optics of sources (1).

The importance of good ion current stability and low electrical noise is self-evident. Table 4-1 summarizes the most salient features of the frequently employed ion sources. These sources are discussed individually in this chapter. Description of the basic principles and experimental designs is followed by a discussion of performance characteristics, merits and limitations, and fields of applications.

II. ELECTRON BOMBARDMENT

A. Principle

The most widely used source of ions in mass spectrometry is the classical electron bombardment or electron impact source. As the name implies, ion formation is based on an exchange of energy during collisions between neutral gas atoms or molecules and energetic electrons. The type of ions formed and their relative amounts depend mainly on the chemical nature of the sample and on the energy of the bombarding electrons. When the electron energy reaches a certain value, characterized by ionization potential, molecular formation commences. The efficiency of ionization increases with increasing electron energy (Fig. 1-7) and reaches a maximum in the 50–90 eV range. Most impact sources are routinely operated at the 70 eV energy level. (For "low voltage" mass spectrometry see Section 10-II-E.) When 70 eV electrons are employed, multiple ionization, dissociative ionization, and fragmentation also occur and the mass spectrum becomes a "fingerprint" of the particular compound. The basis of the widespread use of impact sources in analytical applications is the fact that, as long as experimental conditions are kept constant, the mass spectra of individual compounds exhibit a remarkable constancy even in the presence of other components in a mixture.

The various types of ions that may form and the theory of mass spectra are detailed in Chapter 9. In this section instrumental problems are discussed.

Table 4-1　Summary of Ion Sources

Type	Ion energy spread, eV	Analyzer required	Sample type	Sensitivity	Current stability	Used for
Electron impact	0.1–5.0	Single	Gases, vapors of organic liquids, vapors of solids	High	Good	General purpose
Thermal ionization	0.2	Single	Inorganic solids, salts	Very high or very low	Fair	Isotope abundance; trace analysis
Vacuum discharge	1000	Double	Metals, semiconductors, insulators	Very high	Poor	Trace impurities in solids
Ion bombardment	5–100	Single (double)	Inorganic solids	High	Poor	Surface studies
Field ionization	0.5–5.0	Single	Organic vapors	Varies	Fair	Organics, no fragments; surfaces
Photoionization	0.05–0.2	Single	Gases, vapors	Medium	Good	Molecular spectroscopy
Gas discharge	100–1000	Double	Gases, vapors	High	Poor	Packing fractions; isotope separators

B. Design and Operation

Figure 4-1 illustrates the operation of a simple electron bombardment source. The sample to be investigated is introduced in gaseous form into an evacuated chamber where it is bombarded by a stream of electrons of known energy and intensity. Electrons obtained from a heated filament are accelerated by the potential difference between slit S_2 and the filament, and enter the ionizing chamber through collimator slits S_1 and S_2 which have a diameter of about 1 mm. The electrons continue their flight through the ionization region and leave through slit S_3 to be collected on an anode or trap which is kept at a potential of about 20 V higher than slit S_3 to prevent the escape of secondary electrons. A weak external magnetic field, of the order of 100 G, which is parallel to the electron beam acts as a collimator and keeps the width of the electron beam small.

The positive ions formed are separated from the electrons and removed from the ionization region by the combined effects of the "repeller voltage" and the field penetration from the accelerator section, S_4-S_5. The voltage difference (few volts per centimeter) between the repeller plate and the source case (or S_4) establishes a weak field which directs the ions into the acceleration area where they are accelerated to a few thousand volts by the potential difference between S_4 and S_5. The accelerated ion beam is next focused on the entrance of the mass analyzer. Possible orientations of neutral gas, electron beam, and ion beam are shown in Figure 4-2.

Most present-day structures are based on a design by Nier (2). The main parts of the source are the electron gun, the ionization chamber, and the

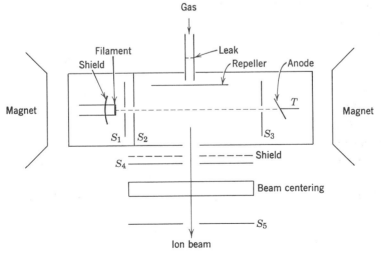

Figure 4-1. Electron bombardment source.

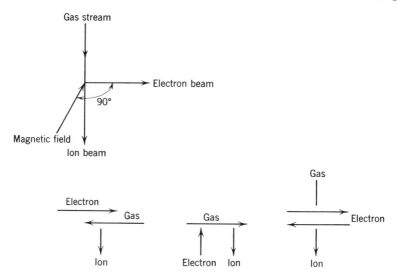

Figure 4-2.   Possible orientations of neutral gas, electron beam, and ion beam in electron impact sources.

ion optical system which forms the beam. The entire source is mounted on the analyzer.

1. Electron Gun

The purpose of the electron gun is to provide a well-defined beam of electrons of known energy and intensity. Electrons are obtained by thermionic emission from a heated filament. The choice of filament material is discussed shortly.

The emitted electrons are collimated and accelerated as they pass through slits S_1 and S_2 (Fig. 4-1), and enter the ionization chamber with kinetic energy equivalent to the potential difference between the filament and the chamber. This *electron accelerating voltage* is normally adjustable up to 100 V, the most frequently used value being 70 V.

Three types of "currents" can be measured in the ion gun in connection with the filament. The *filament current* heats the filament; the presence of this current indicates that the filament is "on," and its magnitude is often a useful guide to the "age" of the filament. The current measurable between the filament and slit S_2 is representative of the total number of electrons emitted and is called *emission current*. Most important is the *ionization current* or *trap current*, which is a measure of the number of electrons reaching the anode. This current is proportional to the number of electrons available for ionization within the chamber and, keeping other experimental parameters constant, the number of positive ions formed is directly proportional to the ionization current. Because of this proportionality, ionization current is normally maintained constant at a predetermined value by electronic regulation. Emission regulators usually operate by comparing the trap current to a reference value and then using the error signal as a feedback to adjust the filament heating supply in the proper direction to increase or decrease emission. Required stabilization is 1 part in 10^4. The linearity between positive ion current (measured at the ion collector) and ionization current usually holds in the 10–150 μA ionization current range.

To obtain an ion beam of small energy spread, the bombarding electron beam must be kept as narrow as possible so that ions are created along an equipotential surface. Collimation is accomplished by a small permanent magnet (50–300 G), aligned with the assistance of small iron blocks, which forces the electrons to describe helical paths. In fully immersed (180° deflection) instruments no auxiliary magnet is necessary. The energy spread of 1–3 eV obtained in conventional electron guns can be reduced by an order of magnitude or more by replacing the collimating slits with an elaborate energy filter or a multiplate pulsing grid system (Section 12-I-C).

The usual source of electrons is a tungsten or rhenium filament heated in vacuum. The electron current, I, which is the number of electrons leaving the filament per second from unit area, is a function of the absolute temperature, T, of the filament according to Richardson's equation

$$I = AT^2 \exp\left(-W/kT\right) \tag{4-1}$$

where W (eV) is the work function of the emitting surface, k is the Boltzmann constant (8.617×10^{-5} eV/deg), and A is a constant with a value of about 60 for clean metals. The process is called *thermionic* emission, and involves the passage of electrons through a potential barrier the "height" of which is represented by the work function.

The choice of the filament material merits some consideration. The filament has to provide an electron current of the order of 10^{-4} A, has to have low vapor pressure at elevated temperature (long life), and must be chemically inert. For many years *tungsten* filaments were used almost exclusively. A typical tunsgten filament has a diameter of 0.2 mm and requires about 4 A filament current to reach the emission temperature of 2200°K. The normal lifetime of a tungsten filament is 6–10 months. A typical filament assembly is shown in Figure 4-3.

It has been known for some time that carbon originating from the cracking of hydrocarbons reacts with the hot tungsten filaments in the ion guns. The tungsten carbide coating which forms causes variable emission characteristics due to the intermittently changing work function.

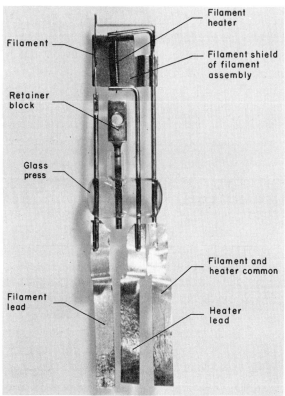

Figure 4-3.   Filament assembly. (Courtesy of Consolidated Electrodynamics Corp.).

The fluctuating electron beam will force the emission regulator to supply more or less current to the filament in its attempt to maintain the ionization current constant. This, in turn, will result in a fluctuation of the temperature of the source which may not be compensated fast enough by the source heater control. The temperature of the source is known to affect sensitivity and cracking pattern reproducibility; thus the formation of tungsten carbide is undesirable. The problem is greatly reduced by *filament conditioning:* an unsaturated hydrocarbon, such as butene or acetylene, is introduced periodically to reform the W_2C layer on the surface of the filament. Conditioning procedures must be rather rigidly adhered to as the carbide thickness depends on the duration of conditioning. The frequency of conditioning depends on the type of samples analyzed. Conditioning reduces another phenomenon, called *interference*, which consists of a change in the cracking pattern and/or sensitivity of a gas in the presence of another gas. This results in poor analytical accuracy. The cause of interference is not well understood, but it is believed to be connected both to changes in the work function of the unconditioned filament and to space charge effects.

In recent years *rhenium* has become a popular filament material (3). It does not form stable carbides, its oxides are unstable at the required emission temperatures, it does not become brittle at high temperatures as does tungsten, and no conditioning is necessary. Filament life is somewhat reduced due to the vapor pressure, which is about 150 times that of tungsten at operating temperature. Rhenium is particularly useful when oxygenated compounds or oxygen are present frequently in the source. Tungsten carbide reacts with oxygen to form tungsten oxide which shortens filament life and carbon monoxide which increases background; rhenium is relatively inert.

The possibility of thermal decomposition of the sample gas on the hot filament surface is reduced by keeping the filament as far away from the ionization chamber as possible and by replacing slit S_2 by a long channel. An alternative is to reduce the operational temperature of the cathode. Thoriated tungsten oxide cathodes or lanthanum boride (on rhenium carrier) provide adequate emission at relatively low temperatures (oxide cathodes cannot be reused after having been exposed to air).

2. Ionization Chamber

The ionization chamber is kept as gastight as possible to minimize required sample size and to reduce the influence of residual gases and variation in pumping speed. It is obvious that only vapors can be analyzed in electron bombardment sources. Elements and compounds with a vapor pressure $>10^{-1}$ mm at room temperature are admitted directly for a sample reservoir through a leak, either molecular or viscous (Section 7-2). Liquids or solids of low vapor pressure may be introduced from a heated inlet system, from a heated internal crucible placed near the source, from a Knudsen cell, or they may be evaporated directly inside the chamber. Problems with such arrangements are fractionation during evaporation, outgassing of the sample holder, memory effects, and difficulties in changing samples. Sample introduction systems are detailed in Chapter 7.

The stability of cracking patterns is influenced by the temperature of the source. Fragmentation, as expected, increases with increasing source temperature. The main source of heat in the ion chamber is the filament, radiation from which results in a temperature of about 200°C. In modern instruments source temperature is maintained at 250°C and is regulated to within a few tenths of a degree.

3. Ion Beam Formation

The formation, shaping, and acceleration of the ion beam and its ejection into the analyzer are accomplished by a number of plates and slits kept at various potentials. Almost every investigator has a favored variant.

In Figure 4-1 the ion beam is removed from the chamber by the combined efforts of a weak repeller field (a few volts per centimeter) and the field penetrating from the first plate outside the chamber. The source magnetic field also helps to separate ions and electrons as it collimates the latter while the heavier ions are only slightly influenced as they move toward S_4. Potential conditions within the source must be selected by considering whether maximum resolution or highest beam intensity is desired.

Ion acceleration is accomplished by providing, from a regulated power supply, a potential difference of a few thousand volts between plates S_4 and S_5. Additional plates and slits may be employed to control intensity, to act as collimators, and/or to provide a means for vertical displacement of the beam. The main acceleration field is often shielded from the ion chamber by a grid attached to S_4.

C. Performance Characteristics

The yield of impact sources can be evaluated by considering the parameters controlling the number of ions that form at a given pressure, P (torr). The theoretically obtainable positive ion current, I^+ (A), is given by

$$I^+ = \beta Q_i S_e I^- P \qquad (4\text{-}2)$$

where β is the efficiency of ion extraction, Q_i (cm^{-1} torr $^{-1}$) is differential ionization, s_e (cm) is the effective path length of electrons, I^- is ionization current, and P (torr) is sample pressure. The differential ionization is the number of ions formed by an electron during 1 cm of its path through a gas pressure of 1 torr. Q_i depends on the chemical nature of the sample and on the energy of the bombarding electrons; its value is in the 1–10 cm^{-1} torr^{-1} range for most gases when 70 eV electrons are employed.

The ionization current is usually limited to about 10^{-4} A due to adverse space-charge effects which result in interference. In the gas-interference test the ion current for a particular gas under given pressure is measured and then an additional amount of some other gas is introduced and its effect on the original ion current measured. For accurate analysis of mixtures, gas interference must be below 3%; this sets the limit for the ionization current at about 10^{-4} A. Interference may also result from excessive sample pressure. Maximum source pressure is 10^{-4} torr, but 10^{-5} torr is more desirable. The electron path is determined by the dimension of the source, its value normally being approximately 1 cm. The efficiency of the ion extraction system depends mainly on the "amount" of collimation required and is usually in the 1–10% range. Under the above conditions the ion current in the source is of the order of 10^{-7} A. Only a fraction of this current will reach the ion collector where the available ion current is normally in the 10^{-8} to 10^{-11} range. Since the lowest measurable current is normally at the 10^{-14} A level with electrometer detection, partial pressure analysis down to 1 part in 10^6 is possible.

Due to the low energy spread, electron impact sources require only single-focusing analyzers. The ion beam is steady and relatively strong, making it possible to employ electrometer amplifiers and achieve excellent reproducibility. The efficiency of ionization is uniform within a factor of 3 for most inorganic gases (helium and neon are notable exceptions), and within a factor of 10 for the majority of organic compounds.

Mass discrimination (4) is caused by the collimating magnetic field in which lighter ions and ions with lower energies are deflected somewhat more than heavy ions or those having higher energies. A compromise must be made: for good electron beam collimation a strong magnetic field is needed, while for lessened discrimination a weak field is desired. The importance of source discrimination is greatest in absolute isotope ratio measurements.

The ionic species produced in impact sources are predominantly singly charged. At 70 eV electron many types of fragment ions form. A large amount of information is available in the literature on the cracking patterns and relative sensitivities of organic compounds (Section 9-I).

Ionized residual gases result in *background* spectra. Peaks most frequently appearing in the background are at m/e 17 and 18 (water), 28 (carbon monoxide and nitrogen), 32 (oxygen), 41 and 43 (hydrocarbon fragments from pump oils), 44 (carbon dioxide), and 198–204 (mercury). Background may be a serious problem in trace studies for particular components, and in those cases where contributions from hot components (crucible source) cannot be avoided. Memory problems may be caused occasionally by decomposition products and compounds which are difficult to pump out. Decomposition products, if deposited, may cause beam distortion.

The normal amount of sample needed in routine analyses is of the order of 1 standard cubic centimeter for gases and a few milligrams for solids. Small amounts of gases can be handled in special instruments, and as few as 5×10^5 atoms of xenon have been analyzed (5). The limit of detection of impurities is of the order of 1 ppm. Special problems concerning trace analysis are discussed in Section 11-II. Precision achievable is $\pm 0.01\%$ in isotope abundance measurements, and ± 0.1 mole $\%$ in the analysis of gas mixtures.

Electron impact sources have many advantages: high yield, good energy homogeneity, and stable ion beam are combined with convenient sample handling and easy operation. They are certainly the most reliable of all types of ion sources, and hundreds of variations are employed in both qualitative and quantitative analysis, in isotope ratio measurements, in studies of electron impact phenomena, and in leak detectors. Many applications are described in Part II.

D. Oscillating Electron Sources

The efficiency of ionization in a simple collimated electron bombardment source depends on, among other factors (eq. 4-2), the average path length of the electrons in the ionizing region. Heil (6) described an ion gun (Fig. 4-4) in which the electron collector is put at filament potential thus becoming a kind of anticathode. The anode is a cylindrical ring, and an axial magnetic field keeps the electrons well oriented in a beam. Electrons emitted by the filament are accelerated toward the anode but are kept to the axis by the magnetic field. As they approach the anticathode, they are decelerated and reflected backwards. The oscillating electrons undergo many ionizing collisions as their effective path lengths become large. Eventually their energy becomes less and less, the effect of the collimating field upon them decreases, and they ultimately become lost on the anode. A source similar to the one above, but employing plane ring electrodes, is described by Finkelstein (7).

The efficiency of such sources may be as high as 80%. Ion energy spread is usually a few electron volts, somewhat larger than in classical sources, but still sufficiently low for single-focusing analyzers. There appears to be a current revival of the oscillating sources, particularly as a primary ion source in ion bombardment studies (Section V in this chapter) and in omegatron residual gas analyzers (Section 13-IV).

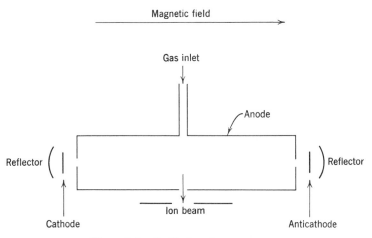

Figure 4-4.   Oscillating electron source.

III. THERMAL IONIZATION

A. Mechanism of Thermal Ionization

The thermal ionization (thermal emission, surface emission) source is based on the fact that when neutral atoms or molecules of (first) ionization potential I (eV) are heated on, or impinge upon, a hot metallic surface of work function W (eV) and temperature T (°K), there is a probability that ions will also evaporate in addition to neutral particles. The ratio of the number of positive ions formed to the total number evaporated, the efficiency of ionization, is given by the Langmuir-Saha equation,

$$n^+/n^0 = \exp\left[e(W - I)/kT\right] = \exp\left[11606\ (W - I)/T\right] \qquad (4\text{-}3)$$

where k is Boltzmann's constant and e is the electronic charge.

To obtain the ratio of the number of negative ions formed, n^-, to the neutral particles, $W - I$ has to be replaced by $A - W$, where A (eV) is the electron affinity of the substances evaporated. Thermal sources can be converted to *electron affinity* sources by merely reversing the polarity of the acceleration voltage. Affinity sources are not widely used.

An examination of equation 4-3 reveals that the thermal ionization source is highly selective. It shows that: (*a*) For higher efficiency ($n^+/n^0 > 1$) the work function of the surface must be larger than the ionization potential of the sample and the temperature should be as low as possible consistent with efficient evaporation. (*b*) For materials where $W - I$ is negative the efficiency of ionization can be very low, and the only hope for adequate sensitivity is to increase the temperature of the hot surface. This, however, is limited by the volatility of the sample which may completely evaporate before the required temperature is reached. Ionization efficiency may be as high as 100%, such as for sodium and potassium on nickel surfaces (8), or may be so low that analysis becomes impossible, e.g., for copper on tungsten. It is concluded that while the thermal source is exceptionally sensitive for the analysis of certain individual elements, it is not suitable for mixture analysis. Since the ionization potential does not differ detectably among the isotopes of an element, thermal ionization is well suited for isotopic analysis. Discrimination problems are discussed shortly.

The applicability of the technique in a given problem can be evaluated by inserting numerical values into the Langmuir-Saha equation. Work functions (and melting points) for a number of metals are tabulated in Table 4-2, while Appendix II lists ionization potentials. From the calculated ionization efficiencies one may estimate, by considering the overall transmission of the instrument, the ion current that is expected on the

Table 4-2   **Work Functions and Melting Points of Some Metals**[a]

Element	Thermionic work function, eV	Melting point, °C
Nickel	4.61	1452
Rhodium	4.80	1985
Palladium	4.99	1552
Tantalum	4.19	2996
Tungsten	4.52	3380
Rhenium	5.12	3180
Platinum	5.32	1769

[a] Data from R. C. Weast, Ed., *Handbook of Chemistry and Physics*, 48th ed., Chemical Rubber Publishing Co., Cleveland, 1967–1968.

collector. Conversely, one may calculate the required operating conditions for obtaining a desired ion current.

B. Design and Operation

Thermal sources are simple: they consist of a filament assembly and a plate–slit system for the acceleration and shaping of the ion beam. It is noted that here the term *filament* refers to a heated strip of metal whose surface is coated by, or bombarded with, the sample material. There are three basic types of filament assemblies: single filament, multiple filament, and specially prepared surfaces.

1. Single Filament

In the simplest thermal sources, a few drops of the sample solution or slurry is placed with a micropipet on a filament ribbon which is about 0.01 mm thick, 15 mm long, and 1 mm wide. The filament is next slowly heated by direct heating until the solvent is evaporated. This is usually performed outside the mass spectrometer. Next, the source assembly is reinstalled inside the mass spectrometer in such a manner that the filament is positioned directly in front of the drawing-out plate. After pumpdown, the temperature is raised and the mass spectrum scanned.

The choice of filament material is straightforward: it should have a high work function, should be a refractory material (low vapor pressure), and should be as pure as possible. Tungsten, tantalum, and rhenium are most frequently used. Platinum has a high work function (Table 4-2) but its low melting point is a disadvantage.

Single filaments are often employed in canoe- and V-shapes (Fig. 4-5a). The higher efficiency of such types is explained by considering them as closed-up triple filaments.

2. Triple Filaments

When $W - I$ is negative, one has to operate at the highest possible temperature for adequate ion intensity. For example, when $W - I = -2\,\text{V}$, the ratio n^+/n^- is only 10^{-8} at 1200°K but increases to the order of 10^{-4} when the temperature is raised to 2500°K. When single filaments are used a compromise must be made between the efficiency of ionization and the length of time for which the sample lasts.

In the sophisticated triple-filament source (Fig. 4-5b), developed by Inghram and Chupka (9), the sample is evaporated from either of two symmetrically located filaments at a relatively low temperature (long sample "life"), and the neutral vapor thus produced is ionized by a central filament which is maintained at high temperature ($\sim$2700°K) for efficient ionization. The main advantage of this arrangement is that sensitivity is increased without increased sample consumption. This technique also permits the use of a greater variety of compounds, e.g., volatile halides.

To further increase sensitivity, Voshage and Hintenberger (10) introduce the samples from a small oven heated externally by electron bombardment. Another recent development is the use of parallel filaments (Fig. 4-5c),

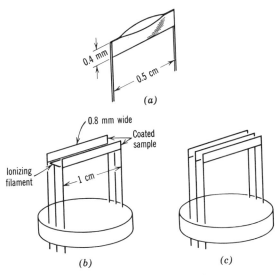

Figure 4-5. Thermal ionization sources: (a) boat-type filament; (b) triple filament; (c) parallel filaments. (After H. W. Wilson and N. R. Daly, *J. Sci. Instr.*, 40, 273 (1963)).

developed by Patterson and Wilson (11), which allows simultaneous analysis of a standard together with the unknown. In addition to increased sensitivity due to the favorable solid angle which reduces geometrical loss, discrimination effects are greatly reduced in this technique.

3. Specially Prepared Surfaces

Several techniques have been developed to increase efficiency by special treatment of the filament surface. These methods are usually restricted to specific applications. It has been reported, for example, that filaments treated with benzene prior to the coating with the sample slurry are often considerably more efficient; this is probably due to the elimination of oxide layers. Cement binders are claimed to increase ion intensity but also contribute to background. The application of fused borax is known to increase efficiency considerably in certain cases. This is attributed partly to such physical processes as wetting and adherence, and partly to chemical combination, e.g., MBO_2^+ peak, where M is the metal in question. In general, all conditioning treatments are aimed to increase the work function of the surface.

Although thermal sources are primarily used for the analysis of very small samples, high capacity and large ion currents may be desired in certain experiments. The so-called Kunsman source (12) delivers ion currents of 10^{-4} A/cm^2 magnitude for hundreds of hours and operates at only 800°C. The source is made by melting together 98% (by weight) Fe_2O_3, 1% Al_2O_3, and 1% of an alkali salt in an oxygen blast, placing a small quantity of the mixture on a platinum foil and sintering by heating to incandescence.

4. Ion Beam Formation

Ion collimation and acceleration in thermal-source mass spectrometers is performed in a more-or-less conventional manner. Figure 4-6 shows the beam extraction system of a multifilament ion source (9). The defocusing slit is typically 1.5 mm wide and 15 mm long. Other dimensions include: discrimination slit 0.75 × 15 mm, first collimating slit 0.3 mm × 15 mm, final collimating slit 0.2 mm × 15 mm, and focusing slit 2.0 mm wide consisting of two separate half-plates, thus also serving for beam centering. Of course, many other designs have been described.

Ions are usually accelerated from high positive potential (filament potential) to ground; this prevents electron bombardment ionization within the source. Filament potential is in the 2000–6000 V range and the collimating slits are normally grounded. A small differential voltage across the

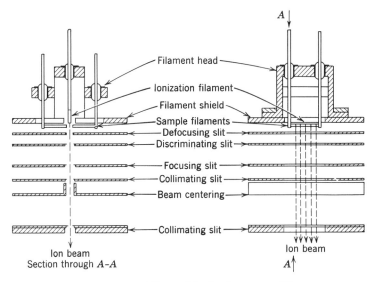

Figure 4-6. Thermal ionization source (9).

deflection plates assists in centering the beam. The resolution obtained is determined by the size of the resolving slits which are usually made adjustable.

Secondary negative ions and electrons which may form on the collimating slit due to positive ion bombardment are accelerated in the opposite direction, i.e., toward the emitting filament, where they cause additional ion emission. These peaks, known as tertiary peaks, would complicate the spectra. They can be suppressed by a suppressor grid placed in front of the collimating slit. In electron affinity sources the use of tertiary suppression is mandatory.

Practical considerations in thermal sources include provisions for easy sample changing (vacuum locks, Section VII-7), wide slits for fast pumping (contrast to the tightness of electron impact sources), proper vacuum practice and cleaning to prevent memory effects, and adequate control for the filament heating current.

C. Performance

Almost every solid element in the periodic table has been analyzed with the thermal source either in elemental or chemical compound form. Exceptions are carbon, silicon, sulfur, rhenium, and a few other elements with high ionization potential. The yield, as is evident from the basic equation, varies widely from low levels to as high as 100%.

The ion beam is nearly monokinetic (0.2 eV energy spread). The ions formed are mostly singly charged, so spectrum interpretation is relatively simple. Electrical noise is minimal; electron multipliers can routinely be employed. Background is usually low as long as good vacuum practice is followed. Sodium and potassium from the filament material are the most bothersome background peaks; hydrocarbon contribution from pump oils is relatively low. Background may further be reduced by proper preheating. Memory effects can be reduced to a minimum by proper cleaning techniques.

The major use of the thermal emission source is in the analysis of selected elements. Extremely high sensitivity and very small required sample size make this technique highly useful in many nuclear applications. Microgram quantities of samples are routinely handled, but even picogram sizes may be analyzed under favorable conditions.

Although widely used in isotope measurements, absolute isotope ratio measurements require the use of reference standards to offset discrimination. Isotope fractionation may be a serious problem, particularly for elements with low atomic weight. Correction for this is difficult, sometimes practically impossible, e.g., in $^7Li/^6Li$ determinations. Discrimination originates from preferential vaporization of the lighter species. This type of discrimination varies with time, while discrimination effects in the analyzer and detector are normally the same throughout the analysis. Thermal sources are widely used in connection with the isotope dilution technique (Section 11-III-A).

IV. VACUUM DISCHARGE SOURCES

A. Vacuum Breakdown

When the potential between two electrodes in vacuum is increased, the current slowly increases and in a *prebreakdown* region the process is reversible. As shown in Figure 4-7 (a voltage–current plot), there is a certain potential (*spark breakdown*) at which the interelectrode potential suddenly drops sharply accompanied by an irreversible current increase. This region is characterized by negative resistance. If the current is not limited by circuit resistance, it continues to increase while the voltage drops to a low value, close to the ionization potential of the vapor: the discharge becomes an arc discharge.

Methods of ion production based on vacuum breakdown in a narrow (<0.1 cm) gap between two electrodes were first developed by Dempster in the 1930's. These designs have recently been revitalized and are now

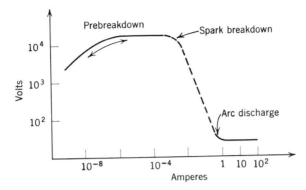

Figure 4-7. Vacuum breakdown.

widely used to analyze impurities in solids. There are three variations: radio frequency (rf) spark, vacuum vibrator, and pulsed dc arc. The basic principle is the same in all three: a potential is built up between the electrodes of the material to be analyzed until a discharge occurs. The rf spark is the most popular at the present time.

B. Design and Operation

In the rf spark source (Fig. 4-8a) an ac potential of 20–100 kV (variable) is generated in the form of short pulses of a few microseconds duration (variable). Repetition rate is also variable from single sparks to 10^4 pulses per second. The mechanism of ion formation in the spark is complex and not well understood. The process probably starts with the appearance of electrons due to field emission, and continues by vaporization of neutral atoms caused by the bomdarding electrons, followed by impact ionization. While the effective temperature in the spark may be as high at 10^4°K, the bulk temperature is kept below 500°K since the pulses are of short duration. Performance characteristics of this source, as discussed shortly, raise special demands for both analyzer and detector.

In the *vacuum vibrator* (Fig. 4-8b) a potential difference of 10–30 V is maintained between the electrodes. While one electrode is stationary the other is mechanically set in vibration and touches the stationary one once in each cycle. As the current-carrying contact is broken, the contact resistance increases and the resulting high temperature melts a small volume of the metal. The liquid bridge that forms will rupture as the electrodes move apart, and an arc will form if a potential difference of about 10 V is kept between the electrodes.

Both electrodes are stationary in the *direct current hot arc* source (Fig. 4-8c). A potential difference between the electrodes is built up by means

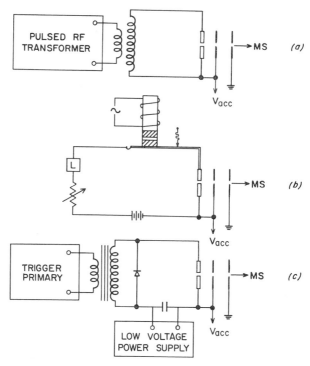

Figure 4-8. Types of vacuum discharge sources: (*a*) rf spark; (*b*) vacuum vibrator; (*c*) low voltage dc arc.

of a condenser connected across them. After spark formation the condenser is discharged and a new cycle starts. Several low-voltage sources have recently been reviewed by Honig (13), who also discusses the mechanism of ion formation in vacuum discharges.

C. Performance

The rf spark source demands an energy filter (double-focusing analyzer) due to its high kinetic energy spread ($\sim 10^3$ eV). The energy passband of double-focusing instruments is 200–600 eV, resulting in relatively low ion transmission. Moreover, according to the investigations of Woolston and Honig (14) and Franzen and Hintenberger (15), various matrix materials exhibit widely differing energy distributions and the distributions are greatly affected by such source parameters as spark gap width and rf spark voltage. To avoid analytical errors, experimental conditions in the spark source must be carefully determined and adhered to.

The energy spread of the vacuum vibrator and pulsed dc arc sources according to incomplete information presently available is at the 100 eV level for small arc currents. Although a double-focusing analyzer is still needed, the higher usable ion current is an attractive feature when high sensitivity is the objective.

In the rf spark source the most intense lines are those due to singly charged ions. The intensities of multiply charged ions (at $\frac{1}{2}$, $\frac{1}{3}$, etc. of isotopic masses) fall by a factor of 3–10 for each degree of ionization. In addition, polyatomic ions are always present in the rf spectra of such semimetallic elements as C, Si, Ge, Se. Relative intensities are normally in the order $X^+ > X_2^+ > X_3^+ > X_4^+$; carbon is an exception. Complex ions, like Fe_2O^+ and AlN^+, and molecular species such as GaP and GaP_2 are occasionally observed, always at low intensities. In the vacuum vibrator and dc arc sources the abundance of doubly and triply charged ions is greater than that of singly charged ions.

An additional feature of spectra obtained in the discharge sources is the presence of charge-exchange lines (Section 9-V) resulting from charge-transfer processes between positive ions and residual gas molecules in both the electrostatic and magnetic analyzers. They appear at odd masses, frequently forming a continuum, and are undesirable from an analytical point of view. They decrease with increased vacuum. Charge transfer lines are more intense in the dc sources due to the presence of an increased intensity of multiply charged ions.

Uniform ionization efficiency, at least within a factor of 3, has repeatedly been reported for the rf spark source. This is understandable since the ionizing energy available is enormously greater than the highest ionization potential of any element present. This property is, in fact, an extremely attractive feature of the rf spark source for trace analysis (contrast with emission spectroscopy). Deviations from uniform ionization efficiency are discussed in Section 11-III-B. There are indications that the efficiency of dc sources may be considerably higher than that of the rf type for certain hard, high boiling materials.

Poor current stability and electrical noise practically necessitates the use of photographic plate detectors which are integrating detectors (Section 5-II). An additional advantage of photoplate detectors in spark source mass spectrography is that it is possible to cover the entire mass range simultaneously. Electrical detectors may also be used for accuracy and convenience, but special circuitry is required and only individual beams can be measured at a given instant.

Memory effects can be reduced to negligible levels by careful cleaning of the ion source and by proper programming of the analyses. The limiting factors in semiquantitative trace analysis are background and interferences lines on the detector, and the level of ion current which can be delivered by the instrument to the detector (source yield, analyzer transmission). Analytical techniques are described in Section 11-III-B.

D. Merits and Limitations

The two principal advantages of spark source mass spectrography are high sensitivity and complete coverage of the elements. The limit of detection is 0.1–0.001 ppm for most elements in most matrices. An overall coverage of all elements in the periodic table is possible in a single analysis, using photoplate detection. Relative ionization efficiencies are almost the same for all elements. There is a generally established linearity with concentration. Both bulk and surface impurities can be studied, and samples as small as a few micrograms can be analyzed.

Major disadvantages are that the instrument required is complex and expensive, photoplate recording is needed, the ion pattern is relatively complex, and quantitative determinations with good accuracy are difficult to achieve. Reproducibility is normally within a factor of 3–10, although precision to 0.1 standard deviation can be obtained when suitable standards are available (Section 11-III-B).

V. ION BOMBARDMENT SOURCE

A. Principles

In the ion bombardment source a solid sample is bombarded by energetic positive ions (20–1000 eV) and as a result of the impacts both neutral and charged particles are removed from the surface. The process is called *sputtering*.* The bombarding ions, even those with relatively high energy, do not penetrate deeply into the surface and it is assumed that the secondary particle emission takes place only from the top few layers. The method is thus nondestructive and permits positive identification of species emanating from surfaces. In a qualitative fashion, the production of positive and negative ions can be predicted from the Langmuir-Saha equation: elements with a low ionization potential produce positive ions, while those of high electron negativity result in negative ions.

While the charged particles that form can be introduced directly into a mass analyzer, sputtered neutral atoms or molecules must first be ionized in a second source of a conventional electron impact type.

---

* Another kind of charged particle–surface interaction phenomenon is *secondary* electron emission, which is utilized in the electron multiplier type detectors. These detectors and the importance of sputtering in ion collector systems are discussed in Chapter 5.

B. Design and Operation

Rare gas ions (usually argon) are used as bombarding particles for their chemical inertness and because they can be removed from the lattice after bombardment. Mass and energy of the impinging ions and the angle of incidence are the defining parameters. Many source arrangements have been designed to obtain an intense *primary ion beam*. High efficiency is achieved in the oscillating electron sources (6,7), in magnetically confined discharges (16), in Penning-type arc sources (17), and in duoplasmatron sources (18). Target bombarding densities of the order of 50–200 $\mu A/cm^2$ are produced in the arc-type sources, and an order of magnitude lower in the oscillating electron types.

A typical source arrangement developed by Smith et al. (19) is shown in Figure 4-9. A bombarding rare gas ion current of 1–3 μA (current density 3–10 $\mu A/cm^2$) can be obtained for 10^{-4} torr gas pressure in the first source (Finkelstein-type electron oscillation). The gas is introduced through a bakeable valve, and the target is mounted behind the ionization region in a holder which is adjustable to permit variation of the angle of incidence. The second source is a modified Nier-type electron bombardment source.

Sputtered ions pass through the ionizing source undisturbed when the electron beam is off and the spectrum consists of secondary ions from surface contaminations on the sample and ions from the target matrix. By switching on the Nier source, both the sputtered and background gases (including reflected neutral particles of the bombarding gas) are ionized and analyzed. To study sputtered negative ions the polarity of the electric and magnetic fields is reversed. Acceleration and shaping of the ion beam is achieved by techniques similar to those in classical impact sources.

C. Performance

The relative number of neutral particles and positively and negatively charged ions (*secondary ions*) can be estimated by the Langmuir-Saha equation (eq. 4-3). Surface temperature T here refers to a highly localized "apparent temperature" which is a function of the mass and energy of the bombarding ions, and is of the order of $10^4 °K$. Sputtering sources are highly selective depending on the values of $W - I$ or $A - W$. Measurable quantities of positive ions are obtained for all elements with an ionization potential below 10 eV, and negative ions for all elements with an electron affinity in excess of 1 eV. Ion energy spread may be 5–100 eV depending on source design.

Sputtering yield, a significant performance parameter, is defined as the ratio of the number of sputtered particles to the number of incident rare

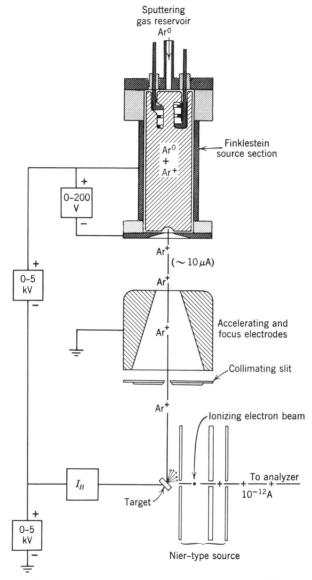

Figure 4-9.   Geometry of sputtering ion source (19).

gas ions. Another way to express the efficiency of a source is to give the ratio of the collector ion current to the bombarding ion current for a given mass at maximum peak height. Many recent studies were directed to measure yield as a function of experimental variables. A good account of present experimental trends is given by Woodyard and Cooper in a detailed description of their experiments and discussion of various yield expressions (16). The number of ions formed is of the order of 0.0001–0.1 sputtered ions per bombarding ion, depending upon the identities of the bombarding ions and the molecules on the surface, and also upon experimental conditions. The number of neutral particles sputtered is 1–10 per bombarding ion.

The resolution of a given analyzer system is considerably reduced, with respect to operation with a "normal" electron bombardment source, when ion bombardment is employed due to scattering and the energy spread of the primary ion beam. The loss may be significant, particularly with arc-type sources. There are cases where double focusing is clearly warranted (17) and recent commercial instruments specifically designed for ion bombardment studies, feature double-focusing analyzers (Section 8-III-E).

Ions produced by this source are predominantly singly charged and usually monatomic. The ion beam is quite stable and electrical noise is low.

D. Merits and Limitations

Vacuum, problems are severe in sputtering sources (20). The target region must be completely separated from the source of primary ions so that a high intensity ion beam can be focused on the target without having an excessive gas pressure caused by an overflow of un-ionized gases. Residual gases will, of course, be ionized together with the neutral sputtered electrons by the ionizing electrons. Background must be determined by a separate measurement in the Nier source with the accelerating voltage of the primary ion beam off. A more sophisticated method with which background peaks can be suppressed is the synchronous source detector (Section 11-III-C).

The strong dependence on the work function of the surface and the uncertainty of the "apparent temperature" make quantitative measurements difficult. Neutral sputtered particles are believed to be somewhat better suited for compositional analysis since discrimination is reduced. The limiting factor is background. Since the spectra consist mainly of adsorbed gases and radicals on surfaces the method is suited for studies of surfaces without significantly increasing the temperature. The technique is nondestructive and gives almost complete coverage in one experiment. Accuracy is poor.

VI. FIELD IONIZATION

Field ion mass spectrometry originated from field ion microscopy, developed by Mueller in 1951. In a field ion microscope a picture of a fine metal tip (radius of curvature $\sim$1000 Å) is produced magnified by a factor of 10^6 on a luminescent screen placed a few centimeters away. The magnified picture is produced by helium ions formed near the metal tip when a field of 10^8 V/cm is applied between the metal tip and the luminescent screen in a helium atmosphere of about 1 μ pressure. A combination of field ion formation with mass spectrometric detection was first employed by Gomer and Inghram in 1955 (21) to study ion desorption from surfaces. The possibility of field ion production from the gas phase, as suggested by Beckey a few years later, opened a new and potentially powerful field in the analysis of complex organic mixtures.

The mechanism of field ionization can only be dealt with briefly here (22). When the metal tip is positive, the strong electrostatic field results in the removal of ions of impurities adsorbed on the surface of the metal, or ions may be produced from impinging gas molecules. This process, which is basically an evaporation of positive ions over a potential barrier on the surface due to the strong field, usually results in small pulses of ions ($\sim$10^{-15} coul). Continuous ion current results when field ionization occurs in the gas phase at or near the emitter. The ion formation results from a strong interaction between the field and the outer electron shell of the neutral molecules. Electron removal (i.e., ion formation) is a consequence of the quantum-mechanical tunnel effect: the potential wall surrounding the molecules becomes narrower as they approach the anode in the strongly inhomogeneous field.

In the earlier instruments, the electric field was set up in the vicinity of a sharp emitter, a metal tip of about 5×10^{-5} cm radius (1.5×10^{-8} cm^2 emitting area), by employing a potential of the order of 10 kV between the anode and the cathode diaphragm. The cathode was placed a few millimeters away from the anode and a central aperture was provided for ion removal. The field strength, E (V/cm) is given approximately by

$$E = [V/r \log (R/r)] \qquad (4\text{-}4)$$

where V is the applied potential, r is the radius of the fine point, and R is the distance between anode and cathode.

More recently (23) Beckey replaced the fine spike anode with a platinum wire of 2.5×10^{-4} cm diameter and 5 mm long (2.2×10^{-4} cm^2 emitting area), placed parallel to the plane of the cathode diaphragm. Also, field ionization and electron impact ionization are now provided in the same source system, with extremely simple conversion from one type of operation into another. Figure 4-10 is a schematic drawing of a combined field ionization–electron impact source. Several commercial versions are now

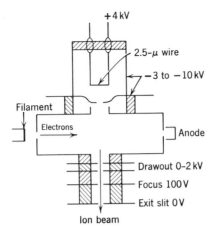

Figure 4-10.   Combined electron impact and field emission source.

available. With the spike-type emitter the ion currents obtainable are at the 10^{-12} A level, necessitating the use of multiplier detection. A factor of 100–1000 is gained with the wire-type anode, permitting the use of conventional electrometers for detection. Another advantage of the wire-type emitter is good ion current stability. Razor blades have also been employed in very simple source arrangements (24).

Since the energy homogeneity of the ions is within 1 eV, single-focusing analyzers are adequate for field ionization mass spectrometry. In field ionization, unlike under electron impact, mainly parent ions are formed since fragmentation occurs only as a secondary process near the anode surface and in ion-molecule collisions in the gas phase (Fig. 10-3). Current investigations show (Section 10-III-F) that fragmentation may be obtained under certain experimental conditions. When fully developed, the technique may be used in organic structure determinations.

The simplicity of the mass spectra under controlled conditions offers obvious applications in the analysis of complex organic mixtures (Section 11-II-F). With the recent availability of commercial field ionization–electron impact sources, the use of field ionization mass spectrometry is likely to become widespread, particularly in the petroleum industry.

VII. PHOTOIONIZATION

Photoionization is based on interaction between light quanta of sufficient energy and neutral molecules by the process

$$XY + \text{photon} \rightarrow XY^+ + e^- \tag{4-5}$$

The required wavelength, λ (in angstroms, $1 \text{ Å} = 10^{-8}$ cm) is given by

$$E = 1.2398 \times 10^4/\lambda \qquad (4\text{-}6)$$

where E is the energy of photons in eV. Since the ionization potential of most elements and compounds is in the 10–25 eV range, the wavelengths required are in the ultraviolet region.

It has been mentioned that the threshold law for single ionization is a linear function of the excess energy for electron impact. The law is a *step function* for photoionization. This is shown in Figure 4-11 where ionization efficiency curves are drawn for argon: the ion current does not increase with increasing photon energy. The photoionization curve thus corresponds to the derivative of the electron impact curve. Photoionization cross sections are one or two orders of magnitude less than those for 70-eV impacting electrons. In addition, the flux density of ionizing particles is usually smaller by an order of magnitude in photoionization sources than in impact sources. As a consequence, the need for electron multiplier detection is almost inevitable.

Photoionization sources are similar, at least in principle, to electron impact sources. The neutral gas is "bombarded" by ultraviolet light in an ionization chamber at reduced pressure (10^{-4} to 10^{-6} torr). Ion withdrawal, acceleration, and analysis is usually done in conventional instruments. A 60 degree sector, 15 cm radius curvature, Nier-type instrument equipped with a multiplier, for example, is adequate.

There are two basic methods for generating the photon beam. The first employs a discharge through krypton or hydrogen at a pressure of a few

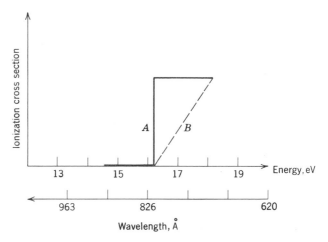

Figure 4-11. (*A*) Photoionization: $\text{Ar} + \nu \rightarrow \text{Ar}^+ + e^-$. (*B*) Electron impact ionization: $\text{Ar} + e^- \rightarrow \text{Ar}^+ + 2e^-$.

mm Hg to provide continuous ultraviolet radiation. The light is introduced into the ionization chamber through a lithium fluoride window (1 mm thick) which separates the low-pressure chamber from the high-pressure discharge tube. Since the transmission of the windows starts to deteriorate at about 1300 Å and falls off completely below 1050 Å, energies above 11.6 eV are not possible.

The second design eliminates the window and employs drastic pumping action (differential pumping) to maintain the pressure in the ionization chamber. Recently, monochromators have become popular in studying photoionization processes. Figure 4-12 is the schematic of one possible design employing a monochromator (25). Provision is also made for electron impact studies in the same source. Ion beam intensity is about 10^{-16} to 10^{-18} A.

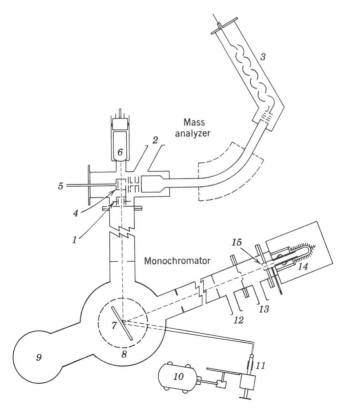

Figure 4-12. Photoionization source (25). (*1*) Secondary electron deflection plates. (*2*) Mass analyzer pump. (*3*) 17-stage Be/Cu electron multiplier. (*4*) Ion source. (*5*) Gas sample inlet. (*6*) Photon detector. (*7*) Grating. (*8*) Grating table. (*9*) Monochromator pump. (*10*) Grating drive motor. (*11*) Micrometer screw. (*12*) and (*13*) Light source differential pumping. (*14*) Light source. (*15*) Inlet slit.

The mass spectra obtained in photoionization sources are always much simpler than those in electron impact sources. The parent peak of diethyl ether, shown in Figure 4-13 (26), appears at about 9 eV, and the only other peak observable is that of the parent ion minus a methyl group at m/e 59. In the electron impact spectrum, which was taken at the same sample pressure and detection sensitivity, a number of intense ionic fragment peaks appear. Reducing the electron energy from 70 eV to 10 eV results in a spectrum essentially similar to that in photoionization but low voltage electron impact sources are limited by stability problems (Section 10-II-E).

The facts that the efficiency of photoionization is finite at the threshold and that the energy of the photon beam can be accurately measured and closely controlled make photoionization sources important tools in the determination of appearance potentials (Section 12-I) and in the study of the ionization process itself.

The simplicity of photoionization mass spectra suggests analytical applications in mixture analysis, particularly when high sensitivity is provided by electron multipliers. Thus far little work has been done to this end.

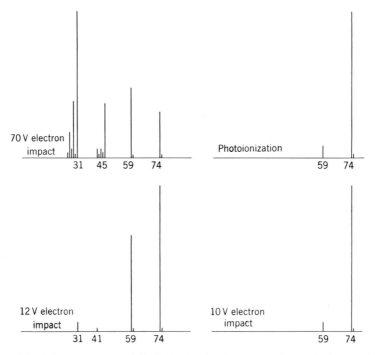

Figure 4-13.   **Mass spectrum of diethyl ether in photoionization and electron impact source at various electron energies (26).**

VIII. GAS DISCHARGE SOURCES

The gas discharge source (Goldstein, 1886) is the oldest and most direct method for ion production. Simple discharge sources were employed in most of the pioneering work in mass spectroscopy, and Aston (27) gives a rather detailed and personalized account of his thirty-five years of experience with these sources.

A *high-voltage glow discharge* source consists of two metallic electrodes placed about 50 cm apart in a cylindrical glass envelope of 5–10 cm diameter. The tube is filled with the sample gas to a pressure of 10^{-1} to 10^{-3} torr. A potential difference of 10,000–60,000 V, depending on construction, between the electrodes maintains a continuous discharge in which ion formation takes place. The cathode has in its center a hole of 0.1–1.0 cm diameter through which ions are withdrawn from the discharge. The escaping ions enter the mass analyzer after passing a collimating slit. Careful adjustments are necessary to keep the ion pencil aligned with the escape channel. Due to intense cathode ray bombardment, the end of the tube opposite the cathode is often pointed to provide better cooling.

To maintain the pressure at the 10^{-6} torr level in the analyzer either a long narrow channel or an intermediate, and separately pumped, chamber must be provided between source and analyzer. The name "canal-ray" originates from early designs employing a long narrow canal for ion withdrawal.

To maintain the discharge, pressure in the tube must be kept constant by continuously admitting the sample gas through an adjustable valve from an outside sample reservoir. Solid samples are made into a paste and placed in slots drilled into the end of the anode. A noble gas supports the discharge. Electron bombardment of the anode evaporates the solid sample which is then ionized in the discharge.

The ion current in this type of source may be as high as several milliamperes. There are many disadvantages: ion energies vary from zero to full discharge voltage, the beam is fluctuating and erratic, there is a need for differential pumping, and the rate of sample consumption is high. Moreover, since many types of collisions take place in the discharge, spectra are complex and not reproducible. The abundance of multiply charged ions is advantageous in packing fraction measurements, and this is the only purpose for which this kind of source is occasionally utilized at the present time.

In the *low-voltage arc discharge* sources, heated filaments are employed to provide an ionizing electron beam. Continuous discharge can thus be sustained by a potential difference at the 100 V level. To obtain high current densities, the arc is confined either by a capillary (Fig. 4-14) or by a magnetic field. The capillary is about 1 cm long with 0.5 cm diameter. Ions are withdrawn through an aperture in the wall of the capillary. The main advantage is high ion intensity: ion currents at the 100 μA level are obtained using an arc current of 0.5 A. Energy spread is 2–10 eV.

Low-voltage arc sources are used extensively in large-scale electromagnetic isotope separators. Capillary types are most useful for gases, while the magnetic confinement types are more practical for solids which are usually evaporated into the arc from an adjacent oven.

The *Penning-type* sources (PIG source, Philips ionization gauge) are variations of the Penning cold cathode gauge (Section 6-VI). The hot filament is eliminated, and discharge is maintained between two plane cathodes by a potential difference of 1000 V. A magnetic field like that in the oscillating electron sources (Fig. 4-4) is employed to increase the effective electron path lengths. Ions are extracted at right angles to the discharge. The cathode material must be selected so as to provide a large quantity of electrons to maintain the discharge and to resist erosion by sputtering. In addition to high ion intensity, advantages are the lack of heated parts, and relatively low gas pressure (10^{-4} mm) required. Ion energies, however, range from zero to full operating potential.

IX. LASER SOURCES

Honig and Woolston (28) investigated the laser-induced emission of ions, neutral atoms, and electrons from solid surfaces using a modified double-focusing spark mass spectrometer. Metals, semiconductors, and sintered insulators were irradiated with an energy input of about 1 joule.

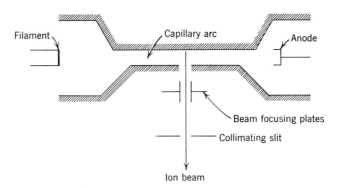

Figure 4-14.   Low-voltage discharge source.

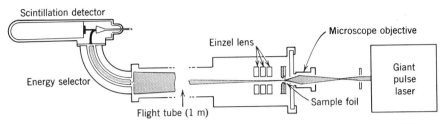

Figure 4-15. Laser source combined with time-of-flight mass spectrometer (29).

Craters formed were 20–100 μ in diameter and up to 1000 μ deep. The emission of thermal ions appeared to have been governed by the Langmuir-Saha formulas, and only singly charged ions were observed. Vaporized neutrals were ionized by electrons generated in a low-voltage arc discharge. Ions thus formed included those from both the target materials and background gases. Ion currents as large as 10^{-5} A (10^{10} ions/laser pulse) were obtained, leading to space charge broadening of the major lines on the photoplate. The authors estimated that impurities to the ppm level can be detected with the proper technique.

The focused beam from a "giant pulse" laser has been used by Fenner and Daly (29) to vaporize and ionize thin samples of solid materials. Since the ions obtained possessed an energy spread of 0–500 eV, an energy filter was needed to obtain a resolution of 30 in a time-of-flight mass spectrometer. The laser produced 6×10^{11} ions when 2×10^{-9} g copper was evaporated. Comparable ionization efficiencies were found for many metals. The experimental arrangement employed is shown in Figure 4-15. (An einzel lens, or equipotential lens, is a special kind of three-diaphragm electrostatic lens. The beam from the lens is focused and there is no net change in energy.)

References

1. Naidu, P. S., and K. O. Westphal, *Brit. J. Appl. Phys.*, **17**, 645, 653 (1966).
*2. Nier, A. O., *Rev. Sci. Instr.*, **11**, 212 (1940); **18**, 398 (1947).
3. Robinson, C. F., *Rev. Sci. Instr.*, **29**, 250 (1958).
*4. Coggeshall, N. D., *J. Chem. Phys.*, **12**, 19 (1944).
*5. Reynolds, J. H., *Rev. Sci. Instr.*, **27**, 928 (1956).
6. Heil, H., *Z. Physik.*, **120**, 212 (1943).
7. Finkelstein, A. T., *Rev. Sci. Instr.*, **11**, 94 (1940).
8. Datz, S., and E. H. Taylor, *J. Chem. Phys.*, **25**, 289 (1956).
*9. Inghram, M. G., and W. A. Chupka, *Rev. Sci. Instr.*, **24**, 518 (1953).
10. Voshage, H., and H. Hintenberger, *Z. Naturforsch.*, **14a**, 216 (1959).

11. Patterson, H., and H. W. Wilson, *J. Sci. Instr.*, **39**, 84 (1962).

12. Kunsman, C. H., *Phys. Rev.*, **25**, 892 (1925); **27**, 249 (1926).

*13. Honig, R. E., in *Mass Spectrometric Analysis of Solids*, A. J. Ahearn, Ed., Elsevier, New York, 1966, Chapter II.

14. Woolston, J. R., and R. E. Honig, *Rev. Sci. Instr.*, **35**, 69 (1964).

15. Franzen, J., and H. Hintenberger, *Z. Naturforsch.*, **18a**, 397 (1963).

16. Woodyard, J. R., and C. B. Cooper, *J. Appl. Phys.*, **35**, 1107 (1964).

17. Beske, H. E., *Z. Angew. Phys.*, **14**, 30 (1962).

18. Liebl, H. J., and R. F. K. Herzog, *J. Appl. Phys.*, **34**, 2893 (1963).

*19. Smith, A. J., D. J. Marshall, L. A. Cambey, and J. Michael, *Vacuum*, **14**, 263 (1964).

20. Barrington, A. E., R. F. K. Herzog, and W. P. Porschenrieder, *J. Vacuum Sci. Technol.*, **3**, 239 (1966).

21. Gomer, R., and M. G. Inghram, *J. Am. Chem. Soc.*, **77**, 500 (1955).

22. Beckey, H. D., *Z. Naturforsch.*, **14a**, 712 (1959); **17a**, 1103 (1962).

*23. Beckey, H. D., in *Advances in Mass Spectrometry*, Vol. 2, R. M. Elliott, Ed., Pergamon Press, New York, 1963, p. 1; also Beckey, H. D., H. Knoppel, G. Meitzinger, and P. Schulze, in *Advances in Mass Spectrometry*, Vol. 3, W. L. Mead, Ed., Elsevier, Amsterdam, 1966, p. 35.

24. Robertson, A. J. B., and B. W. Viney, in *Advances in Mass Spectrometry*, Vol. 3, W. L. Mead, Ed., Elsevier, Amsterdam, 1966, p. 23.

25. Frost, D. C., D. Mak, C. A. McDowell, and D. A. Vroom, *Proc. 12th Ann. Conf. Mass Spectry.*, ASTM Committee E-14, Montreal, 1964, p. 39.

26. Elliott, R. M., in *Mass Spectrometry*, C. A. McDowell, Ed., McGraw-Hill, New York, 1963, p. 93.

27. Aston, F. W., *Mass Spectra and Isotopes*, 2nd ed., Arnold, London, 1942.

28. Honig, R. E., and J. R. Woolston, *Appl. Phys. Letters*, **2**, 138 (1963); also Honig, R. E., *Appl. Phys. Letters*, **3**, 8 (1963).

*29. Fenner, N. C., and N. R. Daly, *Rev. Sci. Instr.*, **37**, 1068 (1966).

Chapter Five

Ion Detectors

Detection and measurement of the positive ions sorted by the analyzer can be accomplished by electrical and photographic means. Electrical recording follows the variation of current with time, while the photoplate integrates the ion current. Sensitivity and speed of response are the defining parameters; one can usually be increased only at the expense of the other. Ion currents in conventional mass spectrometers are in the 10^{-8} to 10^{-16} A range, but when required, ion currents as low as 1.6×10^{-19} A (1 ion/sec) or even lower can be measured by detectors based on secondary electron multiplication. The largest useful ion current is about 10^{-8} A due to space charge effects.

I. ELECTRICAL DETECTION

In "conventional" detectors, the positive ions arriving at the collector are neutralized by electrons arriving from ground after passing through a high ohmic resistor. The potential drop across this resistor is the measure of the ion current. Amplification of the potential drop is facilitated either by a direct current amplifier or a vibrating reed electrometer. In the increasingly popular electron multiplier detectors the ion current is detected and measured by utilizing a phenomenon known as "secondary electron emission." Electrical detection may be broken down to the processes of ion collection, amplification, and recording.

A. Ion Collectors

1. Single Collectors

An ion collector system consists of an ion-resolving slit, the collector itself, and various suppressors and screens.

Resolving Slit. The ion-resolving slit, also known as defining slit, or exit slit, or collector slit, is located at the focal point in magnetic deflection instruments. The width of this slit is critical since this defines the portion of the total ion beam which will strike the collector plate. It has already been pointed out that in order to obtain flat-topped peaks, the heights of which are proportional to the number of ions of the particular mass, the collector slit has to be somewhat wider than the image of the source slit. If the slit is too wide, more than one ion beam strikes the collector; if it is too narrow, i.e., narrower than the focused beam, a certain fraction of the beam intensity is lost. Decreased sensitivity may, however, be accepted as the price for increased resolution.

An ability to vary either the actual or the effective size of the exit slit is obviously advantageous. Slit widths may be adjusted mechanically from outside the vacuum system, through a bellows or a magnetic clutch, and measured by a micrometer. The *effective width* can be controlled by the incorporation of a slit with variable potential (see below). The edges of the various slits are often machined to give a sharp cutoff as the beam moves across. The defining slit is usually kept at the same potential as the collector plate, i.e., at ground potential.

Final Collector. The final collector is a small metal electrode placed in an open-ended reflecting box or cup, known as the *Faraday cage* or cup (Fig. 5-1). The ions impinge on the metal surface which is inclined to the direction of their motion so that ions reflected and secondary electrons ejected (see below) cannot escape from the box. In multiplier detectors the first plate of the multiplier becomes the collector electrode. These will be discussed shortly.

Suppressors. Before explaining the role of the various suppressors and screens, it must be emphasized that the final energy of the ions striking

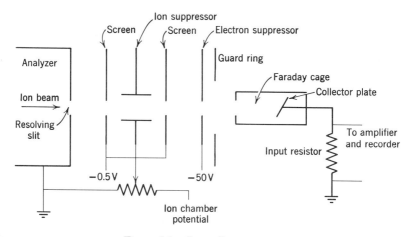

Figure 5-1.   Ion collector system.

the collector plate depends only on the potential difference between the ion source and the collector.

The bombarding of the collector by energetic ions results in the formation of secondary electrons. The number of these secondaries depends not only on the intensity of the ion current but also on the energy and mass of the bombarding particles and the angle of incidence. In order not to change the actual current which is being measured, secondary electrons must be returned to the collector. Similarly, negatively charged particles originating from the bombardment of the defining slit must be prevented from reaching the collector to avoid possible peak height reduction, peak shape distortion, and/or negative peak appearance.

All these effects can be eliminated by the application of either a strong magnetic field around the collector or an additional slit which is negative with respect to the collector plate. In completely immersed instruments (180°) the magnetic field is automatically provided, and secondary electrons are returned due to their small radii of curvature. In other cases a potential of about -50 V is applied on a *secondary electron suppressor* (electron repeller) slit. Leakage current between electron suppressor and collector is prevented by a grounded *guard ring* placed between them.

A second type of suppressor called an *ion suppressor*, which is at or near source potential, is often employed to prevent the passage of metastable ions and scattered ions. Metastable ions spontaneously dissociate during their passage through the mass spectrometer and have kinetic energies smaller than those of normal ions. Formation and characteristics of these ions will be discussed in Section 9-VIII. Scattered ions lose some of their energy in collisions during their passage. Their number becomes large when a relatively high pressure is needed in the source (e.g., in leak detection).

Suppression of both metastable and scattered ions is accomplished by the ion suppressor slit. Since it is kept at source potential a potential barrier is established which can be crossed only by ions having the full energy acquired during acceleration. This barrier slit must be shielded on both sides to prevent field penetration into the analyzer and collector regions. When the potential on the ion suppressor is varied (with respect to ground) it may also be used as virtual slit with a variable width. This may be useful to "select" resolution, or to investigate small peaks at the "skirts" of large ones.

As the presence of suppressor slits may result in harmful beam defocusing effects, the size and spacing of these slits and guard rings should be carefully designed.

2. Double Collectors

In relative isotope abundance measurements where momentary instrumental instabilities can cause serious errors, double collectors are frequently used. Most designs are patterned after the one by Nier et al. (1) shown in Figure 5-2. One electron beam is received by the cup collector while the other is collected on the slit which would be the defining slit in a single collector system. Both collectors are surrounded by their own electron repellers and guard rings. Ion suppressors are not required for multiple collectors as the latter are usually designed for specific isotope ratio applications. Nier's classical design, for example, was used mainly for ^{13}C abundance determinations.

Measurement of the ion current, as discussed shortly, is by a null method which yields the abundance ratio and eliminates source instability.

A dual collector system becomes a simple single collector system when the second collector is grounded. By connecting the two collectors an effective slit-width control is possible; one actually has two "resolution positions."

3. Performance of Faraday Cup Collectors

The Faraday cup type of collector system, combined with a dc amplifier or vibrating reed electrometer, has many distinct advantages compared to the electron multiplier and photoplate types of detection. First of all, the measured and amplified ion current is directly proportional to the number of ions and number of charges per ion. The response is equal, regardless of the energy, the mass, and the chemical nature of the ions. Mass discrimination caused by the mass dependency of released secondary electrons and/or secondary positive ions is practically nonexistent as long as properly

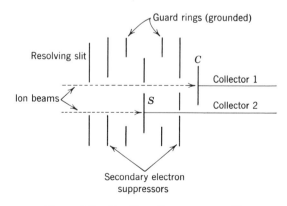

Figure 5-2.   **Double collector system (1).**

designed suppressors are employed. Faraday cup collector systems are constant over long periods of time, can be exposed to air without damage, and are very reliable. In addition, their construction is simple, and they are inexpensive. The electrical noise level of the Faraday cup detector is quite low, so that the lowest detectable current level is essentially determined by the current-measuring device. A definite disadvantage is the relatively long time constant of Faraday cup amplifier systems as compared to multipliers. This problem is discussed later.

4. Collector Systems for Dynamic Mass Spectrometers

In dynamic mass spectrometers the defining slit is usually absent. The beam area in time-of-flight, radiofrequency, and quadrupole mass spectrometers is larger than that in magnetic deflection machines. Often high-transmission grids instead of slits are employed to provide the required suppressor or retarding potentials. Secondary emission effects are reduced in quadrupole and radiofrequency instruments because the ion energies are lower by an order of magnitude or more compared to magnetic instruments.

B. Amplifiers

1. Input Resistor

The ion current is determined from the potential drop across a high ohmic resistor, one side of which is connected to the collector while the other is grounded (Fig. 5-1). In the simple $V = I/R$ relationship, R represents the combined resistance of the calibrated resistor in parallel with the electrometer input resistance. To obtain useful voltages (1 mV to 50 V), the input resistor must be at least 10^{10} ohms. The upper limit is about 10^{12} ohms, set by the requirement that the input resistance of the amplifier be large compared to the value of the input resistor.

For dependable current measurements, variations in resistance due to surface leakage, etc. should be minimized by keeping resistors clean and dry. Carbon-compounded resistors, often in evacuated envelopes, are used for precision work. Polarization or the tendency of the materials composing the resistor to act as voltage sources, due to induced dipoles, appears at a potential difference of about 10 V across the high ohmic resistor. This sets the upper limit to current measurements.

2. Direct Current Amplifiers

Conventional vacuum tube voltmeters, with 10^7 ohms input resistance, cannot be used for ion current measurements since most of the current would be shunted out. The input stage of the dc amplifiers employed to measure the potential drop across the high ohmic resistor consists of special "electrometer" tubes, characterized by low grid current (10^{-15} A) and low anode voltage (<10 V) to prevent residual gas ionization inside the tube. By careful design the impedance between the various electrodes of the electrometer tubes is kept high so that the internal resistance of the vacuum tube voltmeters used in mass spectrometry is of the order of 10^{14} ohms. Additional requirements for the electrometer tube are: low dc noise and relative freedom from microphonics, small size, and as good an amplification factor as possible so that drift and noise in the second stage is minimized. The input or preamplifier section of the electrometer amplifier should be mounted as close as possible to the ion collector assembly to minimize input noise and capacitance caused by the connecting lead. The noise level in dc electrometers is 1×10^{-15} to 1×10^{-14} A.

Since relatively rapid changes in the ion current must be followed, the dynamic behavior of the whole detector system must be considered. The relationship between ion current and voltage at the input of the electrometer is given by Kirchoff's law

$$RC(dV/dt) + V = RI \tag{5-1}$$

where C is the combined capacitance (with respect to ground) of collector, high ohmic resistor, and amplifier feed lines. If the ion current is suddenly changed from I_1 to I_2, one obtains

$$V = RI_2 + R(I_1 - I_2) \exp(-t/RC) \tag{5-2}$$

Figure 5-3 shows this exponential change. RC (ohm/farad) is known as the time constant of the circuit and is measured in seconds. Capacitance cannot be reduced much below 10 pF; thus the time constant for an input resistor of 10^{12} ohms is 10 sec. If the current to be measured is 10^{-15} A, then

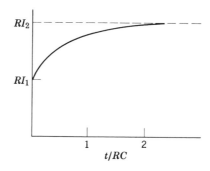

Figure 5-3. Relationship between t/RC and RI.

the time required for the voltage developed to reach 99% of its final value after the current is suddenly changed from zero to some finite value is calculated as $\exp(-t/RC) = 0.01$ and $t = 46$ sec, which is much too high.

Input capacity and consequently the time constant can be significantly reduced by employing *feedback* technique. The principle is explained with reference to Figure 5-4. The output, V_{out}, of a high gain amplifier is fed back through R_0, in series with the input signal. The input voltage to the amplifier is thus

$$V_{in} = V_{signal} - V_{out} = IR - V_{out} \qquad (5\text{-}3)$$

where V_{signal} is equal to the product of the ion current and the high ohmic resistor. Since the gain, G, of the amplifier is the ratio of the output to the input voltage

$$G = V_{out}/V_{in}$$

one obtains

$$V_{in} = IR - V_{out} = IR - GV_{in} \qquad (5\text{-}4)$$

from which

$$V_{in} = IR/(1 + G)$$

and

$$V_{out} = GV_{in} = IRG/(1 + G) \approx IR \qquad (5\text{-}5)$$

assuming $G \gg 1$ (e.g., 10^3). Thus, the impressed and the output voltages are almost exactly equal, meaning that the voltage amplification is unity and the amplifier acts as a current amplifier. Gain depends only on the input and output resistors and gain values of the order of 10^7 are readily obtainable.

Zero stability and current linearity are markedly improved by feedback since the output signal is reasonably independent of fluctuations in gain and supply voltages. In addition, the time constant is reduced because the capacitance, C, is reduced to $C/(1 + G)$, thus affecting the impedance

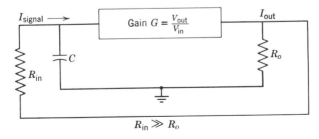

Figure 5-4. Principle of feedback technique.

between input and ground. Time constants as low as 0.1 sec can be achieved this way with a 10^{12} ohm resistor. There are several commercial dc amplifiers available.

3. *Isotope Ratiometers*

In isotope abundance measurements, errors due to source instability are eliminated, or at least reduced, by using a double collector system and measuring the ratio of the two ion beams.   About one order of magnitude increase in accuracy is achievable by employing a method developed by Nier, Ney, and Inghram (1). A schematic is shown in Figure 5-5. The more abundant isotope ($\sim 10^{-9}$ A) is collected on the first collector and the potential drop across the input resistor ($R_1 \sim 10^{10}$ ohms) is amplified by the first amplifier, which employs 100% feedback.

The weaker ion current is collected on the second collector and is amplified without negative feedback. The ouput, V_{02}, of this amplifier can be made zero if fraction x of V_{01} is applied, via switch S, so as to cancel the voltage developing on R_2. One obtains, using equation 5-4,

$$I_2 R_2 = x V_{01} = x I_1 R_1 \, G/(G + 1)$$

or

$$I_2/I_1 = x \, (R_1/R_2) \, [G/(G + 1)] \approx x \, R_1/R_2 \qquad (5\text{-}6)$$

R_1 and R_2 have known values and only x need be determined. As mentioned, accuracy of relative abundance measurements is greatly increased by eliminating source instabilities. For absolute abundance determinations, however, the null method is less accurate than the one based on a single collector.

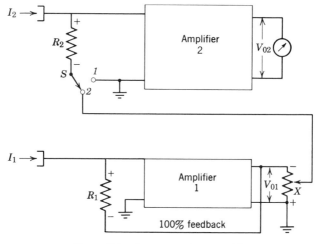

Figure 5-5. Isotope ratiometer (1).

4. Vibrating Reed Electrometer

A rather sophisticated method to measure small ion currents is to convert the dc voltage, generated on the high ohmic resistor, into an ac voltage before amplification (2–4). A gain in sensitivity and an increase in zero stability is obtained because: (*1*) alternating current amplifiers are inherently more stable than dc amplifiers, which must be direct coupled; and (*2*) ac amplifiers may be tuned for maximum gain at the frequency of interruption, resulting in a decrease in noise since background fluctuations, having components over a wide range of frequency, are amplified less efficiently.

Conversion is accomplished electromechanically with the aid of a dynamic capacitor. The dc input is fed to the fixed metal plate of the capacitor. The other (parallel) plate, the so-called reed, is movable and is vibrated several hundred times a second (e.g., at 450 cps) by an electromagnet driven by an oscillator (Fig. 5-6). Since the capacitance is inversely proportional to the separation of the plates, a sinusoidal variation of the separation results in an inverse sinusoidal variation of the capacitance. The strength of the electrical field in the capacitor fluctuates and the alternating voltage developed is directly proportional to the charge on the vibrating reed capacitor. The ac signal is next fed into a conventional, high-gain ac amplifier tuned to the modulation frequency. The output signal is synchronously rectified (rectifier and reed driven by the same oscillator), and is displayed on a recorder. To increase stability sufficient degenerative feedback is provided to the dc side of the vibrating capacitor to cancel the input. If the overall amplifier gain is large (of the order of 10^3) the input

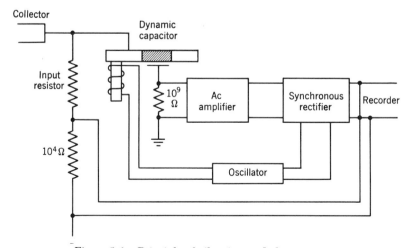

Figure 5-6. **Principle of vibrating reed electrometer.**

current is equal to the output voltage divided by the resistance in the feedback line.

Ion current may also be measured by the so-called rate-of-charge method. Here the resistor is removed from the feedback circuit and the instrument becomes an integrating electrometer. As electrons are collected in the capacitor the charge and, in turn, the ac voltage output increases proportionally to the input. The input current is determined from the slope of the output voltage of the recorder.

Both methods have their respective merits. The rate-of-charge method is more sensitive and more accurate since even small currents result in measurable charges. The "high resistance leak" method, on the other hand, provides a continuous record of current flow; it is used more frequently in mass spectrometry, particularly in the 10^{-6} to 10^{-15} A range.

Sensitivity is limited only by the thermal agitation noise in the input circuit; 10^{-7} A (or 10^{-16} coul) may be measured by the rate-of-charge method. Response speed in current measuring is a function of the value of the high ohmic input resistor. In commercial instruments convenient multiple resistor switches enable one to have several input resistors available; also shorting and open positions are usually provided.

The main feature of vibrating reed electrometers is stability. Zero drift depends only on contact potentials on the condenser surfaces. Changes in amplifier gain and variations in power supply voltages, the usual causes of instability in dc amplifiers, do not contribute to zero drift. With an open circuit input the short period noise is of the order of 10^{-16} coul, the steady drift is about 10^{-17} A. The limiting factor in the accuracy of current measurements is the accuracy of calibration of the input resistors.

5. Signal-to-Noise Ratio

Random and uncontrollable distortions in an amplifier are described as noise. Among the various types of noises (shot noise, flicker noise, etc.) the most important is the thermal or Johnson noise caused by thermal drifting of electrons in resistors. All frequencies are generated since the electron motion is random, but only those handled by the amplifier need be considered. The root-mean-square of the thermal noise voltage in a non-inductive resistor is given by

$$V_{\mathrm{rms}} = \sqrt{kTR\Delta f} = 7.4 \times 10^{-12}\sqrt{TR\Delta f} \qquad (5\text{-}7)$$

where R is resistance (ohm), Δf is the bandwidth of the amplifier (cps), T is the temperature of the resistor (°K), and k is Boltzmann's constant.

For a signal to be distinguishable it must have an amplitude greater than the noise level. A useful concept in the evaluation of amplifier performance is threshold sensitivity, which is defined as the minimum discernible signal that can be detected in a particular application. It is the signal at the input of the amplifier that is equal to the noise output under similar conditions. The signal-to-noise ratio, which is the ratio between the input voltage of the amplifier generated by the ion current on the input resistor, and the noise voltage originating from the same input resistor, is given by

$$\frac{\text{signal voltage}}{\text{noise voltage}} = I \sqrt{\frac{R}{4 k T \Delta f}} \tag{5-8}$$

where I is ion current (A) and R is the resistance of the input resistor.

It is seen from equations 5-7 and 5-8 that while the signal voltage is proportional to R the noise voltage is proportional only to the square root of R. The resistance R should thus be selected to be as high as possible. Its value, however, as mentioned earlier, is limited to about 10^{12} ohms by the requirement that it must be smaller than the grid resistance of the succeeding electrometer tube. Increasing the passed frequency range, on the other hand, would reduce the speed of response. Threshold sensitivity for a 10^{11} ohm resistor and a bandwidth of 10 cps is 1.3×10^{-4} V corresponding to 1.3×10^{-15} A. The practical limit of detection is somewhat higher due to additional noise sources in the amplifier itself (secondary emission noise, ion noise from residual gases, etc.).

C. Electron Multipliers

1. Principle and Design

High detection sensitivity and/or rapid recording (small time constant) may be achieved with detectors based on *secondary electron emission*. The term refers to the phenomenon of the ejection of electrons from surfaces subjected to bombardment by energetic particles. The bombarding particles may be positive or negative ions, electrons (primary electrons), neutral atoms or molecules, and photons. The number of secondary electrons released per incident bombarding particle is the coefficient of secondary electron emission, a dimensionless positive number. The numerical value of the coefficient depends on the mass, charge energy, and nature of the incident particles. Additional variables include the work function and physical "condition" of the surface, and the angle of incidence.

The mechanism of secondary electron emission is not well understood. It involves the excitation of internal electrons, their diffusion to the surface, and passage through a potential barrier. Electron multipliers used in mass spectrometry utilize the phenomenon of secondary electron emission in two ways: First, the ion current is "converted" into an equivalent electron current on the first electrode (*conversion dynode*) of the multiplier. This is secondary electron emission caused by positive (or negative) ion bombardment. The second step in the operation of the multiplier is the *amplification* of the electron beam obtained from the conversion dynode. This is accomplished in many steps utilizing secondary electron emission caused by bombardment of surfaces by primary electrons.

For currents greater than 10^5 ions/sec multipliers are used as pre-amplifiers to conventional dc electrometer amplifiers or vibrating reed electrometers. For currents less than 10^5 ions/sec they serve as pulse amplifiers employing standard scintillation counting. The transit time of electrons is only 10^{-9} to 10^{-8} sec (100 Mc/sec bandwidth) which is desirable in many applications.

In an *electrostatic multiplier* (Fig. 5-7) positive ions leaving the analyzer are accelerated by a potential difference of 2–5 kV and impinge on the first plate of the multiplier. From this electrode (the *conversion dynode*) secondary electrons are released which are accelerated and focused onto a second dynode. The ion current is thus "transformed" into an electron current. This electron current is next amplified as the number of electrons released from the second dynode is greater than the number of electrons impinging upon it. As many as 10–20 stages are customarily used, each stage connected to a successively higher potential by a voltage divider. The final collector (anode) is connected to a conventional amplifier. Most designs are patterned after one by Allen (5).

The obvious requirement of the dynode material is that the values of the secondary electron emission coefficients for ion and primary electron

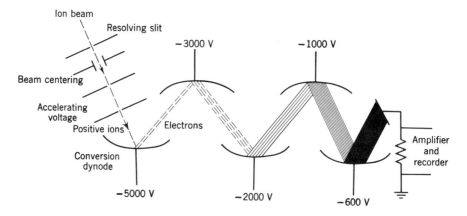

Figure 5-7.   Operation of an electron multiplier.

bombardment be as large as possible. Dynodes are usually made of 2%Be-Cu or 2%Ag-Mg alloy, and are conditioned by heating to 200–400°C under vacuum or in an inert atmosphere (5).

To eliminate the adverse effects of the changing magnetic field as the spectrum is scanned, electron multipliers are either shielded from the fields or immersed in an auxiliary magnetic field. A special magnetic multiplier is described shortly.

In the *scintillation-type* multipliers, ions are accelerated to 10–20 kV before hitting the conversion dynode (aluminum). The secondary electrons released are further accelerated onto a phosphor, resulting in a photocurrent which is then measured in a photomultiplier external to the vacuum system. The resulting voltage pulse is fed to a ratemeter and a pulse height analyzer via a cathode follower and pulse amplifier. Daly (6) reported practically 100% efficiency and small mass discrimination. Measured noise level was 4×10^{-20} A, permitting the measurement of ion beams as low as 10^{-18} A. Using two scintillators back to back, simultaneous measurement of two beams can be made with such accuracy that in the determination of tungsten isotope ratios precision is dependent almost entirely on the statistics of the number of ions counted.

A *magnetic electron multiplier* which utilizes a continuous strip of semiconductor material rather than conventional multielement structure has been developed by Goodrich and Wiley (7) for Bendix time-of-flight mass spectrometers.

Ions striking a stainless steel collector (cathode) release secondary electrons which are forced by crossed electric and magnetic fields to move in successive cycloids. A uniform magnetic field of a few hundred gauss is provided, perpendicular to the plane of Figure 5-8. The multiplier itself consists of two parallel plane glass surfaces $(2.5 \times 5.5$ cm) coated with a semiconducting metallic oxide and spaced 0.5 cm apart. A potential difference of 1700 V is applied along each surface. At the same time, however, the actual potential on the upper surface is 350 V more positive than that on the lower surface. The equipotential lines between the surfaces are thus parallel but slanted and they have directions as shown in Figure 5-8.

As electrons enter the multiplier they continue their cycloidal motion but because of the slant of the equipotential lines they hit the dynode surface before completing a cycle. The impact energy of the electrons corresponds to the difference in potential between the points of origin and impact and is large enough to result in a secondary electron emission yield greater than unity. There are about 40 such stages, resulting in an overall gain of 10^7.

The output may be collected on an anode, but usually a set of crossed-field gates is employed (Fig. 5-9) so that the beam can be directed into any one of six channels which are connected to electrometer amplifiers. Six different masses can be monitored simultaneously by pulsing the appropriate gate electrode at the exact time ions of the particular mass arrive at the channel. The output of the amplifiers is proportional to the ion abundance. One channel may be used to cover the whole spectrum.

Dark current is 0.4×10^{-13} A corresponding to the arrival of 0.02 electron/sec (3×10^{-21} A) on the cathode. The amplifier does not require refrigeration and since no activation process is needed it can readily be exposed to air. Figure 5-10 is a photograph of the Bendix magnetic electron multiplier.

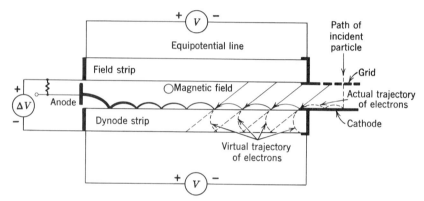

Figure 5-8.   Magnetic electron multiplier (7).

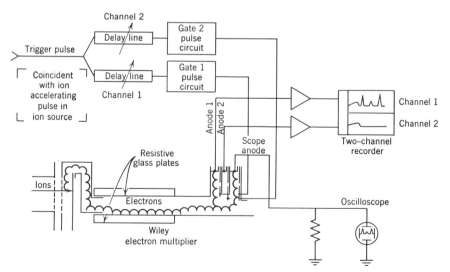

Figure 5-9.   Electron multiplier and cross-field gate system. (Courtesy of Bendix Corp.)

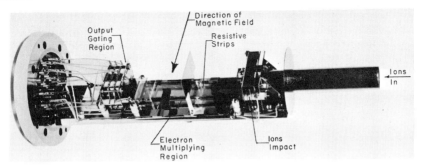

Figure 5-10. Picture of magnetic electron multiplier. (Courtesy of Bendix Corp.)

2. Merits and Drawbacks

The greatest advantage of electron multipliers is extremely high sensitivity. The overall gain, i.e., the average number of electrons collected per ion incident on the first dynode, of a multiplier is given by

$$G = G_1 G_2{}^n \tag{5-9}$$

where G_1 is the ratio of secondary electrons to incident positive ions and G_2 is the multiplication factor of the n remaining dynodes. Since the value of both G_1 and G_2 is about 2, the amplification of a 20-stage structure is more than 10^7. Another often used characteristic is *efficiency*, by which is meant the ratio of the number of output pulses measured to the number of incident ions.

Ion currents as low as 10^{-19} A may be detected with secondary electron multipliers. It is recalled that the arrival rate of 1 ion per second corresponds to 1.6×10^{-19} A. At such ion current levels the accuracy of measurement is limited by statistical fluctuations. The fluctuation about the average in the measurement of a number of particles, N, is equal to $N^{1/2}$. Here N is equal to the average particle current multiplied by the time constant of the overall current measurement and recording system. To achieve a standard deviation of 10%, 100 particles must be counted. At an ion arrival rate of 1 ion/sec, this means a collection time of nearly 2 min. This limitation must be considered when multipliers are used with extremely rapid scanning of the mass spectrum.

The response of electron multipliers depends upon the mass, energy, charge, and chemical nature of the impinging ions. These discrimination effects were first investigated by Inghram et al. (8). Figure 5-11 shows a plot of secondary electrons per ion versus ion energy (eV) for a number of alkali ions impinging at an angle of 45° on a 2%Mg-Ag electron multiplier. Figure 5-12 shows secondary electrons per ion as a function of mass at

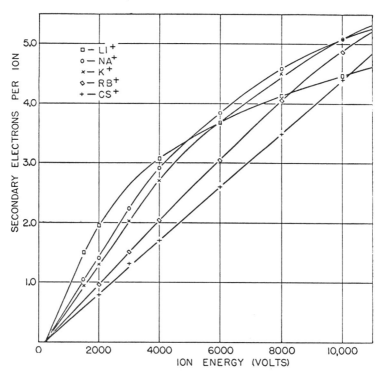

Figure 5-11. Secondary electron emission for alkali ions at 45° incidence on 2%Mg-Ag dynode (8).

constant energy. The two points shown for ^{40}Ca and ^{48}Ca illustrate that a calibration curve taken with different elements is not the same as that obtained using isotopes of the same element. This is very important in isotopic work.

The most commonly used correction for mass discrimination is based on the assumption that ions of equal velocity produce the same number of electrons,

$$(I_1/I_2)_{\text{true}} = (I_1/I_2)_{\text{meas}}(M_1/M_2)^{1/2} \qquad (5\text{-}10)$$

where I is ion current, M is mass (amu), and $M_1 > M_2$. This correction is only an approximation and applies primarily in the higher mass regions.

In practice, multiplier gain variations with ionic mass and other parameters is usually determined by calibration. This is normally only a minor problem which is indeed a modest price to pay for the increased sensitivity. When accurate quantitative measurements are needed, calibration may become more critical.

Two important practical problems with electron multipliers are desensi-

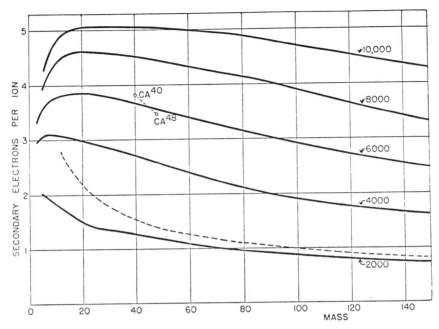

Figure 5-12. Secondary electron emission as a function of mass of constant energy (8).

tization due to very large ion currents and repeated exposures to air. The decrease in gain is usually accompanied by an increase in the "dark current" or electrical leakage. The true or inherent noise of a multiplier refers to noise present at all times. This has nothing to do with statistical fluctuations. Inherent noise is at the 10^{-19} A level in properly operating multipliers. Several techniques have been developed to rejuvenate multipliers. Young (9) suggests baking at 300°C in oxygen at atmospheric pressure for 1 hr. The technique is applicable to both Mg-Ag and Be-Cu dynode multipliers.

A blanking circuit has recently been reported by Haumann and Studier (10) for magnetic electron multipliers in time-of-flight mass spectrometers. Those peaks not pertinent to the experiment are blanked out electronically, resulting in increased linearity, gain, and dynode strip life.

In summary, the principal merits of electron multipliers are extreme sensitivity and fast response. Multipliers are indispensable in the analysis of very small sample quantities and in the study of fast processes. To compensate for various discrimination effects, calibration is often needed. Modern "combination" detector systems, incorporating both conventional Faraday cage-amplifier and electron multiplier detectors are becoming deservedly popular and are featured on many commercial instruments.

D. Recorders

The final stage in a mass spectrometric measurement, as in every instrumental method, is the presentation of the information obtained. The signal emerging from the amplification stage is usually a voltage proportional to the ion current, and it is plotted against the mass scale or a signal proportional to it. Six basic systems are presently used to record mass spectra: (*1*) pen-and-ink recorder, (*2*) recording oscillograph, (*3*) cathode-ray oscilloscope, (*4*) magnetic tape recorder, (*5*) digitizer, and (*6*) photographic emulsion.

Important parameters to be considered when selecting the proper recorder for a particular application are: speed of response, dynamic range, accuracy, and convenience. The required speed of response is determined mainly by the applied scan rate (Section 10-V-C). The requirement of an ability to record large intensity ratios with equal accuracy is understandable when considering trace analytical applications or investigations where accurate cracking patterns are needed. The accuracy of the recorder should, of course, correspond to the overall accuracy of the apparatus. Convenience to the operator and the cost of the recording media should also be considered, particularly in routine laboratories.

1. Pen-and-Ink Recorder

Pen-and-ink recorders (strip-chart recorders) are essentially self-balancing direct current potentiometers. The signal voltage (Fig. 5-13) is compared to a standard reference voltage provided on a precision slide wire. The unbalance signal (error signal) is converted to alternating current by a chopper. The voltage is amplified, and the output of the amplifier is applied to the stator of a phase-sensitive reversible motor which rotates whenever voltage is applied and, hence, whenever unbalance current flows in the potentiometer. The direction and speed of the motor rotation is governed by the polarity and magnitude of the output of the amplifier which, in turn, is determined by the unbalance. The rotor shaft is mechanically connected to the balancing contact of the potentiometer slide wire in such a way that the direction of drive of the sliding contact is always in the sense that will reduce the amplifier output. The error signal thus produces a chain of events resulting in a correction signal that is applied to cancel the error signal: when the balance point is reached the motor stops. A mechanical connection between the slide wire and a pen provides a means to record the potential on a moving paper chart. Thus voltage is recorded against time (constant chart speed).

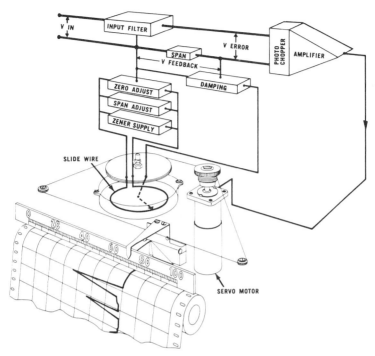

Figure 5-13. Principle of servo potentiometer. (Courtesy of Texas Instruments Corp.)

During the last few years at least two dozen models of potentiometric recorders have become commercially available. Stress is on versatility and convenience. The potential required for full scale deflection ranges from a few millivolts to several volts providing a wide sensitivity range.

Response is specified in terms of time required for the pen to travel to a new balance point across the full scale, rather than by frequency response. An excursion time of 0.25 sec is considered fast, corresponding to 2 cps. Operators sometimes lower recorder gain to reduce apparent noise, without realizing that a recorder operated at low gain may have an effective response time as much as an order of magnitude lower than its rated value. Recorders can follow faster signals of lower amplitude since effective response times are shorter for peaks traversing a smaller fraction of the chart. The price paid for this is a downgraded signal-to-noise ratio, because the smaller time constant (greater band width) is more effective for low amplitude noise impulses than for larger amplitude signals. Input impedance at balance is infinite; external resistance can be as high as 25 kΩ or even higher. Chart speed is often variable from 0.3 in./min to 1 in./hr.

The range can usually be changed by resistance attenuation. This technique becomes cumbersome when the number of peaks is large and when magnitudes change frequently, and becomes almost useless with unknown peaks whose size cannot be anticipated. In such cases, assuming an adequate amount of sample to be available, relative peak intensities may be obtained in a preliminary run performed at such low sensitivity that all peaks appear on scale. Automatic sensitivity selection may be accomplished by an electronic shunt selector which responds instantaneously to the signal voltage and makes the required attenuation before the lagging pen reaches full scale deflection.

Another method, which can be used with mass spectrometers equipped with dual collector systems, operates as follows. The incoming beam strikes the first collector, the ion beam is amplified, and the output is used by means of a range-selecting circuit to preselect the appropriate range for the main collector–amplifier–recorder system. Range changing is accomplished in milliseconds and the speed with which recording can be done is limited by the recorder. The double collector cannot be used with electron multipliers and its use is limited to low masses. The selected sensitivity range is automatically marked on the chart either by a second pen or by means of some other device.

Automatic attenuation is conveniently accomplished with multichannel potentiometric recorders (up to four pens). Sensitivity ranges are simultaneously plotted on a single chart paper using different color pens.

Precision of recording potentiometers is about 0.3% of full scale voltage. In what follows, it will become obvious that the galvanometer and, whenever available, peak digitizing technique, is much more convenient to display mass spectra consisting of numerous peaks.

2. Recording Oscillograph

The recording oscillograph (flying spot detector) acts essentially as a photographic strip-chart recorder. The output of the amplifier, representing the ion current reaching the collector, is fed into a number of mirror galvanometers (often four or five), each having a different sensitivity. The deflections, which are proportional to the impressed signal, cause a spot of light, focused by an optical system, to move across the recording paper (Fig. 5-14). The displacement of the light beam is proportional to the ion current. The light source is often a high pressure mercury vapor lamp with a maximum output in the ultraviolet region; since modern recording papers are highly sensitized to ultraviolet, records are instantly usable ("latensifying," direct print). When a permanent record is desired, traces can be fixed

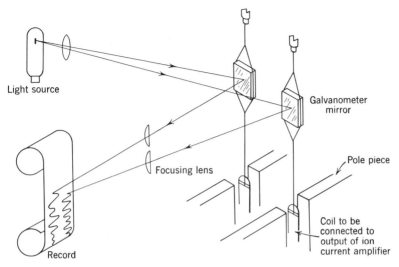

Figure 5-14. Principle of optical recording oscillograph.

photographically. Horizontal grid lines for amplitude reference (0.1 in. apart), timing line systems, trace identification, and numbering are useful accessories aiding data interpretation.

The sensitivity of the galvanometers is preadjusted, by resistors in series, to cover the full voltage range (often 0.1–10 V) in several steps. The objective is to put *every* peak on scale with a height of not less than 30% of full scale deflection. This assures accuracy in amplitude measurements and also provides a wide *dynamic range*. A dynamic range, for example, of $10^5 : 1$ for largest peak to smallest peak is often desirable. Several spectra taken on oscillographs are shown later in the text, e.g., Figure 10-5.

Another advantage of oscillographs is their relatively high frequency response, from dc to 1000 cps. Still another advantage in multichannel recording is the possibility to include on the same record such experimental parameters as electron and total ion current, and electron accelerating voltage. Chart speed is usually variable by pushbutton-controlled transmission from 0.1 in./sec to 50 in./min. Accuracy in oscillographic recording is 0.5–1.0%.

3. Cathode-Ray Oscilloscope

The main advantage in displaying mass spectra on a cathode-ray oscilloscope is, obviously, speed of response. It is possible to obtain 10,000 or more spectra per second. The method is normally used with time-of-flight, quadrupole, and other dynamic instruments, but can also be employed with direction-focusing mass spectrometers. Requirements for the cathode-ray oscilloscope for viewing and/or photographing mass spectra are fast rise time (e.g., 15 nsec) and accurately calibrated triggered sweeps. An oscilloscope trace of xenon is shown in Figure 5-15.

The multiplier output of time-of-flight instruments is transmitted to the vertical deflection plates of the oscilloscope. The horizontal sweep is synchronized with and triggered by the ion acceleration pulses. The m/e value is thus measured from the position of the peak on the time axis, while the relative number of ions is proportional to the area under the corresponding peak or, approximately, to the peak height. The portion of the spectrum to be studied is selected by the horizontal sweep controls and by varying the delay between the ion accelerating pulse and the oscilloscope trigger.

The cathode-ray oscilloscope technique is chiefly used where transient phenomena are investigated and fast qualitative identification is required, e.g., in explosion studies, gas chromatographic peak identifications, and continuous control of the behavior of ultrahigh vacuum systems (momentary gas surges).

The speed of the oscilloscope is more and more often coupled to the sensitivity of electron multipliers. The price one has to pay for rapid

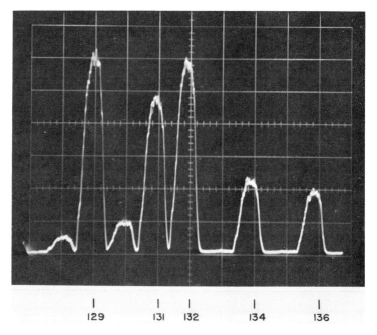

Figure 5-15. Xenon spectrum on an oscilloscope. The spectrum was taken in 7 msec. (Courtesy of Finnigan Instruments Corp.)

scanning is less accuracy in peak height measurement, which is of the order of 5% for cathode-ray oscilloscopes. Another inconvenience is the inability to display a wide range of peak heights at a given time.

4. Magnetic Tape Recording

Magnetic tape recording is the newest method for data acquisition in mass spectrometry (11,12). Commercial recorders with various degrees of sophistication (and cost) have recently become available. Tape speeds range from $1\frac{7}{8}$ to 60 in./sec. Accuracy of speed is better than 0.5% with short-term deviation (flutter) of $<1\%$. As many as 14 channels are available so that a wide dynamic range can be covered by multiple track recording, similar to oscillographic recording. In addition to peak heights, other instrumental parameters such as magnetic field strength and acceleration voltage, as well as an accurate time signal (from a pulse generator), can also be recorded simultaneously.

Magnetic tape recording is a versatile method, most useful for recording large quantities of digital data, such as in the analysis of chromatographic effluents. Data may later be reproduced at faster or lower speeds, facilitating the use of complex retrieval systems. Before permanent recording from the tape the information can be examined on an oscilloscope or a recording oscillograph to select important portions. Final display may be made either in graphic or digital form.

A mass spectrum of hexadecane is shown in Figures 5-16a and 5-16b (12). The upper trace is a direct oscillographic record (chart speed 8 in./sec); the lower one is the same spectrum recorded on magnetic tape (tape speed $7\frac{1}{2}$ in./sec). Displacement between the spectra represent the time required for the tape to move from the record head to the reproduce head. The spectrum was taken from the electrometer output of a Bendix TOF instrument with a scan rate of 6 sec for the m/e 12–200 range.

The importance of magnetic tape recording in high resolution mass spectrometry is further discussed in Section 10-IV.

5. Digitizers

Accurate and rapid measurement of peak heights has always been of concern in mass spectrometry. Peak heights are usually read with the aid of grid lines provided on the recording medium. The smallest distance measurable with a ruler-comparator is about 0.01 in. The method is burdensome and peak measurements may take hours on a sample which was run in a few minutes. The economic importance of digitization of mass spectra will be appreciated when it is considered that the number of routine

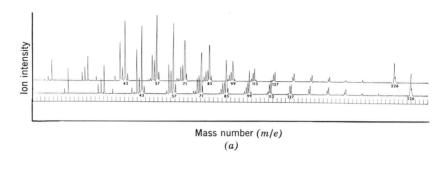

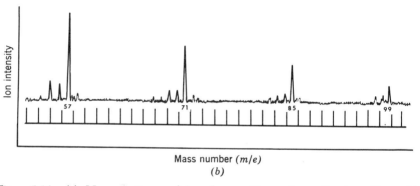

Figure 5-16. (*a*) Mass spectrum of hexadecane. Upper trace direct oscillographic record; lower trace spectrum recorded on magnetic tape. (*b*) Mass spectrum of hexadecane (*m/e* 57–59) recorded on magnetic tape at 7.5 in./sec and reproduced on an oscillograph (12).

samples may be as high in 10,000 per year in a petroleum industry laboratory. With the inception of high molecular weight mass spectrometry where several hundred peaks may be required to be evaluated in every analysis, advanced data reduction has become a necessity. Digitizers of varying complexity have been described in the literature, and at least three commercial models are available.

The inputs of a mass spectrum digitizer system are two analog voltage signals. One represents the ion current, the other identifies the mass. The voltage signal which is proportional to the ion current originates from the amplifier system. The voltage signal which is proportional to the mass is normally the ion acceleration voltage (for voltage-scanning instruments). The two signals are accepted by the digitizer in an analog form, are converted in two subsystems (in parallel) to digital signals, necessary calculations are performed, and the data are presented in digital form ready for further calculations by the operator. The acceleration voltage readout may, of course, be calibrated in amu units.

Thomason (13) describes a solid state digitizer, an improved version of which is not available through Consolidated Electrodynamics Corp. The sequence of logic control in the signal identification can be described with reference to Figure 5-17 (13). Logic state I represents the base line: the circuit is "looking for" a signal that is greater than the zero signal. As soon as an ion signal appears (state II) the input signal is increasing. The top of the peak is identified by a reading that is smaller than the previous one. A digital filter, consisting of five additional readings, is provided at this point (state III) to prevent false identification of noise peaks. In logic state IV the acceleration voltage input is read. This voltage—in the 0.000–9.999 V range—is a known fraction of the ion acceleration voltage for a given magnet current setting and is inversely proportional to the mass number. In state V command is given to print out the information stored in the digital voltmeters (peak height and mass number). In logic state VI the system is looking for a signal equal to the previous reading. The peak is decaying, the base line is reached and command is given to reset all registers and prepare for a new peak. The entire sequence of events occurs in milli-seconds. State VIII indicates a positive signal which is not analyzed. The operation of the digitizer is controlled by two clocks, operating at 60 cps and 100 kc, respectively.

Available readout devices include digital printers (1–40 peaks/sec), paper tape punches (20–100 characters/sec), and magnetic tape recorders. The sampling interval is usually adjustable from 1 to 1000 msec, and the dynamic range is $10^5:1$.

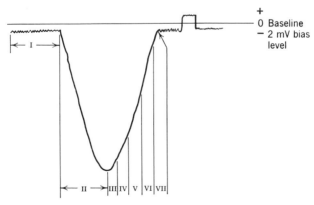

Figure 5-17. Signal identification in digitizers. Sequence of logic control (13).

II. PHOTOGRAPHIC DETECTION

A. Principle

The oldest detection technique involves the use of screens made of materials that fluoresce when bombarded by ions. Indeed, positive ions were discovered by Goldstein (1886) on the basis of the fluorescence of glass under their impact. Thomson used zinc sulfide scintillation screens, and many other materials (e.g., willemite) had been employed. In 1910 Koenigsberger and Kutschewski discovered the ion-sensitive character of photographic plates, and after that use of scintillation detectors was practically non-existent until the recent revival of scintillation counters (6).

The usual photographic emulsion for the detection of positive ions consists of a suspension of silver bromide crystals in gelatin. This suspension is spread as a thin layer of film on a glass plate. The plates are normally 2×15 in. or 2×10 in.

A photosensitive emulsion is a composite collector-transducer-recorder. The bombardment by an energetic positive ion beam (keV range) results in a latent image on the silver bromide grains which are suspended in gelatin. The halide crystals always contain within the crystal lattice a number of free positive silver ions and also silver sulfide specks ("sensitivity specks"). The impinging ions release electrons inside the halide crystals and the diffusing electrons transfer their charge onto the sensitivity specks. The negative charge thus acquired by the silver sulfide specks is next neutralized by the free positive silver ions which precipitate as metallic silver and serve as the nucleus in the process of developing. The developing agent performs what amounts to chemical amplification, and the final negative image consists of reduced silver grains around the sites of the latent image. The unreduced silver halide is removed by the fixing bath.

The main role of gelatin is to act as the agent in which the silver halide grains are mechanically suspended; the size of the grains is 0.5μ or less, and there are about 10^9 grains/cm^2. Since the penetration range of kV-energy ions in gelatin is short, the gelatin coating must be quite thin; the proportion of silver in the gelatin is higher then in conventional emulsions. The nature of ion detection with photographic emulsions is discussed by Owens (14).

B. Emulsion Response to Ion Bombardment

The two most important performance characteristics of ion-sensitive photoplates are sensitivity and the shape of the response curve. Sensitivity has been defined in various ways. Fundamentally, sensitivity is the number of blackened silver bromide grains per incident ion, for ion density approaching zero (15). Since this definition does not include grain size, it is better to consider sensitivity as a fractional area blackened per incident ion density, for vanishingly small ion densities (16).

A number of methods have been proposed for emulsion sensitization: reduction of the gelatin content ("Schumannizing"), treatment with various chemicals (sulfides, iodides), preexposure, etc. In present-day mass spectrography, commercially available photoplates are employed almost exclusively, and most of the development work is done by plate manufacturers on the basis of suggestions made by mass spectroscopists.

At this time there are five types of emulsion available for positive ion detection. Their properties are described by Rudloff (17) and McCrea (18). The five emulsions are: three kinds of Ilford Q plates (Q1, Q2, and Q3), Eastman Kodak's SWR emulsion, and the Schumann plates, manufactured by Agfa. The Ilford Q emulsions are by far the most widely used, the Q2 type being the most reliable for analytical applications because of its intermediate granularity and contrast and relatively high sensitivity. The Q1 emulsion has low speed, fine grains, and high contrast, while the Q3 is just the opposite.

The sensitivity of photographic emulsions, within batches of the same type or of different types, may be compared by determining the minimum ion charge density necessary to attain a certain blackening. Owens reports (19) a sensitivity of 10^5 ions/mm^2 or about 10^{-14} coul/mm^2 for Q2 plates when they are bombarded by ions of mass 100 accelerated to 15 keV. Since the area of a line is about 0.1 mm^2 the minimum detectable line involves about 10^4 ions. Similar values were found by a number of other authors, and the value is also about the same for SWR plates. More recently, the sensitivity of Q2 plates has been increased by improved manufacturing; however, no numerical values are available at this time.

Reproducibility of the blackening, assuming constant exposure, depends on the uniformity of the emulsion, both from plate to plate and over the surface of single plates, and on the process of developing. Errors caused by uneven development can be reduced by rigidly controlled developing (20,21) in a mechanical developer. A widely applied method for correction is the use of standard exposures of a known sample (tin with its many isotopes) on each plate. Much research work is being done presently to advance the production of uniformly sensitive plates.

The sensitivity of emulsions exhibits ion mass dependence and also ion energy dependence. Owens has shown (22) that the mass dependence of Q2 plates may be expressed by the function $\sqrt{M_{Pt}/M}$, where M is the mass of the ion under test and M_{Pt} is the mass of the reference ion ^{198}Pt$^+$. Es sentially the same $M^{-1/2}$-type dependence was also found by Burlefinger and Ewald (15), who determined absolute plate sensitivities for various ion masses by counting grains, and by Rudloff (17), who studied the dependency of blackening upon ion mass of Q1 plates using a relative method, resulting in continuous blackening curves. Owens' *mass* dependency curve (22) is shown in Figure 5-18. The *energy* of impinging ions is largely determined by the acceleration voltage. As shown in Figure 5-19, the ion energy dependence of the photographic response is linear for Q2 plates in the 3–15 kV range. The energy range in which photographic response is linear with ion energy was shown to extend to 45 kV (22). When evaluating multiply charged ions it must be remembered that ion energy is the product of ion charge and accelerating potential.

The variation of photoplate sensitivity as a function of ion mass ($\sim M^{-1/2}$) and ion energy ($\sim M$) has been considered as a result of the presence of the gelatin. Hunt (23) investigated the properties of vapor-deposited silver bromide plates which do not contain gelatin. These investigations have recently been extended by Honig et al. (24) and Woolston et al. (25). These and other studios currently in progress will undoubtedly lead to the development of photoplates with optimal properties for ion detection in mass spectrographs.

C. Use of Photoplates

Photographic plates have been used since the early days of mass spectroscopy for precision mass measurements. Since no resolving slit is needed, the highest possible resolving power is obtained for a given analyzer geometry. Photographic detection is essential with spark source instruments

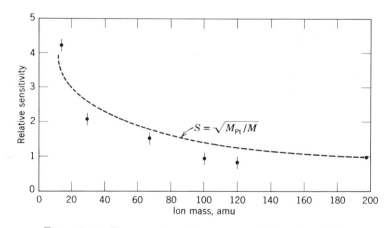

Figure 5-18. Ion mass dependency curve of photoplates (22).

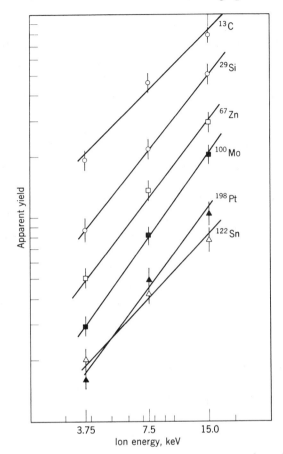

Figure 5-19.   Ion energy dependency curve of photoplates (22).

for the analysis of trace impurities in solids because the spark source is unstable, has a low useful yield, and produces electrical noise. The photographic plate, which is an integrating detector and does not involve electronics, is thus particularly suited to spark sources. In addition, a complete spectrum over a wide mass range can be obtained in the Mattauch-Herzog geometry (1:36, e.g., from Li to U) and can be recorded simultaneously. This is a distinct advantage both in exploratory searches for trace impurities and when only small samples are available. Problems of quantitative measurements are discussed in Section 11-III-B. It is becoming increasingly popular to employ photographic recording in high resolution instruments in connection with organic structure determinations. The merits and limitations of both electrical and photoplate recording in high resolution work are evaluated in Section 10-IV.

Light sensitivity, plate introduction and positioning (to within 0.001 in. of the correct focal plane), and requirements for prepumping for low background in the analyzer are basically engineering problems which have been successfully handled in commercial instruments. The limited dynamic range of photoplates (50:1 in one exposure) can be overcome by making as many as 15 exposures on a single plate, increasing the exposure each time by a factor of $10^{1/2}$, thus covering an intensity range of 10^7 to 1.

The principal advantage of photoplates is the ability to record simultaneously complete spectra, up to hundreds of ion beams, at high resolution. This is important when the sample is available only for a limited period of time, e.g., the effluent of a gas chromatograph. Other advantages include good precision for mass measurement, data availability for later retrieval and use, and no problems with complex electronics. Disadvantages are: relatively low sensitivity (10^4 ions to produce detectable lines), no possibility to obtain data on rates, difficulties in quantitative measurements due to nonlinear response, expensive auxiliary equipment for automatic data handling, and problems in experimental procedures (storage, light protection, outgassing).

References

*1. Nier, A. O., E. P. Ney, and M. G. Inghram, *Rev. Sci. Instr.*, **18**, 294 (1947).

 2. Palevsky, H., R. K. Swank, and R. Grenchik. *Rev. Sci. Instr.*, **18**, 298 (1947).

*3. Williams, W. R., and R. C. Hawes, Instruments & Control Systems, Nov., 1963.

 4. Inghram, M. G., R. J. Hayden, and D. C. Hess, *Phys. Rev.*, **72**, 349 (1947).

 5. Allen, J. S., *Rev. Sci. Instr.*, **18**, 739 (1947).

*6. Daly, N. R., *Rev. Sci. Instr.*, **31**, 264 (1960); **34**, 1116 (1963).

*7. Goodrich, G. W., and W. C. Wiley, *Rev. Sci. Instr.*, **32**, 846 (1961).

*8. Inghram, M. G., R. J. Hayden, and D. C. Hess, "Mass Spectrometry in Physics Research," *Natl. Bur. Std. (U.S.), Circ.*, **522**, 1953, p. 257.

 9. Young, J. R., *Rev. Sci. Instr.*, **37**, 1414 (1966).

10. Haumann, J. R., and M. H. Studier, *Rev. Sci. Instr.*, **39**, 169 (1968).

11. Dorsey, J. A., R. H. Hunt, and M. J. O'Neal, *Anal. Chem.*, **35**, 511 (1963).

*12. Merritt, C., P. Issenberg, M. L. Bazinet, B. N. Green, T. O. Merron, and J. G. Murray, *Anal. Chem.*, **37**, 1037 (1965), also P. Issenberg, M. L. Bazinet, and C. Merritt, *Anal. Chem.*, **37**, 1074 (1965).

13. Thomason, E. M., *Anal. Chem.*, **35**, 2155 (1963); also, Bulletin Model MSD4, Mass Spectrometer Digitizer, Consolidated Electrodynamics Corp., Calif., 1966.

*14. Owens, E. B., in *Mass Spectrometric Analysis of Solids*, A. J. Ahearn, Ed., Elsevier, Amsterdam, 1966, Chapter III.

15. Burlefinger, E. and H. Ewald, *Z. Naturforsch.*, **16a**, 430 (1961).

16. Franzen, J., K. H. Maurer, and K. D. Schuy, *Z. Naturforsch.*, **21A**, 37 (1966).
17. Rudloff, W., *Z. Naturforsch.*, **17a**, 414 (1962).
18. McCrea, J. M., *Appl. Spectry.*, **20**, 181 (1966).
19. Owens, E. B., *Rev. Sci. Instr.*, **32**, 1420 (1961).
20. McCrea, J. M., *Appl. Spectry.*, **19**, 61 (1965).
21. Kennicott, P. R., *Anal. Chem.*, **37**, 313 (1965); **38**, 633 (1966).
22. Owens, E. B., *Appl. Spectry.*, **16**, 148 (1962).
23. Hunt, M. H., *Anal. Chem.*, **38**, 623 (1966).
24. Honig, R. E., J. R. Woolston, and D. A. Kramer, *Rev. Sci. Instr.*, **38**, 1703 (1967).
25. Woolston, J. R., R. E. Honig, and E. M. Botnick, *Rev. Sci. Instr.*, **38**, 1708 (1967).

Chapter Six

Vacuum Techniques

I. REVIEW OF KINETIC THEORY

In the gaseous state molecules move freely and occupy uniformly the total volume of the vessel into which they are introduced. The bulk behavior of ideal or nearly ideal gases can be described by the basic gas laws which can be derived and interpreted on the basis of the kinetic theory. Only the most pertinent concepts are reviewed here.

Vapors are substances in the gaseous state below their critical temperatures, e.g., water vapor and mercury vapor at room temperature. They may be compressed to liquid or solid. The *vapor pressure* of a pure liquid or solid is the pressure of the vapor which is in equilibrium with the substance at a given temperature. Since a system with one component and two coexisting phases has, according to the phase rule, only one degree of freedom, the vapor pressure of pure substances is uniquely determined by the temperature.

Pressure in a gas is defined as "the net rate of transfer of momentum in the direction of the positive normal to an imaginary plane surface or specified area, located in a specified position in the gas, by molecules crossing the surface in both directions, momentum transmitted in the opposite direction being counted as negative, divided by the area of surface" (1). On the basis of this definition, the units of pressure are the microbar ($= 1$ dyne/cm²) and the pascal ($= 1$ newton/m²) in the cgs and mks systems, respectively. Other units, such as kg-force/m², or pound-force/in.², are used only for pressures greater than atmospheric. In engineering practice psig (pounds/in.² gauge) denotes pressure measured above atmospheric.

Pressure may also be defined on the basis of the statics of perfect fluids applied to the equilibrium of a liquid column in a Torricelli experiment. Here the pressure is expressed by multiplying the height of a liquid column by the density of the liquid, and then multiplying the product by the acceleration of free fall at the location of the experiment. The normal atmosphere used to be defined as the pressure exerted by a column of mercury 760 mm high at 0°C at a location where the acceleration of free fall was 980.665 cm/sec². More recently, 1 atm is taken to be exactly equal to 101,325 N/m². On this basis, 1 mm of mercury, which was earlier proposed to be called 1 torr, became equal to 1.000,00014 torr. For all practical purposes,

$$1 \text{ torr} = 1 \text{ mm Hg} = 1/760 \text{ atm} = 1333 \text{ mbar} = 133 \text{ pascal}$$

Pressure between 1 and 760 torr is usually called *rough* vacuum. The pressure range of 10^{-3} to 10^{-6} is normally called *high* vacuum, while the adjectives for the 10^{-6} to 10^{-9} and 10^{-9} to 10^{-13} torr ranges are *very high* and *ultrahigh*, respectively. Pressure in the 10^{-1} to 10^{-5} torr range is often described in terms of fractions of multiples of the *micron*, 1 μ being equal to 10^{-3} torr.

According to kinetic theory, pressure is given by

$$P = \tfrac{1}{3} mnv^2 = \tfrac{1}{3} \rho v^2 = \tfrac{2}{3} E_{\text{kin}} \tag{6-1}$$

where m is the mass of the molecule, n is the number of molecules per unit volume, ρ is density, v^2 is root-mean-square velocity, and E_{kin} is the total kinetic energy. For a given mass of gas an increase in pressure from P_1 to P_2 is accompanied by a corresponding decrease in volume from V_1 to V_2, and vice versa, as long as the change is isothermal ($T = $ constant). This is *Boyle's law*, $P_1 V_1 = P_2 V_2$. When the pressure is kept constant, the volume of gases varies directly with changing temperature, $V_1/V_2 = T_1/T_2$ (*Charles' law*). Similarly, the pressure of a given mass of gas varies in direct proportion with the absolute temperature when the volume is maintained constant, $P_1/P_2 = T_1/T_2$. The well-known *general gas law* is a combination of the above laws, yielding for a gram mole of gas $PV = R_M T$, where R_M is the molar gas constant which is the same for all gases ($R_M = 8.31$ ergs/°K/mole $= 1.987$ cal/°K/mole). The *quantity of gas*, in important concept in mass spectrometry, refers to a PV product. For example, the quantity of gas in a 1 liter volume at 1 micron pressure is 1 micron-liter. It is noted that 10^3 micron-liters $= 1$ mm-liter.

According to *Avogadro's law*, equal volumes of all gases at the same pressure and temperature contain equal numbers of molecules. One mole of any gas occupies 22.41 liters at 0°C and 1 atm, and the number of molecules, N_A (Avogadro's number), has been determined by several methods to be 6.023×10^{23} molecules/mole. If M is molecular weight and m is mass (in grams) of a molecule, then by definition

$$M = m \times N_A \tag{6-2}$$

The sample reservoir of mass spectrometers is usually about 3 liters. The number of molecules in this volume at a pressure of 1 torr is about 10^{20}. The pressure in mass spectrometer analyzers is normally kept at $\sim 10^{-8}$ torr. One liter of gas at this pressure contains about 3×10^{11} molecules.

If several gases are present at the same time, the total pressure of the mixture is the sum of the pressures that would be exerted if each gas were present separately. This is *Dalton's law*; it is valid as long as the gases do not react chemically with each other. The pressure of each gas is described by the term *partial pressure*. Thus, total pressure is the sum of the partial pressures.

Three more useful concepts may be derived from the kinetic theory: mean free path, number of collisions, and the rate at which molecules strike a surface. The *mean free path, L,* is "the average distance that a particle travels between successive collisions with other particles in the gas phase" (1). In terms of the number of molecules per unit volume, n, and the effective molecular diameter, d, the mean free path is given by

$$L = 1/\sqrt{2}\pi n d^2 = \text{const.}/P \qquad (6\text{-}3)$$

The term $\sqrt{2}$ enters on account of the Maxwell-Boltzmann distribution of molecular velocities. Note that the mean free path is inversely proportional to the pressure (at $T = $ constant). A useful practical form of the above equation is $L = 5/P$, which gives an easy-to-remember means to estimate the mean free path of air or nitrogen. Here the pressure is measured in microns and mean free path in centimeters (at room temperature). Gas kinetic data for air are shown in Table 6-1. Mean free path values are different for different gases, due to the d term in the equation; however, the molecular diameter for most inorganic gases is in the 2–4×10^{-8} cm range.

Table 6-1   Gas Kinetic Data for Air

Pressure, torr	Molecular density, molecules cm^{-3}	Mean free path, cm	Molecular incident rate, molecules cm^{-2} sec^{-1}	Time to form monolayer, sec
760	2.5×10^{19}	7×10^{-6}	3×10^{23}	3×10^{-9}
1	3.5×10^{16}	5×10^{-3}	4×10^{20}	2×10^{-6}
10^{-7}	3.5×10^{9}	5×10^{4}	4×10^{13}	20

Average molecular diameter: 3.7×10^{-8} cm
Thermal conductivity: 0.057×10^{-3} cal/cm/sec/°C (at 0°C)
Viscosity: 180 micropoises (at 15°C)

For ions the mean free path is given by $L_{ion} = \sqrt{2}\,L$, while for electrons moving through a gas $L_{electron} = 4\sqrt{2}\,L = 5.66L$, where L always refers to the mean free paths of the neutral molecules. It is emphasized that both L and $L_{electron}$ are mean values assuming uniform distribution of molecules within a volume. In reality, because of the chaotic nature of molecular motion, the free path of individual molecules varies from zero to infinity. The number of molecules, n_m, able to traverse a distance l in a gas without collision is given by

$$n_m = ne^{-l/L} \tag{6-4}$$

where n is the total number of molecules in unit volume, L is the mean free path, and e is the exponential base, 2.71828. In a typical ion source, at a pressure of 10^{-5} torr, the fraction, f, of the number of electrons able to travel a distance of, say, 2 cm without collision with nitrogen molecules is given by

$$f = n_m/n = e^{-2/2830} = 0.993$$

This means that only the fraction $1 - 0.993 = 0.007$ is available for ionization.

The *collision frequency*—the total number of collisions, Z, per second between molecules in 1 cc volume—is derived as

$$Z = \tfrac{1}{2}\sqrt{2\pi}\,(nN_A)^2 d^2 v_{av} \tag{6-5}$$

where v_{av} is the average velocity of the molecules. Since nN_A is of the order of 10^{19} molecules/cc at STP, the molecular diameter is about 10^{-8} cm, and the values of v_{av} are of the order of 10^5 cm/sec, one obtains Z in 1 cc of gas at STP as approximately 10^{28} sec^{-1}. The collision frequency per molecule (the average number of collisions per second suffered by one molecule) is called the rate of collision.

The *rate at which molecules strike a surface*, i.e., the number of molecules, ν cm^{-2} sec^{-1}, striking a unit area per unit time is given by

$$\nu = \tfrac{1}{4}nv_{av} = 3.5 \times 10^{22}\,P(\text{torr})/\sqrt{MT} \tag{6-6}$$

where M is molecular weight. At 10^{-8} torr there are about 4×10^{12} impacts cm^{-2} sec^{-1} for air. The ratio of the number of molecules which condense on a surface to the number incident (in unit time) is given by the condensation or sticking coefficient, β. The time, t_m, required for a clean surface to become covered with a monolayer of gas is obtained as

$$t_m = n_m/\beta\nu \tag{6-7}$$

where n_m is the number of sites per monolayer. The value of β is in the 0.1–1.0 range for most gases on metals. Some data for air are given in Table 6-1.

II. VACUUM REQUIREMENTS IN MASS SPECTROMETERS

Vacuum in mass spectrometers is usually maintained at the 10^{-7} to 10^{-8} torr level except in the immediate region where the ions are formed. At this pressure the number of molecules per cubic centimeter is about 3×10^8. The operation of mass spectrometers requires a vacuum environment because of two adverse effects of residual gases: ion scattering and "background."

When an ion passes a gas molecule so that the distance between particles is comparable with atomic dimensions, interaction occurs and the ion is *scattered*. The amount of scattering may be measured experimentally or calculated on the basis of geometrical and interaction potential considerations (2). While in electromagnetic separators ion scattering produces contamination, in mass spectrometry it results in *peak broadening* as ions are diverted from their prescribed trajectories. Scattering in the analyzer, for example, reduces the ultimate sensitivity of photoplates due to diffuse darkening. The appearance of charge-exchange lines in spark-source instruments is also a result of collisions between (multiply charged) ions and residual gases (Section 9-V). To minimize the number of collisions between ions and residual gas molecules, as the ions travel from source to collector, pressure must be kept at a level such that the mean free path of the neutral molecules is large. Since the ion path length in most mass spectrometers is of the order of 0.5–2 m, it is concluded that 10^{-7} torr, at which the mean free path is of the order of 500 m, is adequate to prevent serious peak broadening.

Residual gases in the ionization chamber are ionized together with the sample material. All spectra must therefore be corrected for *instrument background*. High resolution may be required to separate superimposed background peaks from those of interest (Section 8-II), and background often constitutes the limiting factor in attainable detection limits in trace analysis. In this respect, the existing partial pressure of particular components is more important than the total pressure. The type and amount of residual gases contributed by the various components of the vacuum system are discussed later in this chapter; residual gas analyzers are described in Sections 8-III and 13-IV.

III. CONSTRUCTION MATERIALS

A. Gas Load

The equilibrium pressure, p_e, in a vacuum system, after a sufficiently long pumping, is given by

$$p_e = r\tau = Q/S \qquad (6\text{-}8)$$

where $r(t) = Q(t)/V$ is the rate of rise of pressure in the absence of any pumping, $\tau = V/S$ is the time required to pump the system to $1/e$ (38.2%) of its original value (time constant), Q is the total rate of influx of gas, S is the pumping speed of the whole system, V is the volume, and t is time. In other words, the ultimate pressure in a vacuum system is determined by an equilibrium between the rate of removal of gas (pumping) and the rate of inflow of gas from various sources (gas load).

Apart from backstreaming and/or backdiffusion from pump fluids and influx from "real" leaks, there are four sources of gas inside a high vacuum system: (a) desorption of adsorbed gases, (b) volatilization, (c) diffusion of gases from the inside of solids, and (d) gas permeation. The release of gas molecules adsorbed on surfaces determines pumpdown characteristics, while the other three, particularly volatilization, determine the ultimate pressure. The amount of gas released into the vacuum system by any of these mechanisms depends strongly on the construction materials (see later).

Gas Desorption. When gaseous molecules or ions strike a bare surface *in vacuo*, there is a probability that they may become sorbed, i.e., retained by the solid substance. When the molecules are retained on the surface the process is called *adsorption*, while the term *absorption* refers to the case when the molecules enter the interior of the solid by diffusion and/or solution. Classification may also be made on the basis of whether the binding occurs through weak van der Waals forces or by chemical compound formation. Often it is difficult to determine the exact nature of the process. When gases are released from the surface, the process is known as *desorption*. Eventually an equilibrium is reached in which the number of molecules arriving at the surface equals the number leaving the surface. This equilibrium involves several monolayers, and the equilibrium pressure in the system is in some respects analogous to a vapor pressure. During pumping, and immediately after introducing a sample gas, the equilibrium is disturbed. In fact, in most practical mass spectrometry problems there exists a continuous exchange of gas with the walls of the instrument through adsorption and desorption. This phenomenon is usually difficult to control. In trace analytical applications, for example, a "conditioning" of the instrument is often needed to prevent the loss of minute components and/or to eliminate unwanted contributions from previous samples.

Removal of adsorbed molecules can be speeded significantly by increasing the temperature, i.e., by *baking*. Water is removed at 150°C; however, to remove other adsorbed gases and to achieve vacuum lower than 10^{-7} torr, baking at 300–450°C is practically mandatory. Since the whole system, including valves, gauges, etc., must be heated, all construction materials have to withstand the degassing process. Degassing rates (torr-liter/sec-cm^2) for various materials can be found in vacuum technique monographs. Degassing rate multiplied by surface area yields the gas load which must be removed by the pumps. On the surface of stainless steel, for example, there are about 50 monolayers of air at 1 atm, corresponding to about 1×10^{-3} torr-liter/cm^2, and it takes about 4 hr of pumping to decrease the degassing rate by a factor of 10. Surface roughness (welded joints) may increase the number of monolayers by a factor of 10^3. It is noted that one monolayer contains about 3×10^{-5} torr-liter/cm^2 gas at 20°C.

Vapor Pressure and Outgassing. The requirement of low vapor pressure at the material's highest operating temperature (which is usually the baking temperature) does not require further explanation. Vapor pressure data for solid, liquids, gases, pumping fluids, and sealing materials at various temperatures are given by Honig and Hook (3). In studying their tables it becomes obvious, for example, that brass cannot be used in baked systems because the vapor pressure of zinc is several microns at 300°C.

The residual pressure after prolonged pumping and baking is usually determined by the outgassing characteristics of the construction materials. Gases inside solids may reside as entrapped gas, as gas in solution, or in the form of chemical compounds, such as hydrides and nitrides. The original source of the gas is either the surrounding atmosphere during melting, transferring, or processing, or the containers, mold materials, and fluxes with which contact is made. Gases generally evolving are hydrogen, carbon monoxide, carbon dioxide, nitrogen, and water. Hydrogen, which has a high solubility in metals, and nitrogen (low solubility) are usually present in interstitial solution, while the evolution of carbon monoxide and carbon dioxide is often the result of a reaction between carbon diffusing to the surface and oxide impurities on the surface. Gas solubility, C, as a function of pressure, p, is given by

$$C = C_0 p \qquad \text{in nonmetals} \qquad (6\text{-}9a)$$

$$C = C_0 p^n \qquad \text{in metals} \qquad (6\text{-}9b)$$

where the dimensionless C_0 is the quantity of gas in cubic centimeters at STP in 1 cc of solid at 1 atm pressure; n is usually about 0.5. The diffusion of gases into the surface obeys Fick's law* and diffusion coefficients may be determined experimentally. In carbon monoxide evolution, where a chemical reaction also takes place, relations are more complex.

Gas Permeation. The permeation rate (molecules/sec) through metals is given as a function of the pressure differential, $p_2 - p_1$ (where $p_2 > p_1$), across the material by

$$F = KA(p_2 - p_1)/d \qquad \text{gases through nonmetals} \qquad \text{(6-10a)}$$

$$F = KA(p_2 - p_1)^n/d \qquad \text{gases through metals} \qquad \text{(6-10b)}$$

Here K is the permeation constant, i.e., the quantity of gas in cc at STP passing through 1 cm^2 area (A) of a wall of 1 cm thickness (d) when a pressure difference of 1 atm exists across the wall; n is again about 0.5. Table 6-2 summarizes the general features of the permeation of gases through various materials (4). Mass spectrometers are normally continuously pumped systems, so the slight gas penetration through walls at room temperature is not significant, except in special cases, e.g., in the Reynolds type of static instrument (Section 8-III).

Table 6-2   General Features of the Permeation of Gases through Various Materials (4)

Glasses	Metals	Semiconductors	Polymers
H_2, D_2, He, Ne, O_2, Ar	No rare gas through any metal	H_2 and He through Ge and Si	All gases permeate all polymers
Measurable through SiO_2	H_2 permeates most, especially Pd	Ne, Ar not measurable	Water rate apt to be high
Vitreous silica fastest	O_2 permeates Ag H_2 through Fe by corrosion, electrolysis, etc.		Many specificities
All rates vary directly as pressure	Rates vary as pressure	H_2 rate varies as pressure	All rates vary directly as pressure

Note: In all, rate is an exponential function of temperature for true permeation.

* Fick's law states that the rate of diffusion is inversely proportional to the thickness of the material at constant pressure difference and temperature.

B. Glass

The well-known advantages of glass as a construction material are the many available shapes and sizes, easy fabrication, low vapor pressure, chemical inertness, transparency, excellent electrical insulation, etc. The silica content of glasses (50–100%) is responsible for their three-dimensional structure. The high melting point of silica is reduced in *soft* glasses with lime, soda, and other oxides. In the *borosilicate* glasses boron oxide is used as flux to increase chemical stability and heat resistance by reducing the alkali content. The composition and properties of many glass types are given by Corning (5). In vacuum systems the most frequently employed borosilicate type is Pyrex (Corning 7740) with the composition of 80.5% SiO_2, 12.9% B_2O_3, 3.8% Na_2O, 2.2% Al_2O_3, and 0.4% K_2O. Pyrex can be heated to 500°C and baked at 450°C. For temperatures up to 1000°C, or where severe thermal shocks are expected, high silica glass, such as Vycor (Corning 7900, 96% SiO_2, $<3\%$ B_2O_3, $<1\%$ oxides) is used.

In early home-made instruments, glass was used extensively as construction material. When the mass spectrometer tube itself is made of glass, a conducting coating can be made (6) by passing a mixture of $SnCl_2$ vapor and water vapor through the heated tube, thus creating a tin suboxide semiconducting film. Conducting transparent glass is now commercially available. The use of glass in modern mass spectrometers is restricted mainly to sample introduction systems (Chapter 7) and special experimental attachments. Small all-glass systems are extensively used in ultrahigh vacuum studies, and a number of residual gas analyzers, such as omegatrons, have all-glass vacuum envelopes. All-glass systems are usually smaller than 10 liters in size due to mechanical reasons.

When glass inlet systems are employed, one must be aware of the fact that glass releases large amounts of gas when heated. The principal gas evolving from the surface is water; it may amount to 50 monolayers. Carbon dioxide is second in quantity, corresponding to 5 monolayers, while carbon monoxide and other gases are removed with relative ease at 450°C. After removal of the surface gas, water appears again on prolonged baking. This water originates in the interior of the glass and diffuses to the surface (Fick's law is obeyed). The bulk diffusion of water may be accompanied by slow decomposition of the glass. Gas permeation through glass is usually negligible as far as normal mass spectrometry is concerned; in ultrahigh vacuum studies permeation may constitute the limiting factor in attaining the desired vacuum. The rate of permeation for helium through 1-mm thick vitreous silica at 25°C is about 5×10^{-11} cc (STP) per second for every square centimeter surface when the pressure difference is 1 cm across the wall. For other gases the value is orders of magnitude lower.

C. Metals

In addition to the general vacuum requirements of chemical inertness and low vapor pressure at temperatures up to bakeout, metallic components must also be nonmagnetic. The most frequently employed stainless steel is the austenitic-300 type containing 15–20% Cr, 6–13% Ni, <0.15% C, and <2% Mn. Copper of OFHC grade (oxygen-free high-conductivity) is used extensively in mass spectrometer tubes because copper oxide is semi-conducting. In the ion optical system the presence of insulating surfaces is undesirable because they may become charged when hit by the ion beam and the charged surfaces cause beam deflection which may be of considerable magnitude. Metallic surfaces can become insulating by surface oxidation or by oil contamination. To prevent oxide film formation, internal surfaces are sometimes gold or silver plated and highly polished. An example is the electrostatic sector in Mattauch-Herzog type double-focusing instruments.

Nickel is used in both pure and alloy forms: Nichrome V, Monel, Inconel, etc. are frequently used in ion sources. Tungsten finds applications in ionization gauges and ion sources; oxygen reacts with the carbon content of tungsten (Section 4-II). Rhenium is often used as filament material. Molybdenum and tungsten are employed where mechanical strength is needed at high temperatures, as in Knudsen cells (Section 7-VI-B). The outgassing of metals has been treated extensively in treatises on vacuum technology. The gas content, as mentioned already, greatly depends on composition, mechanical treatment, and thermal history. The most frequently evolved gases that contribute to background are hydrogen and carbon monoxide. Gas permeation, apart from hydrogen through palladium and nickel, and oxygen through silver is usually negligible.

D. Ceramics and Elastomers

Ceramics containing alumina (Al_2O_3), steatite ($MgSiO_3$), beryllia (BeO), and thoria (ThO_2) are noted for their resistance to thermal shock. Their electrical properties depend on composition and applied temperature. Ceramics are used mainly in seals, terminals, and special applications. A pure form of alumina (known as sapphire) is extensively used for *spacers* in ion sources.

Elastomers are chiefly used in demountable seals. Their drawback is outgassing which limits their use, particularly at higher temperatures. Synthetic rubbers, though many variations are available, are usually not recommended for use in seals because they contain volatile plasticizers and oils. Recently, fluorinated elastomers, such as Viton A, Kel-F, and Fluorel, have come into widespread use due to their favorable outgassing properties. More will be said about gaskets shortly.

IV. PUMPS

A. Performance Characteristics

Pumps produce vacuum by establishing a lower molecular density in themselves than exists in other parts of the system. This results in a net movement of molecules from the system into the pump. Gas molecules are removed permanently either by rejecting them through a no-return path or by transferring them into a condensed phase inside the pump. Depending on the mechanism by which the gas removal is accomplished, pumps may be classified into five groups: (1) positive displacement, (2) vapor stream, (3) getter-ion, (4) sorption, and (5) cryogenic.

Pump performance is characterized by three parameters: pumping speed, ultimate pressure, and limiting forepressure. The *pumping speed* (exhaust speed) of a pump, S_p, is defined as the volume, V, of gas removed by the pump per unit time, t, at the pressure, P, prevailing at the throat of the pump,

$$S_p = \left| \frac{dV}{dt} \right|_P \qquad (6\text{-}11)$$

Pumping speed changes considerably with varying inlet pressure. The term *preferential pumping* refers to the fact that pumping speed is also a function of the kind of gas being pumped (see later). The unit of pumping speed is liters/sec (1 liter/sec = 2.12 cu ft/min). A pump with a speed of 1 liter/sec evacuates a volume of 1 liter to a pressure of 0.37 times its initial pressure in 1 sec, assuming no wall effects. Pumps are available with speeds from a few liters/sec to thousands of liters/sec. The pumping speed employed in mass spectrometer vacuum systems ranges from 75 to 300 liters/sec.

Ultimate pressure (final pressure, ultimate vacuum) refers to the condition when further pumping will result in only negligible pressure reduction in a particular system. Assuming no leaks present, the ultimate pressure is determined by the characteristics of the pump, the type of pumping fluid used, and the outgassing of the construction materials. In modern mass spectrometers an ultimate vacuum of 10^{-8} torr is normally achieved. Often it is important to know what the ultimate pressure is "made of," i.e., the partial pressures of the components. The contribution of the various pump types to instrument background is considered in the description of their operation.

The *limiting forepressure* of a pump is the maximum allowable pressure of the forevacuum side of the pump. When this value is exceeded, direct communication is established between the forevacuum and high-vacuum sides of the pump, and effective pumping ceases. The positive displacement pumps, commonly known as mechanical pumps, the simple water-jet pump, and the sorption pumps operate against atmospheric pressure. These are normally used as *forepumps* (or backing pumps) in connection with the other pumps which begin to operate only below a certain pressure level, usually below 10^{-3} torr.

B. Mechanical Pumps

Mechanical pumps are used for initial pumpdown ("roughing") from atmospheric to about 10^{-2} torr (single-stage) or 10^{-3} torr (two-stage), and also as forepumps for vapor stream and other pumps. All *rotary pumps* consist of a solid cylindrical rotor which rotates within a cylindrical housing, the stator. The rotor sweeps the gas in the volume between it and the stator from the inlet to the exhaust port. The various makes of rotary pumps differ mainly in the method of sweeping the volume between rotor and stator. Light paraffin oil (high vapor pressure fractions removed) is used for sealing and lubrication of the accurately machined pump parts (tolerance is 0.0001 in. at the point where rotor and stator are closest together).

Figure 6-1 shows a *double-vane* mechanical pump. The rotor is mounted concentrically on the pump shaft, while the shaft is eccentrically located in the stator. As the rotor revolves at 200–500 rpm, the crescent-shaped gas space is swept twice each revolution and the gas molecules are forced out into the atmosphere through a pressure-operated exhaust valve. In the *single-vane* type of pump the vane is mounted in the stator between inlet and exhaust ports and is spring-loaded against the surface of the rotor. The rotor is eccentrically mounted on the pump shaft and the shaft is centrally located in the stator. Here the vane provides the stationary seal, and, as it turns, the rotor sweeps the crescent-shaped volume. In the *rotary-piston* mechanical pump the shaft is centrally located in the stator

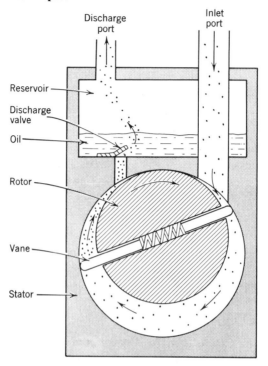

Figure 6-1. Double-vane mechanical pump. (After W. F. Brunner and T. H. Batzer, *Practical Vacuum Techniques*, Reinhold, New York, 1965, p. 27).

and the rotor is eccentrically mounted on the pump shaft. Here the piston does not turn. Instead, the piston sweeps the volume between it and the stator as the rotor turns with the shaft. There is a vane-like extension on the piston ("slide") which moves up and down in an oscillating seal, called the slide pin.

In the commonly used *two-stage* pumps (compound pumps) two chambers are arranged in series within the same housing and are connected by a channel. The exhaust of the first stage is coupled to the inlet of the second stage. The pumping speed of mechanical pumps is given by the manufacturers in terms of free air displacement which refers to the volume of air passed per unit time through the pump when there is 1 atm pressure on both the intake and output sides. Numerical values are quoted in cubic feet per minute; to convert to liters per minute multiply by 28.3. Required pumping speed in mass spectrometer forepumps is usually of the order of 20 liters/min.

Because the rotary pump is actually a compressor, condensible vapors such as water and organics may be liquefied during pumping. To minimize the loss in pumping speed caused by the mixing of condensation products with the pump oil, pumps are often featured with *gas ballast* (vented exhaust). The effective compression ratio of the pump is reduced from the normal 10^3:1 ratio to 10:1 by admitting a calculated amount of air at atmospheric pressure after the pump chamber is shut off from the space to be evacuated. When the exhaust valve opens, as the pumping cycle proceeds, air and noncondensed vapor are released together into the atmosphere or into a trap. Gas displacement in mass spectrometers is usually small and the backing pressure required for diffusion pumps can readily be provided even with full gas ballast. Gas ballast is normally not needed in mass spectrometer vacuum systems.

Mechanical pumps contribute water and hydrocarbon peaks to the basic mass spectrometer background. Higher hydrocarbons may be reduced by applying a liquid nitrogen or zeolite trap in the foreline, but total removal is difficult (cold-cathode discharge may be used to polymerize hydrocarbons). For the production of completely oil-free forevacuum, adsorption pumps (see later) have been used.

Two other types of mechanical pumps are becoming increasingly popular in vacuum technology, although they have not yet been used in mass spectrometers, mainly because of price considerations. In the *Roots-type* mechanical booster pumps two figure-eight-shaped rotors are counter-rotating at high speed. Clearance between the lobes and between lobes and the casing is 0.01 in. There is no friction and no oil lubrication is needed. Advantages are high pumping speed in the region (1–500 μ) where oil-sealed rotary pumps and vapor pumps are not very efficient, and oil-free operation. The *turbomolecular drag pump* (7) resembles a turbine: both rotor and stator consist of several slotted disks. The slots are obliquely machined, the rotor slits being in reverse to that of the stator slits. At a speed of 16,000 rpm, momentum is transferred to the molecules by molecular drag, resulting in compression in the axial direction. Oil-free ultimate vacuum is as low as 10^{-10} torr.

There is a version of a mercury piston pump, known as the *Toepler* pump, which is often used in connection with mass spectrometers for the collection of gas samples. Figure 6-2 shows the operation of an automatic Toepler pump (8). The gas to be pumped is allowed to flow into the piston chamber through the input dip tube, D. When air is admitted to the reservoir chamber by slowly opening stopcock F, mercury is forced to enter the lower dip tube, and the rising mercury pushes the gas in the piston chamber past the mercury-filled output valve B. At the same time, input dip tube D becomes closed. When the mercury reaches valve B, electrical

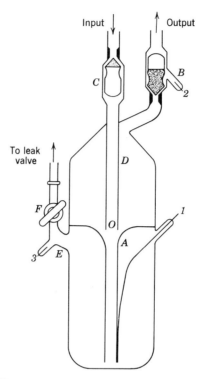

Figure 6-2.   Automatic Toepler pump (8).

contact is established between tungsten wires *1* and *2*, and a relay is closed. The relay contact stops the entering of air and simultaneously turns on an auxiliary pump to evacuate the reservoir chamber. This results in the lowering of the mercury level, and the closing of valve *B*. (The seat of the valve remains covered with mercury, providing positive closure.) As the chamber is evacuated, another portion of the gas is introduced in small bubbles through the opening in the input dip tube. The mercury level falls in the piston and rises in the dip tube until it reaches point *A*. The electrical contact between *1* and *3* shorts a relay which, in turn, turns off the evacuating pump and permits air to enter the reservoir chamber again. This cycle of operations continues until all gas is pumped into a reservoir chamber connected to the output. More recent improvements concern the replacement of the tungsten electrodes by photoelectric sensors (9).

C. Diffusion Pumps

In diffusion or vapor stream pumps a fluid, mercury or special oil, is evaporated from a boiler (Fig. 6-3) and the vapors stream out of a jet in

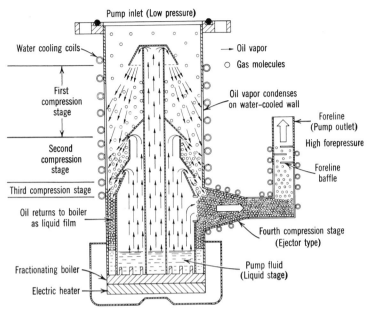

Figure 6-3.   Diffusion pump. (Courtesy of Norton Corp., Vacuum Equipment Division).

downward direction with supersonic velocities, hit the cooled walls of the pump, condense, and flow back to the boiler. Pumping action is based on the transfer of momentum from the high-speed vapor to entrained gas molecules which are forced to move toward the forevacuum. A pressure gradient is thus created between the space above the jet (system side) and the forevacuum (discharge side). The process can only take place after pressure has been lowered to molecular flow level ($<10^{-2}$ torr) by a mechanical backing pump. Two or three stages are normally employed in series.

Pump fluids must meet many requirements: low vapor pressure, low viscosity (return to boiler is by gravity), molecular weight in the 200–500 range (to increase "dragging" efficiency), thermal stability, and chemical inertness. Both mercury and oil diffusion pumps have their respective merits. Mercury pumps were favored for many years in mass spectrometer vacuum systems, but recently high-speed oil pumps with very low vapor pressure fluids are becoming popular.

Mercury is advantageous in mass spectrometers because it does not decompose on hot filaments (source, ion gauges), it is chemically inert, and its background in the m/e 198–204 (Hg^+) and 99–102 (Hg^{2+}) range does not interfere in analysis. Indeed, these background peaks are often used as "mass markers" for the m/e scale. Disadvantages are the necessity of a refrigerated trap and sensitivity to air inrush (oxidation). Trapping is absolutely necessary since the vapor pressure of mercury at the temperature prevailing at the "top" of the pump ($\sim$20°C) is about 1 μ. At this pressure many metals form amalgams. Dry Ice (-78°C) reduces the vapor pressure of hydrogen to the 10^{-9} torr level. Water, however, still has a vapor pressure of almost 1 μ at Dry Ice temperature, and for this reason liquid nitrogen traps (-196°C) are commonly used with mercury pumps. Trap designs are described in the next section. A small hydrocarbon background is often found with mercury systems; it originates either from forepump oils or from insufficiently cleaned parts.

Oils employed as pumping fluids can be divided into four major categories: (1) *Apiezon* oils consisting of various hydrocarbons obtained in petroleum refining, (2) *phthalic and sebacic esters*, particularly octyl phthalate (Octoil) and octyle sebacate (Octoil S), (3) *silicone oils* (methyl polysiloxanes), and (4) *polyphenyl ethers*. Vapor pressure and some other data for the most popular types of oils is given in Table 6-3. Octoil S has been a popular all-purpose pumping fluid in mass spectrometers on account of its good pumping speed. The silicone oils are distinguished for their high resistance to thermal decomposition and oxidation (air inrush). The DC-705 oil and the polyphenyl ethers are becoming

Table 6-3   Diffusion Pump Fluids

Trade name	Composition	Vapor pressure in torr at 25°C	Specific gravity at 25°C
Apiezon A	Hydrocarbon mixture	2×10^{-5}	0.873 (15°C)
Apiezon B	Hydrocarbon mixture	4×10^{-7}	0.871 (15°C)
Apiezon C	Hydrocarbon mixture	1×10^{-3}	0.880 (15°C)
Convalex 10	Polyphenyl ether	1×10^{-10}	1.2
Convoil 20	Hydrocarbons	5×10^{-7}	0.865 (15°C)
Narcoil 10	Ditrimethyl hexyl phthalate	1×10^{-7}	0.973
Octoil	Di-2-ethyl hexyl phthalate	3×10^{-7}	0.980
Octoil S	Di-2-ethyl hexyl sebacate	3×10^{-8}	0.910
Silicone DC 704	Semiorganic single molecule	2×10^{-8}	1.07
Silicone DC 705	Semiorganic single molecule	3×10^{-10}	1.09

Vapor pressure of mercury at various temperatures:

-180°C	negligible	20°C	1×10^{-3} torr
-78°C	3×10^{-9} torr	100°C	2×10^{-1} torr
0°C	2×10^{-4} torr	200°C	17 torr

increasingly popular because of the possibility of obtainingoil-free vacuum without traps.

Oil vapors decomposing in the ion source contribute to nearly every peak in the mass spectrum providing an undesirable (often fluctuating) instrument background which severely reduces instrument sensitivity. Pyrolysis of the oils inside the pumps is reduced by the use of multistage fractionating pumps. Here the vapors reaching the first stage (i.e., the jet nearest to the system being pumped) are purged of the more volatile impurities, resulting from decomposition, by partially condensing and refluxing the vapor at the various stages in such a way as to force volatile components to concentrate in stages closest to the forevacuum. Even with fractionating, however, pressure cannot normally be lowered below 10^{-7} torr without trapping. The use of polyphenyl ethers has first been suggested by Hickman (10). Vacuum as low as 10^{-9} torr has been claimed without any trapping. These oils are now commercially available and their use is becoming widespread. Overall performance, however, is still not documented adequately. Background peaks such as CH_4, C_2H_6, and CO appear to indicate some decomposition inside the source.

D. Traps and Baffles

A *trap* is a system of cooled walls or plates placed near the inlet of a vacuum pump to condense vapors that migrate from the pump (or from the vacuum system). *Baffles* are defined similarly, except that here the condensate is returned from the baffle to the boiler to reduce fluid consumption. There are two types of fluid migration in vacuum systems: *backstreaming* refers to the movement of oil or mercury molecules in the "wrong" direction, i.e., toward the high vacuum side. It is normally caused by scattering and the effect can be reduced by the use of baffles and proper jet design. *Back-migration* is the reevaporation of vapors from baffles and traps.

A widely used baffle type is the *chevron baffle* which consists of a series of V-shaped, water-cooled metal plates arranged in such a manner as to provide an optically dense system of surfaces. For better efficiency liquid nitrogen cooling may also be employed. Figure 6-4 shows three cold-trap designs. In the *reentrant* type trap the trap itself is immersed in liquid nitrogen which is held in a Dewar flask; such traps are commonly used to protect vacuum gauges. Efficiency is increased by employing a copper foil insert which provides a constant condensing surface irrespective of the level of the refrigerant due to the thermal conductivity of copper. The second design shown in Figure 6-4 is the *internal reservoir* type into which

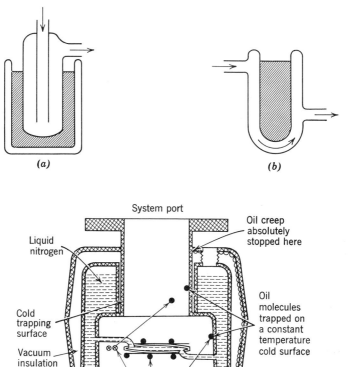

Figure 6-4. Three cold trap designs: (a) reentrant trap; (b) internal reservoir trap; (c) optically tight metal cold trap (courtesy of Granville-Phillips Corp.).

the refrigerant is poured. Suitable insulation of the outside walls cuts down on loss of liquid nitrogen. The positions of inlet and outlet can be varied. A rather sophisticated cold trap design is shown in Figure 6-4c. The trap offers minimum impedance to the diffusion pump, but at the same time the optically tight design virtually eliminates the entry of oil vapors into the vacuum system. A constant low temperature of all cryo-pumping surfaces prevents pressure bursts resulting from the reevaporation of condensates. In addition, the trap is bakeable to 450°C.

Many ingenious designs have appeared in the literature for automatic refilling of liquid nitrogen traps. In one design, a methane-filled glass tube submerges into liquid nitrogen; when the liquid level falls, methane evaporates and pushes a mercury column to make an electrical contact which, in turn, activates the refilling mechanism. In other designs thermistors are used (Fig. 6-5). The probe is submerged in the liquid and when the liquid level falls the thermistor is exposed only to the vapor above the liquid and the large difference in electrical resistance is utilized to signal for refilling.

Although the use of liquid nitrogen traps with oil pumps reduces the background significantly, light hydrocarbons still have a chance to "creep" into the ionization chamber. To further reduce the partial pressure of oil inside the system, adsorption traps and copper foil traps may be used. Both were developed in connection with ultrahigh vacuum research. In the adsorption trap of Biondi (11), artificial zeolite or activated alumina is used to remove oil; the trap is bakeable to 450°C. The Alpert-type (12) copper foil trap appears to function by a gettering action. It is actually a continuous spiral of corrugated OFHC copper sheet placed between the inner and outer tubing in a reentrant type trap. This trap operates at both room and liquid nitrogen temperatures and is an effective oil remover, particularly after activation by repeated oxidation (air) and reduction (hydrogen) at 450°C. With Octoil S, pumping fluid pressure of $\sim 10^{-10}$ torr can be maintained for a few days, after which saturation effects appear.

It should be kept in mind that cold traps act like pumps for condensable vapors such as water and carbon dioxide which, on heating, may reenter the ion source. Also, the conductance of traps and baffles must be considered when determining required pumping speed.

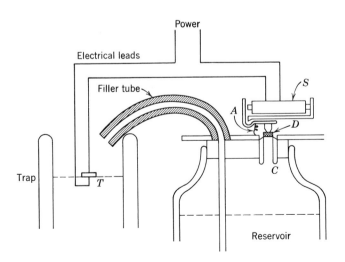

Figure 6-5. Liquid nitrogen filler using thermistor. (*T*) Thermistor; (*S*) solenoid; (*C*) tube to permit air to escape into atmosphere; (*D*) neoprene disk; (*A*) spring. (After A. Guthrie, *Vacuum Technology*, Wiley, New York, 1963, p. 342.)

E. Getter-Ion Pumps

In *chemical* or *getter* pumping an evaporated active material such as titanium or barium reacts with the gas molecules and removes them permanently in the form of low vapor pressure componds. In *ion* pumping the gas to be removed is ionized by a strong electric field or by electron bombardment and the ions are permanently removed by driving them into a surface. The two mechanisms are effectively combined in the getter-ion pumps. Here chemically active gases, such as oxygen, hydrogen, and nitrogen, are removed by an evaporated getter, while inert gases are ionized and the ions are buried in a subsequent getter film deposition.

Getter-ion pumps may be divided into two broad classes: hot-cathode and cold-cathode pumps. In *hot-cathode* pumps electrons from a heated filament are accelerated by a positive grid, and inert gases are ionized by electron impact in the area between the grid and the walls. In the meantime, titanium wire is fed onto a heated post and the evaporated metal reacts with the active gases that wander in from the vacuum chamber. The chemical compounds that form deposit on the cold walls, and the fresh metal film also buries the positive ions of the inert gases that are reaching the walls. In the *cold-cathode* pumps, flat-plate titanium cathodes and a hollow metal anode with a shell structure are mounted between the poles of a permanent magnet. When a high voltage is applied between cathode and anode, electrons are emitted. These electrons ionize neutral gas molecules, and the efficiency of ionization is good because of the oscillatory motion of the electrons due to the magnetic field. The positively charged ions travel toward the cathodes and become buried as they strike it. At the same time, the bombarding ions knock out titanium atoms from the cathode in a process known as *sputtering*. The sputtered titanium is deposited on the pump walls and anode cylinders and acts as a getter for the chemically active gases inside the pump, resulting in chemical compound formation. Thus both ion and getter pumping are accomplished.

There are several commercial getter-ion pumps available and improved designs appear frequently. Figure 6-6 shows a *triode* structure where auxiliary hollow-mesh cathodes are employed to increase sputtering by providing a means for the incident ions to strike at a grazing angle instead of normally, and also to acquire an increased energy during acceleration (13).

Getter-ion pumps start to operate at a pressure of about 20 μ and operate most effectively at pressures below 10^{-5} torr. In diffusion pump systems the pump must be isolated during roughing, and the pump must be connected to the mechanical pump continuously during operation. In getter-

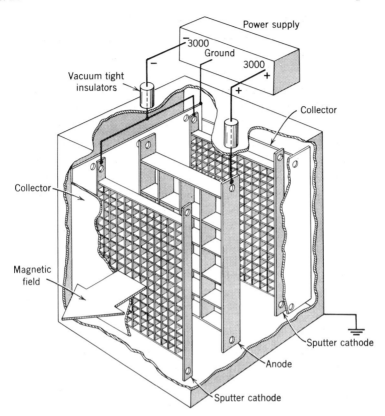

Figure 6-6. Structure of triode pump. (Courtesy of Consolidated Vacuum Corp.)

ion pump systems after the initial evacuation there is no further need for mechanical pumps since the pumped gases are in effect digested by the pump. However, an auxiliary pump (mechanical or sorption) is needed to reduce the system pressure initially to about 20 μ where the ionic pumping action will commence. Getter-ion pumps operate most effectively at pressures below 10^{-5} torr, and ultimate vacuum as low as 10^{-10} torr may be achieved. Advantages include oil-free vacuum, no need for refrigeration, economical operation, and long life. (Pump life is inversely proportional to pressure.)

Getter-ion pumps exhibit preferential pumping to a considerable degree. Since the pumping mechanism is different for different gases, the pumping speed for various gases depends on the chemical activity, ease of ionization, and the size of the molecules. Sputter-ion pumps are fastest for hydrogen and slowest for argon, the ratio of speeds for these gases being about 300. It is recalled that the argon content of air is about 1%. In addition, certain diode-type pumps show "argon instability," i.e., bursts of previously buried argon may occur. Thus, when noble gases are involved in a mass spectrometric study (such as in gas analysis), the use of getter-ion pumps is not recommended. Also, there is a tendency to release pumped gases (memory effect) which may be important in certain applications. It is recommended that getter-ion pumps should not be mounted in the line of sight with the mass spectrometer source region or collector assembly because ion pumps are basically glow discharge reactors and thus interfere with source and/or detector operation. Several studies appeared on the application of getter-ion pumps with mass spectrometers with the conclusion that, apart from special applications, there are no inherent reasons against the use of such systems, even in combination with diffusion pumps. Of course, proper design with respect to pumping speed, capacity, valving, etc. is essential for successful operation.

F. Sorption and Cryogenic Pumps

In addition to getters, other materials such as zeolites, activated charcoal, and activated alumina may also be used to reduce pressure by removing neutral molecules from the gas phase. Some of these pumps operate at room temperature, but liquid nitrogen is often used to increase efficiency. A promising application of sorption pumps is their use as oil-free forepumps, particulary in connection with sputter-ion pumps. Activated charcoal, for example, was found (14) to remove about 10 times its own volume of air at liquid nitrogen temperature. Pressures less than 10 μ could be achieved in a few minutes; this is adequate to start getter-ion pumps.

In cryogenic pumping, molecules are removed from the gas phase by condensation on a cold surface; the resultant pressure is the vapor pressure of the condensate. In mass spectrometry cryopumping is mainly applied in cold traps and in special applications (Section 11-II). Cryopumping is extensively used in space simulators and other large vacuum systems. Table 6-4 shows the melting and boiling points of a number of substances and their vapor pressures at liquid nitrogen and liquid helium temperatures. It is seen, for example, that at 20°K the vapor pressure of all common gases, except H_2, Ne, and He, is negligible. Recent availability of liquid hydrogen and helium suggests many applications.

G. Basic Vacuum Equations

The properties of vacuum systems can be described in a way analogous to electrical circuits by utilizing laws derived from the kinetic theory. A *vacuum circuit* consists of a series or parallel combination of tubings,

Table 6-4   **Melting and Boiling Points and Vapor Pressures of Various Gases**

Gas	Melting point, °C	Boiling point, °C	Vapor pressure, torr	
			$-190°C$	$-268.8°C$
Hydrogen	-259	-252.7	—	10^{-6}
Helium	-272	-268.8	—	760
Water	0	100	10^{-22}	—
Neon	-249	-245.9	—	10^{-26}
Nitrogen	-210	-195.8	760	10^{-81}
Carbon monoxide	-207	-192	760	10^{-94}
Oxygen	-219	-183	350	10^{-104}
Carbon dioxide	—	-78	10^{-7}	—
Mercury	—	—	10^{-32}	—

forepumps, diffusion pumps, traps, baffles, valves, and other components in addition to the volume that is being evacuated. The quantity of gas pumped is usually measured in pressure times volume (PV) units instead of mass units. Gas flow or *throughput*, Q, is defined as the quantity of gas flowing across a specified open cross section in the vacuum circuit per unit time at the prevailing pressure, P, at a given temperature,

$$Q = PV/t \quad \text{torr-liter/sec} \tag{6-12}$$

Pumping speed, S_p, may now be redefined (eq. 6-11) as the volume of gas removed from the system per unit time at a pressure P prevailing at the throat of the pump,

$$S_p = Q/P \tag{6-13}$$

Flow resistance or *impedance*, Z, is next defined as the pressure difference per unit throughput,

$$Z = (P_1 - P_2)/Q = 1/C \tag{6-14}$$

where P_1 and P_2 are measured at the high and low sides, respectively, of the particular component. The reciprocal of impedance is called conductance, $C = 1/Z$. Both pumping speed and conductance are expressed in volume per unit time. Conductance implies "passive" components such as tubings and bends, while pumping speed refers to pumps. For a vacuum system with components in parallel one obtains $C_{total} = C_1 + C_2 + \ldots$, while for components in series $1/C_{total} = 1/C_1 + 1/C_2 + \ldots$, where the subscripts refer to individual components. As in the electrical analog, if one of the series conductances is much smaller than the others, the overall conductance will be nearly equal to the small conductance.

The quantity of gas striking an area A per unit time is given by

$$Q = PV = AP(kT/2\pi M)^{1/2} \tag{6-15}$$

where M denotes molecular mass. When molecular flow prevails (Section 7-I), the net flow of gas across the aperture is determined by the pressure difference, $P_1 - P_2$ and

$$Q = A(P_1 - P_2)\,(kT/2\pi M)^{1/2}$$

For the conductance one obtains

$$C = Q/(P_1 - P_2) = A(kT/2\pi M)^{1/2}$$

or, when $P_1 \gg P_2$

$$C = Q/P_1 = S_p \tag{6-16}$$

This equation determines the maximum speed of any pump operating on the basis of kinetic flow of gases through an orifice, and shows that it varies inversely with the square root of the molecular weight of the gas being pumped; for oxygen $C = 11.1A$ liters/sec.

The overall pumping speed of a system (S) is determined by the combined speeds of the pumps $(S_p = S_1 + S_2 + \ldots)$ and the combined conductances (C_{total}) of all components

$$1/S = 1/S_p + 1/C_{\text{total}} \tag{6-17}$$

This relationship is utilized in calculations involving *differential pumping*, i.e., maintaining a pressure differential between two regions inside a vacuum system by using apertures of low conductance to connect the different regions. An example is the electron bombardment ion source (Section 4-II-B), where the pressure in the region of ion formation is several hundred times greater than in the rest of the system. This is accomplished by a "tight" source structure providing only low conductance paths toward the pumps.

Conductance and pumping speed calculations and examples for system designs are detailed in many vacuum monographs and are beyond the scope of this chapter. Flow conditions are discussed in Section 7-II.

V. VACUUM PLUMBING

A. Seals

In addition to joining components together, seals are also used in vacuum systems to bring in electrical power or signals, to provide access ports, to introduce mechanical motion, and occasionally to allow observation of the inside of a component. Seals and valves are important components in all vacuum systems and the need for familiarity with the many types available cannot be overemphasized. A detailed discussion of sealing techniques is given by Roth (15) in a recent text.

According to the degree of permanency, seals are classified into permanent, semipermanent, and demountable seals. Special forms of demountable seals are used to introduce translatory or rotatory motion. According to the sealing technique used, sales may be grouped into welded and fusion seals, brazed or soldered seals, wax and resin seals, ground seals, liquid seals, and gasket seals. A third way of classification is suggested by the types of materials joined together, such as metal-to-metal seals and glass-to-metal seals. To obtain vacuum lower than 10^{-7} torr, seals must be bakeable to several hundred degrees.

1. Permanent Seals

Techniques for permanently joining separate parts of a vacuum system require much skill and experience and specialized equipment. Permanent seals are discussed according to the types of materials involved.

Metal-to-Metal Seals. In *welding* an intimate union is made between two materials by heating them until a molten or plastic state is reached, with or without the application of pressure, and with or without the use of filler metal. The filler, if used, must have a melting point about the same as that of the base metal. Among the welding techniques, electric arc welding—preferably in a protecting inert atmosphere to avoid porosity—is best to make vacuum-tight joints and seams in stainless steel, copper, or aluminum parts. Electric resistance welding (spot welding) is often used in joining wires and sheets of metals (e.g., in ion sources). Rather precise welds can be made by electron beam welding.

In *brazing* and *soldering*, coalescence is produced with the aid of another (nonferrous) metal or alloy with a melting point considerably lower than that of the base metal. The filler metal seals as it becomes distributed between the fitted surfaces by capillary action. Brazing is done with silver and copper alloys (e.g., Ag:Cu eutectic) above 400°C, whereas in soldering, a low melting alloy (<300°C) of tin and lead is used (soft solder). Brazing gives good mechanical strength and permits high temperature bakeout. When possible, brazing should be performed in a reducing furnace, thus eliminating the need for fluxes and providing cleaning at the same time. When brazing is done in air (by torch or hf induction heating) and also in soldering, a flux must be used for good adhesion. Fluxes decompose into acidic substances and corrode into the parent metal. Brazing fluxes are usually composed of borax with boric acid, while soldering fluxes consist of mineral grease, wax, and resins with zinc chloride. In electronic soldering, rosin is often used as flux. Parts should be carefully cleaned after brazing to remove all flux.

Dirt and occluded gas can be extremely troublesome in welded or brazed joints. Virtual leaks caused by released gas are difficult to find and may apparently last forever.

Metal-to-Glass Seals.   Classification into "matched" and "unmatched" seals indicates if the seal partners have nearly equal thermal expansion coefficients. The seal between glass and metal is due to an oxide layer on the metal which forms just before the seal is made. This layer partially diffuses into the glass when the parts are brought together under slight pressure. The coefficient of linear expansion, when plotted as a function of temperature, results in a straight line for metals but curves somewhat for glasses as their melting point is approached. Direct seals can be made as long as the coefficients are nearly equal. For example, platinum can be sealed to Corning 0010 glass, and tungsten to Corning Type 7720 (Nonex) glass. Good seals are light brown.

Copper can be joined to Pyrex glass using the Housekeeper seal (15) in spite of the difference in thermal coefficients. Glass is sealed to the very thin edge (feather edge) of an ultrapure (OFHC) copper tube, and the ductility of the copper assures that the glass will not crack under stress during cooling. Housekeeper seals, developed in 1923, have largely been superseded by *Kovar-to-glass* seals. Kovar is an inexpensive Fe-Ni-Co alloy (54:29:17) which can readily be joined to Corning 7052, 7040, and 7053 and Philips 28 glasses. Since the thermal expansion curve for Kovar almost coincides with that of the 7052 and 28 glasses, good mechanical strength and vacuum tightness is obtained. Kovar can easily be machined and copper-brazed. Many commercial Kovar–glass seals are now available including tubing, windows, single terminals, electrical feedthroughs, etc. A good seal has a light gray color. Connection between the 7052 glass and Pyrex 7740 is made through a "graded" glass seal to compensate for the difference between the thermal expansion of soft and hard glasses. When used to support electrometer collectors or similar arrangements, the insulating resistance of normal Kovar–glass seals can be increased from 10^{12} ohm to 10^{15} ohm by incorporating a highly insulating, high alumina glass.

Metal-to-Ceramic Seals.   Ceramics possess higher mechanical strength, withstand higher temperatures, have better dielectric properties, and are less permeable to gases than glass; their use in vacuum systems is increasing. Large "crunch" (ram) seals are made by pushing the ceramic cylinder into the beveled end of a copper-plated steel cylinder at high temperature and pressure. In the sintered metal process, powders of refractory metals are applied in a binder to the surface of the ceramic, followed by sintering in hydrogen firing. Next the metallized ceramic is copper or nickel plated and brazed to the metal. Sapphire windows can be sealed in Kovar rings using an active alloy process, where the ceramic surface is treated with titanium or zirconium hydride. Silver-copper eutectic (or nickel) is next placed on top, followed by another layer of hydride and the metal part. Firing in a vacuum oven results in decomposition of the hydride, the hydrogen is pumped away, and the titanium, alloyed with silver, acts as the binder.

Electrical Lead-Throughs. High current, low voltage lead-throughs are used primarily for heating elements, while connections for the various electronic devices normally carry low current at high voltages. The insulating materials used in the seals must have low vapor pressure, be mechanically strong, and have the proper resistance. Kovar, Houskeeper, and iron-nickel alloy seals are readily available commercially in many sizes and shapes. They are usually connected permanently to the system, but demountable forms are also available, e.g. the well-known Stupakoff Kovar seal (16).

2. Demountable Seals

There are dozens of techniques to make demountable seals. Although the required sophistication depends on the specific problem, with emphasis on maxium operating temperature and ultimate vacuum required, there are two basic considerations: sealing materials must have good outgassing properties, and continuous molecular contact must be established between the mating parts. Most demountable joints are made with stainless steel flanges with various gaskets providing the sealing. Many commercial types are available, and vacuum components such as pumps and traps are normally designed for flange connections with grooves for O-ring type sealing. Other flanges are made flat with carefully controlled surface roughness, or have sharp ridges. Flanges must have sufficient thickness to withstand permanent deformation and a sufficient number of bolt holes for uniform sealing.

Where extreme cleanliness is desired or where bakeout is necessary, *metallic gaskets* are employed. There are two basic types. Application of *gold O-ring* gaskets in all-metal mass spectrometers was first suggested in 1949 by Hickam (17). An O-ring is formed from gold (or aluminum or nickel) wire, 0.5–1.0 mm diameter, by fusing the ends together with a small torch, followed by annealing. The O-ring is clamped between flat, highly polished stainless steel flanges or in the corner of a stepped flange. Seal is accomplished by compressing the O-ring with bolts to about 50% of its original diameter. The seal can be baked to 450°C and can be reused repeatedly; after use the gold can be salvaged. Figure 6-7a shows an O-ring seal. When gold O-rings are employed, flanges must have backoff holes because it is often difficult to separate mating flanges after bakeout. Indium and lead are also good gasket materials, particularly at lower temperatures (18).

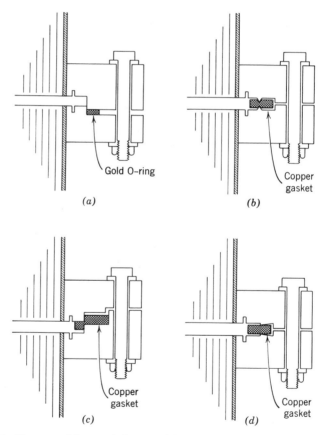

Figure 6-7. Demountable metal flanges: (*a*) metal O-ring seal; (*b*) knife edge seal; (*c*) step seal; (*d*) "Conflat" seal. (After R. W. Roberts and T. A. Vanderslice. *Ultrahigh Vacuum and Its Applications*, Prentice Hall, Englewood Cliffs, N. J., 1963, p. 63.)

The second basic type of demountable seal employs flat gaskets, usually made of OFHC copper, and knife edge flanges. Three variations are shown in Figure 6-7. In the *knife edge* seal (19) sharp edges are forced into the copper gaskets. The radius of the tip is normally 0.125 mm, and the wedge angle is 30–45°. In the *step seal* a flat copper gasket is sheared between two concentric steps on circular flanges. This seal (20) is easier to machine than the knife edge seal, and large pressures can be developed in the small area of the gasket determined by the 0.25 mm overlap in the flanges. In 1962 Wheeler and Carlson described (21) a seal design which later became commercially available. A flat copper gasket and conical sealing surfaces are used. The sealing edges intrude about 0.40 mm into the copper gasket which is 2 mm thick (Fig. 6-7d).

O-ring seals are often made using elastomer gaskets. The term *elastomer* refers to both natural and synthetic materials that are vulcanized into a state in which they can be exposed to and recover from extreme deformation. There are many types, and they come under various trade names. Elastomer gaskets are either ring-shaped or, occasionally, rectangular. The cross section may be circular (O-rings), square, or rectangular. Among the rubber types Bune N (nitrile butadiene rubber) and neoprene (chloroprene rubber) are the best known. Natural and even synthetic rubber gaskets are not recommended for vacuum below 10^{-6} torr. Outgassing is impossible, and rubber is permeable to atmospheric gases. Recently, fluorinated elastomers, such as Kel-F, Viton A, and Fluorel have become popular due to favorable outgassing properties. Viton A (DuPont) is the most recommended at the present time. It can be used up to 250°C and, as shown in Figure 6-8, only small quantities of water, carbon monoxide, and carbon dioxide evolve to contribute to the background (22). Vacuum as low as 10^{-9} torr has been achieved with Vitron A O-rings.

Vacuum *greases, cements, and waxes* are often used in demountable vacuum connections. Vacuum greases are normally employed at room temperature and should be applied sparingly since they absorb gases and organic vapors. Their vapor pressure should be less than 10^{-5} torr. Most popular are the Apiezon, Celvacene, and Vacuseal brands, each of which has various types for different purposes. Among the waxes the Apiezon W types are employed perhaps the most frequently for semipermanent joints. Apiezon W is a hard black wax which is applied at about 100°C. Once solidified it is safe up to 80°C. Both the surface and the wax must be heated; overheating will cause bubbling and charring. The W-type waxes are soluble in xylene and other solvents. The famous deKhotinsky cement (shellac plus 20–40% wood tar) has been employed in mass spectrometry

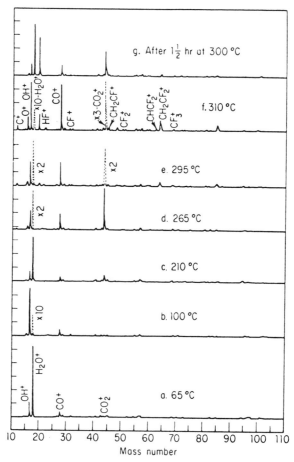

Figure 6-8. Mass spectra obtained during heating of Viton fluoroelastomer (22). Ordinate is ion current, 1 div = 10^{-12} A.

for many years, but its use is now declining; its vapor pressure is 10^{-3} torr at room temperature. Glyptal (Glycol phthalate) is known as a "friend in need" for many mass spectroscopists. It is often used (with xylol solvent) to temporarily seal small leaks by painting with a small brush. Decomposition products of all these sealants give peaks at practically every mass; therefore they should not be used at elevated temperatures. It is best to avoid them altogether.

3. Flexible Connection and Motion Seals

It is a standard practice in all-glass systems to incorporate flexible connections at critical points to avoid stresses. Often a simple bend is adequate;

in other cases glass bellows are used. Another well-known example of the need for flexible connections is in the connection of mechanical pumps. To prevent the transmission of mechanical vibrations to other parts of the vacuum system, heavy-walled rubber tubing is normally employed between the low-vacuum side of the diffusion pump and the high-vacuum side of the mechanical pump. Such connections are usually adequate, although it is better practice to use all-metal lines with flange connections, incorporating stainless steel bellows.

Bellows connections are used whenever mechanical movement must be transmitted through a vacuum boundary. Bellows can transmit both rotation and limited reciprocation. *Bellows* are pipes having their walls bent to form consecutive parallel rings in such a manner that the corrugated shape allows a considerable amount of axial compression or bending. Bellows are made from rubber, Teflon, bronze, stainless steel, and glass; many types are available commercially. Connection of the ends of the bellows to the rest of the system may be made through brazing, welding, elastomer O-rings, etc. Application of bellows in mass spectrometer systems include: mechanical adjustments to position and align mass analyzer tubes in the magnetic field, positioning of ion sources, moving of sample electrodes in spark sources, opening and closing of adjustable source and collector slits, retrieving of Faraday collectors to expose electron multipliers, and the introduction of samples through vacuum locks (Section 7-VII).

Other techniques to transmit motion include the use of diaphragms, ground seals, gasket seals (O-rings, lip seals such as the Wilson seal, friction shaft seals, etc.), magnetic coupling through thin nonmagnetic walls, and the use of liquid metal seals such as gallium, indium, and tin around a shaft going through the walls of the vacuum system. For details the reader is referred to reference 15.

B. Valves

Valves are devices for adjusting the rate of flow of a fluid or for completely stopping the flow. Obvious requirements for high vacuum valves are: freedom from leakage both from the atmosphere (14.7 psi) and from the "low" vacuum side it is supposed to isolate, minimum flow resistance (maximum conductance), chemical inertness, low outgassing of construction materials, and often bakeability. Details of the many available valve types and their respective merits are available in many texts (15), and only the most frequent designs are outlined here.

Most high vacuum valves have been adopted from conventional designs by improving the sealing. Figure 6-9 shows several types of metal valves.

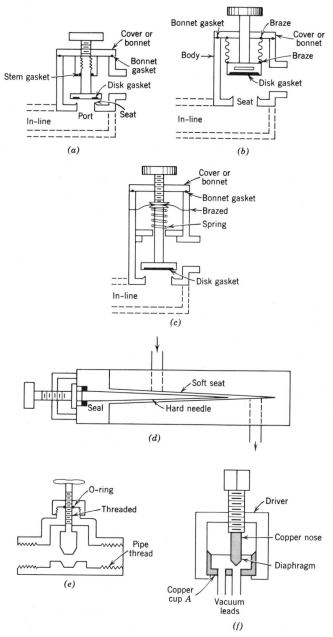

Figure 6-9. Various types of valves: (*a*) disk valve with stem O-ring; (*b*) disk valve with bellows; (*c*) disk valve with diaphragm; (*d*) principle of needle valve; (*e*) blunt-nosed needle valve; (*f*) principle of Alpert valve. (After A. Guthrie, *Vacuum Technology,* Wiley, New York, 1963, pp. 354, 357, 364.)

The main parts of all valves are the *body* or housing containing the external openings, the *bonnet* through which the motion is transferred and which provides the seal and support for the stem, the *stem* which transfers the motion and closes or opens the *port* of the valve by pressing a disk against a *seat*. The port is the opening within the valve body through which the gases flow, and the seat is the section of the valve surrounding the port against which the seal is made. According to Roth (15), any valve consists of a closing system, a sealing system, and an operating system. The *closing system* may be based on liquid or molten metal seals, ground joints, diaphragms, or gasket seals. The ground joints may be plane, tapered, or spherical. The many types of gasket seals include plates (flaps), plugs, cones, gates (slides), and balls. The *sealing system* may be based on packings, or it may be packless, using bellows, diaphragms, or magnetic motion transmissions. The *operating system* is usually mechanical, although many types of pneumatic, magnetic (solenoid), and thermal expansion types have also been described.

Operating below 10^{-7} torr normally requires bakeout at 300–450°C, and elastomers can no longer be used as construction materials. Most all-metal, bakeable, ultrahigh vacuum valves are based on a design by Alpert (23). As shown in Figure 6-9*f*, the Alpert valve consists of two sections, a valve assembly sealed into the vacuum system and a detachable driver mechanism. Sealing is accomplished by forcing a hard polished conical nose (Kovar, Monel) into a soft copper cup. The driver works by a differential screw mechanism and several tons of pressure is exerted. Various modifications have been largely directed to the design of the driver assembly and the use of materials of construction. These metal valves may also be used as variable gas leaks (conductance from 1 to 10^{-14} liter/sec) by controlling the sealing torque with a torque wrench.

VI. VACUUM GAUGES

Most vacuum gauges are based on the measurement of the change in: (*1*) the height of a liquid column, (*2*) the distortion of a solid surface, or (*3*) some kinetic molecular property of the gas (heat conductance, viscosity, ionization, etc.). The first two types are differential methods, although frequently used for the indication of absolute pressure with good approximation, while those in (*3*) actually indicate molecular concentration and thus depend on the nature of the gas. Required pressure range, desired precision, and the nature of the gas present are the major considerations in selecting vacuum gauges. The presence of a proper gauge in strategically located places is exceptionally useful in both the operation and maintenance of a mass spectrometer. Figure 6-10 shows the pressure ranges of many vacuum gauges.

A. Liquid Column Gauges

U-tube mercury manometers are widely used in sample inlet systems to measure pressures in the 1–200 mm Hg range. They are employed both as absolute manometers with one end closed, and as differential manometers with one leg exposed to a reference pressure. When oil is used as the manometric liquid for pressures <0.5 mm Hg, conversion into mm Hg is accomplished by multiplying the difference in column heights by the oil/mercury density ratio. The accuracy of liquid manometers is about 0.1%.

The well-known *McLeod gauge* is a true absolute device although experimentally the height of a liquid column is observed. The gauge operates on Boyle's principle: a known volume of the gas is trapped at the unknown pressure, followed by a compression into a new smaller volume at which the pressure can be measured by an ordinary manometer. The operation of a simple McLeod gauge is explained with reference to Figure 6-11a. The first step in a measurement is to raise the level of mercury in tube D by admitting air into the mercury reservoir through the air bleed valve. When the mercury level reaches LL' a known volume of gas is trapped in volume V. Note that V is connected to the vacuum chamber through tube C. The pressure at which the gas sample is trapped in V is the one to be determined.

mm Hg	10^{-14}	10^{-13}	10^{-12}	10^{-11}	10^{-10}	10^{-9}	10^{-8}	10^{-7}	10^{-6}	10^{-5}	10^{-4}	10^{-3}	10^{-2}	10^{-1}	10^{0}	10^{1}	10^{2}	10^{3}
Mercury manometer													●	●	●	●	●	●
Bourdon gauge														●	●	●	●	●
Radioactive ionization gauge												●	●	●	●	●	●	
Bimetallic strip													●	●	●	●	●	
Thermistor											●	●	●	●	●	–		
Thermocouple											●	●	●	●	●	–		
Capacitance manometer											●	●	●	●	●	●	●	
Quartz fiber									●	●	●	●	●	●	●			
Cold cathode gauge								●	●	●	●	●	●	●	–			
McLeod gauge									●	●	●	●	●	●				
Pirani gauge										●	●	●	●	●	●	–		
Quartz membrane							●	●	●	●	●	●	●	–				
Molecular gauge							●	●	●	●	●	●	–					
Knudsen gauge							●	●	●	●	●	●						
Thermionic ionization gauge (no magnetic field)						●	●	●	●	●	●	●	–					
Cold cathode ionization gauge with magnetic field			●	●	●	●	●	●	●	●	●	●	●	–				
Thermionic ionization gauge with magnetic field				●	●	●	●	●	●	●	●							
Mass spectrometer	●	●	●	●	●	●	●	●	●	●	–							

Figure 6-10. **Chart showing ranges of pressures for different types of vacuum gauges.** (After S. Dushman, *Scientific Foundations of Vacuum Technique*, 2nd ed., Wiley, New York, 1962, p. 350.)

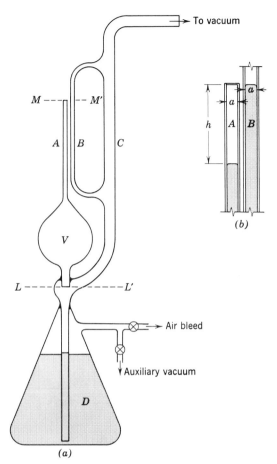

Figure 6-11.   Principle of McLeod gauge.

Next, the mercury is raised further until its level in capillary B reaches MM', which is the top of capillary A (Fig. 6-11b). Capillary B is always exposed to the unknown pressure. The gas trapped in bulb V is compressed into capillary A as the mercury is rising. The amount of compression is determined by the ratio of the total volumes of bulb V and capillary A to the volume of capillary A after compression. What one actually measures is the difference, h, between the mercury levels in A and B. The unknown pressure P (torr), is obtained from

$$P = ah^2/(V_0 - ah) \approx ah^2/V_0 = kh^2 \qquad (6\text{-}18)$$

where a (mm^2) is the cross-sectional area of capillary A, and V_0 (mm^3) is the volume of the gas originally trapped. The volume of the compressed gas is ah, a value very much smaller than V_0. The ratio a/V_0 is known as the gauge constant, k. The pressure is thus proportional to the square of h. Typical values for a and V_0 are 0.1 mm^2 and 200 cc (2×10^5 mm^3). Using a total length of 10 cm, the lower reading limit is around 10^{-6} torr. The quadratic scale is an advantage because it results in increased sensitivity at lower pressures. (McLeod gauges may also be operated using a linear scale method; this technique is useful for pressures of several torrs.)

The principal use of McLeod gauges is for calibration of other gauges. It is noted in connection with the practical uses of McLeod gauges that: (1) they do not indicate the vapor pressure of their own mercury; (2) although measured pressure does not depend on the nature of the gas, condensable vapors (e.g., water) cannot be measured; the presence of condensables can be tested by successively raising the mercury to different levels, measuring h_1 and h_2 for each level and checking $h_1 (h_1 - h_2)$ for constancy; (3) when the gauge is connected to the vacuum system through a liquid nitrogen trap, not only are condensables removed but also contamination of the system by mercury is avoided (vapor pressure of mercury at room temperature is about 1 μ); (4) no continuous pressure indication is possible with McLeod gauges; (5) they are not bakeable.

B. Mechanical Gauges

Mechanical gauges depend upon the distortion of a solid surface when unequal pressures are exerted on its two sides. Such gauges may be used as either absolute or differential manometers. Mechanical gauges based on the Bourdon tube principle are used mainly for pressures over 1 atm. Bellows, capsules, and diaphragms are increasingly more sensitive than Bourdon tubes. In the Wallace & Tiernan absolute pressure indicator, which is often used in sample introduction systems, the unknown pressure is admitted into the sealed instrument where it exerts a pressure on a rather flat, evacuated and permanently sealed capsule. The movement of the pressure-sensitive element is transmitted by a lever system to a pointer calibrated to indicate absolute pressure above zero. Pressure may be measured in the 0.1 mm to 1 atm range with accuracy of 0.3% of full-scale range. In the Ruska Instrument Corp. gauge the pressure-sensitive element is made of quartz and has a very small volume.

In *micromanometers* (Fig. 6-12) the displacement of a flexible membrane caused by a pressure difference is measured electrically. Micromanometers are widely used in mass spectrometer sample systems for pressure measurement in the 1–500 μ range. Here the pressure-sensitive diaphragm constitutes one plate of an electrical condenser which, in turn, forms one arm

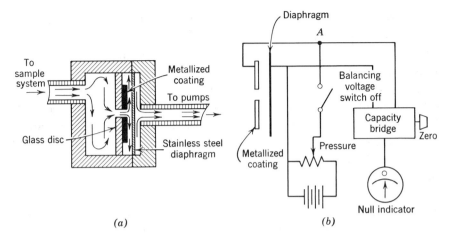

Figure 6-12.   Micromanometer. (*a*) Pressure-sensitive capacitor (not to scale). (*b*) Before sample is introduced, vacuum exists on both sides of diaphragm. The applied sample pressure deforms the diaphragm and changes its capacity; thus a balancing voltage is needed to regain the null reading. (Courtesy of Consolidated Electrodynamics Corp.)

of a capacitance bridge. The capacitance change caused by the pressure-sensitive plate bending away from the fixed plate is compensated by adjusting a dc balancing voltage. The magnitude of the restoring voltage is proportional to the unknown pressure, and direct pressure readings are obtained after proper calibration. Accuracy is of the order of $\pm 1 \mu$ at 100μ pressure, while sensitivity is about 0.1μ. Up to about 200μ, accuracy is independent of sample composition. At higher pressures the dielectric constant of the sample gas must also be considered, and temperature variations become important. Micromanometers are often bakeable to 450°C, with some loss in accuracy due to increased zero drifting.

C. Thermal Conductivity Gauges

Thermal conductivity gauges operate on the principle that the heat loss from a filament heated in vacuum is pressure dependent. Since the rate of heat loss from a given filament material depends on the thermal conductivity of the surrounding gas, one has to be careful in evaluating gauge readings, particularly with gas mixtures. Hydrogen, for example, has a thermal conductivity coefficient almost an order of magnitude higher than nitrogen (0.416×10^{-3} vs. 0.057×10^{-3} cal/cm/sec/°C at 0°C). The useful pressure range is limited to 10^{-3} torr at the low side by the increasing contribution to the heat transfer by radiation and conduction through the leads, and to about 10^2 torr on the upper pressure side because the heat loss becomes independent of pressure.

In practice, a filament with a large temperature coefficient of resistance (tungsten, nickel, or thermistor) is heated by a constant electric current in a glass or metallic envelope attached to the vacuum system. The variation of the filament temperature with pressure is measured either in terms of the change in the resistance of the wire in a Wheatstone-bridge arrangement (*Pirani* gauge, Fig. 6-13*a*), or directly, by means of a thermocouple attached to the wire (*thermocouple* gauge, Fig. 6-13*b*). The unbalance voltage, or the power required to maintain constant filament temperature, in the Pirani gauge, and the developed thermocouple voltage (of the order of millivolts) in the thermocouple gauge is measured. Calibration is usually furnished by the manufacturer and scales are calibrated against a McLeod gauge using dry nitrogen or air. Thermal gauges are often used in pairs, one of the pair acting as a reference to correct for changes in wall temperature. Several thermocouple units may be used in series, as in thermopile gauges. Thermal conductivity gauges are quite rugged but not very accurate.

D. Ionization Gauges

Ionization gauges measure the number of ions formed in collisions between residual gas molecules and bombarding electrons, or alpha particles in the case of the Alphatron gauge. Since the number of ions formed is directly proportional to pressure below 10^{-3} torr, the measured current can be translated into pressure units. Ionization gauges actually measure particle densities. Since the number of ions produced by electrons of given energy is a function also of the ionization cross section (eq. 4-2), ionization gauges depend on the nature of the gas to be measured. Of course, mass spectrometers may, and indeed have been, thought of as fancy ionization gauges that measure partial pressures. Two basic types of ionization gauges can be distinguished: hot filament (thermionic emission) and cold cathode (vacuum discharge) gauges.

Figure 6-14*a* shows the basic triode circuit of a *hot filament* gauge. Electrons emitted from a heated tungsten filament are accelerated toward the grid, pass through its open wire structure, and ionize residual gas molecules in the region between the grid and the collector. Positive ions are collected on the collector, which is negative with respect to both filament and grid; electrons are collected on the grid. The positive ion current is measured with a microammeter calibrated in pressure units.

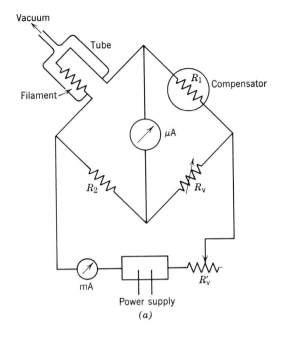

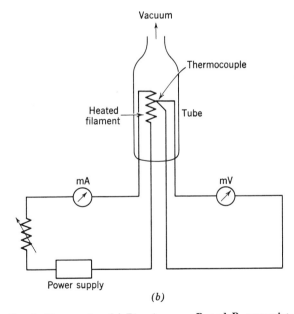

Figure 6-13. Circuit diagram for: (*a*) Pirani gauge; R_1 and R_2 are resistors with resistance close to that of the filament; R_v is variable resistor to balance the bridge; $R_v{'}$ is variable resistor to adjust current from rectifier power supply. (*b*) Thermocouple gauge.

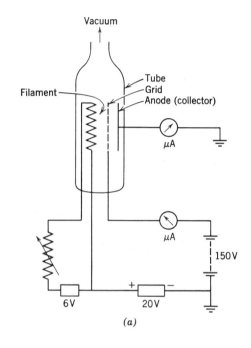

(a)

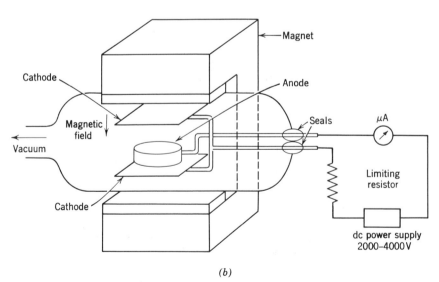

(b)

Figure 6-14. (a) Hot filament ionization gauge. (b) Cold cathode gauge. (After A. Guthrie, *Vacuum Technology*, Wiley, New York, 1963, p. 177.)

In the classical sensing tube structure a central filament is surrounded by a concentric grid and outer collector. The lowest measurable pressure with such a structure is 10^{-8} torr. This limit is set by the x-ray effect (24): soft x-rays produced by electrons hitting the grid irradiate the collector and release photoelectrons. Removal of an electron from the collector is equivalent, electrically, to the arrival of a singly charged positive ion. Thus, when at a low pressure the ion current becomes lower than the photocurrent, pressure measurement is no longer possible. In 1950 Bayard and Alpert (25) adapted an "inverted" structure: the filament is outside the cylindrical grid structure and the ion collector consists of a fine wire located centrally inside the grid. The area of the collector exposed to x-ray radiation is thus significantly reduced, resulting in a reduction of the residual current background (dark current). At the same time the efficiency of both ionization and ion collection is increased. This technique extends the lower limit of pressure measurement into the 10^{-10} torr region. A Bayard-Alpert inverted gauge is shown in Figure 6-15. Two filaments are usually provided so that the gauge does not have to be replaced when a filament burns out.

Calibration of ionization gauges is performed using a McLeod gauge. This permits calibration down to 10^{-5} torr pressure below which values are extrapolated assuming linearity. Gauge sensitivity, S, is given by the measured collector current, i_+ (μA) for a given electron emission, i_- (mA) for a given pressure change, P (torr),

$$S = i_+/i_-P \qquad (6\text{-}19)$$

The value of S varies with the gas used. Relative sensitivities are given in Table 6-5.

Table 6-5 Typical Relative Sensitivities of Hot-Cathode Ionization Gauges

Gas	Relative sensitivity[a]
Hydrogen	45
Helium	15
Water	200
Neon	25
Nitrogen	110
Air	100
Oxygen	85
Argon	160
Mercury	270
Carbon monoxide	110
Carbon dioxide	120

[a] Air is taken as standard

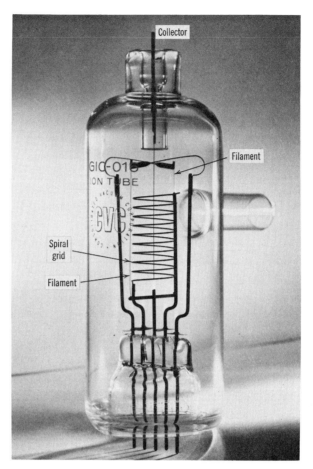

Figure 6-15.   Hot filament ionization gauge tube. (Courtesy of Consolidated Vacuum Corp.)

In cold cathode gauges, such as the Penning and Phillips gauge, electrons are produced by a high-voltage discharge, and their effective path length is increased by forcing them with a magnetic field to spiral as they move toward an anode (Fig. 6-14*b*). The high-voltage discharge current is thus increased to the point where it can be read on a simple meter. The discharge ceases at about 5×10^{-6} torr. The upper pressure limit is 50 μ.

Discharge gauges are simple, usually in an all-metal form with simple measuring circuitry. There is no filament to burn out. A disadvantage is that organic vapors decompose and form a carbonaceous coating on the walls, producing erratic readings. When a cylindrical anode is used instead of the usual loop, the discharge is more concentrated on the walls helping to keep the walls clean longer. At the same time sensitivity is also increased because of more efficient trapping of electrons inside the cylinder.

VII. LEAK DETECTION

Next to problems created by failures in electronic circuits, leak hunting is the most unpleasant and time-consuming task in trouble-shooting mass spectrometers. Like any vacuum system, mass spectrometers are subject to leaks the origin of which may range from mechanical failures to spurious outgassing of entrapped cleaning solvents. Leaks occur most often during initial pumpdown of instruments after they have been exposed to the atmosphere for repairs, modifications, or sample changes such as in spark instruments. It is good practice to leak check all new components before assembling; valves, gauges, welds, glass-to-metal seals, etc. are relatively easy to test individually but difficult to test in the final assembly. Metal parts can be rough-tested by pressurizing with oil-free nitrogen to 1.2–1.4 atm and spraying the surface with soap solution. Leaks as small as 10^{-4} torr-liter/sec can be found by this method. For best results, individual parts should be leak tested with a mass spectrometer leak detector.

The term "leak" has several connotations. A *leak* is a hole or porosity in the wall of an enclosure capable of passing gas from one side to the other under the action of a pressure differential existing across the wall. When the hole or porosity is in the walls of the vacuum enclosure, the continuous passage of gas from the outside into the vacuum system is obviously undesirable, and such leaks are called *real leaks*. Often, however, a capillary or porous wall leak, usually in a glass or metal tube, is used as a device to introduce gases at a specified rate into a vacuum system. Such leaks are known as *standard or calibrated leaks* because the conductance is selected so that leakage occurs at a specified rate. Such leaks are usually also specific with respect to the gas they permit to pass. An example is the use of helium gas to calibrate mass spectrometer leak detectors. The leak rate of these leaks ranges from 1×10^{-7} to 5×10^{-10} atm-cc/sec. There are also metal valves available whose conductance is accurately adjustable by a simple driver mechanism. Such *variable leak valves* have many uses in mass spectrometry.

While the term "leak" refers to the physical orifice, the actual flow of gas is known as *leakage*. The *leak rate* is defined as the quantity of gas, in PV units, flowing per unit time into the system from an external source. For real leaks the external source is the atmosphere; for the various calibrated and adjustable leaks the external source is a gas reservoir. The high-pressure side of leaks is usually 1 atm, although a variable leak may be connected to a gas cylinder with relatively high pressure. Leak rates are expressed in units of micron-liter/sec = lusec,

1 lusec = 1μ/sec pressure rise in 1 liter volume
= 12.7 min time for 1 cc (STP) gas inflow
= equivalent opening of a rectangular slit with 1 cm width, 0.1 mm height, and 1 cm depth

Conversion of lusec into other units can be accomplished by considering the following conversion factors,

7.9×10^{-2} cc (STP)/min = 4.2×10^4 cc (STP)/year
= 1.32×10^{-3} atm-cc/sec
= 4.74 atm-cc/hour
= 1×10^{-3} torr-liter/sec
= 1.5×10^{-3} g/sec (air)

Virtual leaks are caused by the slow release of sorbed or occluded gases from within the vacuum system. Most frequent sources are water desorption from the walls, slow evaporation of entrapped cleaning solvents in valves, under gaskets, etc., and gas release from elastomers. The following rule of thumb may be used to evaluate outgassing rates from construction materials: 10^{-4} lusec/cm^2 for metals, 10^{-3} lusec/cm^2 for ceramics, and 10^{-2} lusec/cm^2 for elastomers.

Real and virtual leaks may be distinguished by the pressure rise method. After the best attainable vacuum is reached, the system is closed off and the pressure rise is plotted against time. As shown in Figure 6-16, the pressure rises linearly with time when real leaks are present, while a saturation effect is the result of a virtual leak. In curve *3*, the slope of the straight part shows the leak rate per unit volume.

For leak hunting in glass systems the *Tesla coil* is a useful aid. A high-voltage high-frequency alternating current of very low amperage is supplied by a Tesla transformer. As the electrode is drawn close to the walls of the glass vacuum system, luminous phenomena appear. As the pressure decreases from atmospheric to 10 torr, small sparks appear inside the system which slowly change to a thread-shaped glow discharge. The color changes to red or violet in the 10 to 10^{-1} torr range when nitrogen, hydrogen, or hydrocarbons are present. A greenish fluorescence appears opposite the electrode as the pressure is further lowered to 10^{-2} torr. The pink or violet luminescence remains but slowly fades as 10^{-3} torr is reached, and all luminous color disappears at pressures below 10^{-3} torr. When the system is "painted" with acetone or carbon tetrachloride, the presence of a leak is

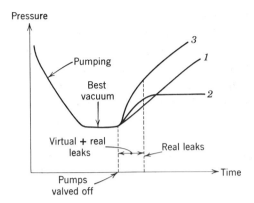

Figure 6-16. Pressure versus time curves to distinguish between real and virtual leaks; (*1*) only real leak; (*2*) only virtual leaks; (*3*) real and virtual leaks together.

indicated by the color of the glow discharge, which changes to bluish. One has to be careful because too strong a spark might produce pinholes in the glassware.

Thermocouple and/or ionization gauges strategically located throughout the mass spectrometer system can be of great assistance in locating leaks. When subsequent sections are exposed to the vacuum pumps, these gauges permit leak checking from 10^{-1} to 10^{-5} torr using the pressure-rise technique. When a jet of hydrogen or helium is directed against a suspected leak, the reading on the ionization gauge will decrease due to the low ionization cross section of the gas. Argon and acetone, on the other hand, will cause an increase in the reading (Table 6-5). In another technique, a hose is connected to an auxiliary vacuum pump and its free end is pressed against the chamber wall. The gauges should indicate a reduced pressure when the area contacted includes a leak. The pressure differential across the leak may be increased by blowing a sharp air jet from a small nozzle over the suspected area.

When the leaks are so small that the mass spectrometer can be pumped down to operating level, the instrument itself may be used as its own leak detector (provided it has an electron bombardment source). Hydrogen, helium, or argon may be used as the test gas. The smallest detectable leaks are of the order of 10^{-10} torr-liter/sec.

The most sensitive leak detector presently available is the mass spectrometer leak detector; many commercial models are available. These instruments are usually magnetic deflection instruments permanently tuned to helium. There are no problems with resolution and all design parameters are optimized for maximum sensitivity for helium alone. The helium ion current is amplified to be displayed on a calibrated meter. Often audio indication is also provided. When mass spectrometer systems are leak tested with a mass spectrometer leak detector, the instrument is connected to the leak detector and the suspected areas are sprayed with helium. Often a plastic bag is used to isolate sections and to build up sufficient helium pressure. Helium leak detectors are extremely sensitive. The detection of one part of helium in 10 million parts of air is possible at a source pressure of 0.2 μ.

Leak detection has often been described as an art. Everybody working with mass spectrometers is painfully aware of the ever-existing possibility of leaks, and is faced with the problem of finding them in order that they can be repaired. Although it is desirable to learn as much as possible about the "theoretical" aspects of leak detection (26,27) practical experience is indispensable.

References

*1. American Vacuum Soc., *Glossary of Terms Used in Vacuum Technology*, Pergamon Press, London, 1958.

2. Menat, M., *Can. J. Phys.*, **42**, 164 (1964); **43**, 1525 (1965).

3. Honig, R. E., and H. O. Hook, *RCA Rev.*, **21** (No. 3), 360 (1960); also Honig, R. E., *ibid.*, **23** (No. 4), 567 (1962).

4. Norton, F. J., *Natl. Symp. Vacuum Technol. Trans., 8th, 1961* (pub. 1962), Vol. 1, p. 8.

5. "Properties of Selected Commercial Glasses," and "Engineering with Glass," Corning Glass Works, Corning, New York, 1962; Hutchins, J. R., and R. V. Harrington, "Glass," in *Kirk-Othmer Encyclopedia of Chemical Technology*, Vol. 10, 2nd ed., A. Standen et al., Eds., Interscience, New York, 1966, pp. 533–604.

6. Gomer, R. *Rev. Sci. Instr.*, **24**, 993 (1953).

7. Becker, W., *Vacuum*, **16**, 625 (1966); also Outlaw, R. A., *J. Vacuum Sci. Technol.*, **3**, 352 (1966).

8. Urry, G., and W. H. Urry, *Rev. Sci. Instr.*, **27**, 819 (1956); also Wheeler, E. L., *Scientific Glassblowing*, Interscience, New York, 1963, p. 372.

9. Bufalini, M., and J. E. Todd, *J. Chem. Ed.*, **44**, 425 (1967); Rice, D. A., and Roach, J., *J. Sci. Instr.*, **44**, 473 (1967).

10. Hickman, K. C. D., *Nature*, **187**, 405 (1960).

11. Biondi, M. A., *Rev. Sci. Instr.*, **30**, 831 (1959).

12. Alpert, D., *Rev. Sci. Instr.*, **24**, 1004 (1953).

13. Brubaker, W. B., *Natl. Symp. on Vacuum Technol. Trans., 6th, 1959* (pub. 1960), p. 302.

*14. Jepsen, R. L., S. L. Mercer, and M. J. Callaghan, *Rev. Sci. Instr.*, **30**, 377 (1959).

*15. Roth, A., *Vacuum Sealing Techniques*, Pergamon Press, New York, 1966.

16. DeVilliers, I. W., *Rev. Sci. Instr.*, **29**, 527 (1958).

*17. Hickam, W. M., *Rev. Sci. Instr.*, 20, 291 (1949); also Hawrylak, R. A., *J. Vacuum Sci. Technol.*, 4, 364 (1967).

18. Seki, H., *Rev. Sci. Instr.*, **30**, 943 (1959).

19. VanHeerden, M. M., *Rev. Sci. Instr.*, **28**, 726 (1957).

*20. Lange, W. J., and D. Alpert, *Rev. Sci. Instr.*, **28**, 726 (1957).

21. Wheeler, W. R., and M. Carlson, *Natl. Symp. Vacuum Technol. Trans.*, *8th 1961* (pub. 1962), Vol. 2, p. 1309.

22. Addis, R. R., L. Pensak and N. J. Scott, *Natl. Symp. Vacuum Technol. Trans., 7th, 1960* (pub. 1961), p. 39.

*23. Alpert, D., *Rev. Sci. Instr.*, **22**, 536 (1951); also Bills, D. G., and F. G. Allen, *ibid.*, **26**, 654 (1955); also Thorness, R. B., and A. O. Nier, *ibid.*, **32**, 807 (1961).

24. Nottingham, W. B., *National Symposium on Vacuum Technology Transactions*, Pergamon Press, New York, 1955, p. 76.

*25. Bayard, R. T., and D. Alpert, *Rev. Sci. Instr.*, **21**, 572 (1950).

26. Turnbull, A. H., *Vacuum*, **15**, 3 (1965).

*27. Santeler, D. J., D. H. Holkeboer, D. W. Jones, and F. Pagano, "Vacuum Technology and Space Simulation," NASA SP-105, U.S. Govt. Printing Office, Washington, D. C., 1966, Chap. 11.

Chapter Seven

Sample Introduction Systems

I. SAMPLE VOLATILITY AND QUANTITY

The function of sample introduction systems is to introduce a sufficient quantity of the sample into the ion source in such a way that its composition accurately represents that of the original sample under investigation. In certain types of ion sources, such as the spark source, the method of sample introduction is self-evident. In thermal ionization sources sample vaporization and ionization take place in a single process inside the ion source. Here the sample introduction system is a vacuum lock which serves mainly for convenience; several arrangements are discussed later in this chapter. Sample introduction systems must, however, be considered as subsystems in connection with electron bombardment sources. Several questions concerning the physical and chemical properties of the samples and the objectives of the investigation must be considered before a proper sample introduction system can be selected. These questions include:

- Is the sample a pure substance or is it a mixture?
- What kind of information is needed: qualitative (identification), semi-quantitative, or quantitative?
- What is the volatility of the sample?
- What is its thermal stability?
- How much sample is available? Is there a need for sample recovery?
- What kind of container is the sample in, how difficult is it to transfer the sample?
- Is the sample reactive? Will metals catalyze its decomposition?

It is recalled that ion sources are normally operated at a pressure of 10^{-5} to 10^{-6} torr. The maximum tolerable pressure of about 10^{-4} torr is set primarily by nonlinearity effects due to space charges and shortened filament life. (It is noted that sources are operated at the millimeter pressure level in special applications; examples are given in Part II.) When the sample is a gas, the main problems are the reduction of the initial pressure (normally atmospheric) to a predetermined level in the sample reservoir and the introduction of the sample into the ion source under controlled flow conditions. In the majority of the cases, the sample container is connected to the ion source through a "leak" of low conductance; this permits the use of conveniently measurable sample pressures within the container, maintaining at the same time the required low ion source pressure. Since it often takes 10–25 min to obtain a complete mass spectrum, there must be enough sample available in the reservoir to prevent sample depletion. Normally, for reasons discussed shortly, the sample reservoir has a 3–5 liter volume, and the sample pressure is of the order of $10\ \mu$ (10^{-2} torr). Flow conditions and the construction of leaks are discussed later in this chapter.

Liquids and solids must be heated to achieve the required 10^{-2} torr vapor pressure. An approximate estimation of the vapor pressure of most organic compounds at any temperature may be made as long as their boiling point is known. According to Trouton's law, the molar entropy of vaporization is the same for all nonpolar liquids at their boiling points at 1 atm pressure; the constant is about 30 cal/deg at 300°C. The value of the saturated vapor pressure as a function of absolute temperature is given by the Clausius-Clapeyron relationship. Tables and graphs giving vapor pressure as a function of temperature are available for a variety of organic compounds. From these the temperature required to obtain a vapor pressure of about 1 mm Hg can be estimated; for lower pressures extrapolation is needed.

Liquids and solids are normally heated in a system external to the source and the vapors obtained are introduced into the source in a manner similar to that used for gases. Many organic compounds have a vapor pressure of the order of 10^{-2} torr at temperatures of 150–350°C and heated inlet systems are usually designed to provide this temperature range. Sometimes the sample, itself not volatile, may be converted into a derivative which retains the important structural features of the original compound but now has sufficient vapor pressure.

Methods have recently been developed to obtain vaporization of solids directly inside the ion source. Sample pressure of only 10^{-6} torr is required when such "internal sampling" is employed. Such techniques are most useful when the thermal instability of the sample or available sample quantity prevent the use of more conventional systems.

Table 7-1 summarizes the sample quantities required by the commonly used inlet systems in connection with electron bombardment ion sources. These values are only guidelines since required sample quantity varies widely according to the type of information sought and analytical technique employed. In general, an increase of 1 to 2 orders of magnitude above the detection limit is required for identification.

It is important to realize that only a small fraction of the sample introduced will be ionized and eventually detected. For example, when the ion current detected corresponds to 100 ions, the sample quantity that reached the collector amounts to only 10^{-20} g. The rest of the original sample introduced is lost in three major steps: (*1*) When a batch inlet is used, only a few per cent of the original sample enters the source; (*2*) due to the efficiency of ionization, only a fraction of the sample becomes ionized; and (*3*) due to the dispersion factor of the analyzer, only a fraction of the ions introduced will be collected. The minimum quantity of material that must be introduced from the ion source into the analyzer is 10^{-8} to 10^{-12} g/sec. This wide range is due to the range of the required resolution. The higher the resolution needed, the more sample is required. For example, in a direct evaporation inlet one might need a sample size of 0.2 μg when resolution of 10,000 is required and mass measurement to 10 ppm is sought. A full order of magnitude less sample would be adequate for a resolving power of 1000 and mass measurement to 100 ppm. This problem is further discussed in Section 10-V.

Table 7-1   Sample Quantities Required by Various Inlet Systems

Inlet system	Typical sample, μg	Detection limit, μg
Standard batch for gases, ambient	10–100	0.01–0.1
Heated inlet; gallium inlet	10–100	0.1–1
Direct insertion; direct evaporation	5–20	0.01–0.1
Gas chromatography effluent	0.1–1	0.01–0.1

After a brief discussion of the laws of gas flow and the construction of leaks, the frequently used types of introduction systems for gases, liquids, and solids are described. The introduction of gas-chromatographic eluents and the problems involved in fast scanning are discussed in Section 10-V.

II. LAWS OF GAS FLOW

Inghram and Hayden (1) list four conditions for the ideal operation of a sample introduction system: (*1*) the composition of the gas mixture in the ionization region should be identical with that of the sample; (*2*) the composition should not change with time; (*3*) the partial pressure of each component should be independent of other components present; and (*4*) the gas flow rate should remain constant during the analysis. These requirements cannot all be met in any single introduction system, and an understanding of the flow conditions is necessary to avoid serious errors in quantitative measurements.

Gas flow in vacuum systems can be divided into three regions: molecular, viscous, and intermediate. *Molecular flow* prevails when the mean free path of the molecules is so large with respect to a characteristic dimension of the enclosure (e.g., radius of a tube) that intermolecular collisions can be disregarded and only collisions with the walls of the vessel are responsible for the flow resistance. Expressed in terms of pressure, molecular flow prevails when the product of the pressure in microns and the characteristic dimension in centimeters is less than 5. When the product is greater than 500, the flow is *viscous*. Here the mean free path of the molecules is small compared to the diameter of the pipe or enclosure, and intermolecular collisions are more important than collisions with the walls. Conditions in the intermediate or *transition* range are complex and difficult to express mathematically.

Recalling the equations of Section 6-IV-G, it is clear that the gas flow from the sample inlet system (high pressure) into the ion source (low pressure) is determined by the conductance of the interconnection between the two chambers. In *molecular inlet* systems, the connection is made by a cylindrical pipe or tube in series with a small aperture, called the *leak*. The molecular flow conductance, C_p, of a circular pipe of diameter D (cm) and length l (cm) is given by

$$C_p = 3.81 \sqrt{T/M} \ (D^3/l) \tag{7-1}$$

where C_p is in liters/sec, M is the molecular weight of the gas, and T is temperature. The conductance, C_a, of a circular opening of diameter D is given by

$$C_a = 2.86 \ D^2 \sqrt{T/M} \tag{7-2}$$

Since the diameter of the leak is almost always smaller than 0.02 mm, the conductance of the connecting pipes may be disregarded as long as the length of the tube does not make the conductance of the pipe the same order of magnitude as that of the leak. This is normally not a problem in mass spectrometer inlets.

There are three important consequences of the molecular flow conditions which must be considered in the analysis of mixtures. First, the flow of gas out of the reservoir into the ion source is such that the pressure in the reservoir falls off exponentially with time. Second, the ratio of the partial pressures inside the ion source is always the same as that in the inlet reservoir. Third, the lighter gases leak more rapidly from the reservoir than the heavier ones.

It is evident that although the rate of gas flow for each component is independent of other components present, it is inversely proportional to the square root of M, resulting in a fractionating effect. In a two-component mixture where the partial pressures of the components are p_1 and p_2 the numbers of molecules n_1 and n_2 passing through a leak are given by

$$n_1/n_2 = (p_1/p_2) \, (M_2/M_1)^{1/2} \qquad (7\text{-}3)$$

Taking an extreme example, the composition of a 1:1 hydrogen–xenon mixture changes considerably after only a few minutes of pumping through a leak. To prevent the change in composition with time, a relatively large volume (3–5 liters) is required for the sample reservoir. Moreover, when calibrating the instrument with pure gases for mixture analysis, rigid adherence to a predetermined time schedule must be maintained for accurate analysis (2,3). Of course, filling up a large sample reservoir requires a relatively large amount of gas. When only a small quantity of sample is available, the large reservoir may be valved off. With careful calibration and fast analysis, semiquantitative results can usually be obtained without difficulty.

In viscous flow, the flow rate, Q_v, in micron liters/sec, is obtained from Pouseuille's equation,

$$Q_v = K_v(p_1 - p_2)^2/\eta \qquad (7\text{-}4)$$

where K_v is a term determined by geometrical considerations of the constriction and the temperature. The molecular weight does not appear in the above formula; thus no fractionation occurs in mixtures. The flow, however, is now controlled by the viscosity (η, poise), a characteristic of the mixture. Moreover, the effective coefficient of viscosity depends in an entirely unpredictable manner on the relative abundance of the components. Two conclusions may be made: (1) in viscous flow the partial pressure of a component in the source is not proportional to its mole fraction in the sample and is not independent of other components present; (*2*) concentration ratios do not change with time (in contrast to molecular flow).

In viscous flow an additional problem is presented by the fact that the flow is necessarily molecular at the ionization source side of the viscous leak and that an *intermediate* flow region exists along each side.

III. CONSTRUCTION OF LEAKS

Molecular leaks usually consist of 4–6 parallel pinholes 0.005–0.02 mm in diameter in thin foils of gold about 0.25 mm thick. Many ingenious methods have been developed for the construction of small holes in metals and glass. Marks (4) developed a method in which a sharp tunsgten carbide point is pressed into the metal until it almost breaks through; the hole is opened by grinding. Other techniques include spinning fine needles and various types of pins. For glass and quartz, sparks from high-frequency Tesla coils may be used. In addition to pinholes in diaphragms, molecular flow conditions may also be provided by sintered glass disks and porous plugs with small capillaries.

The number of molecules ν of mass M striking an area of 1 cm^2 per second at a pressure of P torr is given by equation 6-6. Using a hole 0.02 cm in diameter, and having a pressure of 100 μ in a 3-liter sample reservoir, the value of ν for nitrogen at room temperature is about 5×10^{19} molecules/cm^2-sec, resulting in an effusion rate of about 0.025 micron-liter/sec. Conversion to mg/hour is made by dividing micron-liter/sec values by 0.2.

To obtain viscous flow conditions the simplest method is to use a metal or glass capillary tube 10–15 cm long with an internal diameter of 0.1 mm. The use of such capillaries in isotopic and analytical work is described by Mattraw et al. (5). Many types of adjustable needle valves providing viscous flow have been described; some of these are based on micrometer action, others use lever systems. A thermal expansion leak where the flow is controlled by the variable thermal expansion of two materials is described by Nester (6). A gold plug is inserted into a glass capillary and they are drawn together to form a fine point. On heating the leak is closed; on cooling it is open because the thermal expansion of gold is much greater than that of glass. Several other designs are described by Harrison (7).

Pure gases may be introduced utilizing the selective diffusion of certain gases through membranes. Such "leaks" include quartz for helium diffusion, a palladium–25% silver alloy for hydrogen diffusion, and silver for oxygen diffusion.

IV. INTRODUCTION OF GASES

The introduction of gases into the ion source is relatively simple. Samples are usually available in glass bottles or metal cylinders which can be connected through stopcocks or metal valves to the inlet system. If needed, commercial pressure reducers may be applied. Very small gas samples are often taken in small ampoules with one end drawn into a capillary. The ampoule is placed into an adapter which is connected to the inlet system. After initial evacuation, the capillary is broken by means of a turning stopcock or a glass-enclosed metal slug, followed by sample expansion into the reservoir.

A multipurpose gas inlet system is shown schematically in Figure 7-1. Similar systems are offered with most commercial mass spectrometers. These systems may be made of glass with stopcocks, or magnetically operated glass ball valves, or of stainless steel with metal bellows valves. The latter type is usually equipped with heater coils for baking. Each of the two reservoirs shown has a volume of about 3 liters, and the sample expansion from the bulbs into the reservoirs through a standard volume of about 3 cc is self-evident. Such double-inlet systems reduce the time interval between analyses and permit fast comparisons with reference sam-

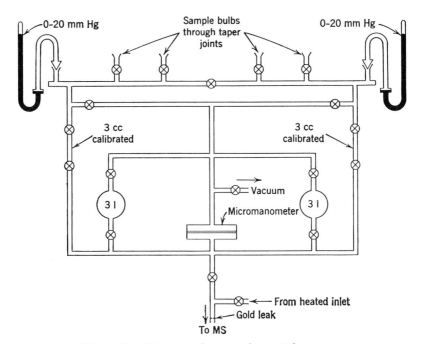

Figure 7-1. Schematic drawing of a gas inlet system.

ples. The pumping system is conveniently connected so that each section may be pumped (or leak tested) separately. For small samples a Toepler pump (Section 6-IV-B) may be used to produce adequate sample pressure behind the leak.

In the measurement of isotope ratios it is more important that the flow rate of the isotopes should not depend on mass than it is that the ion intensities should be proportional to the partial pressures in the reservoir. Accordingly, the inlet systems employ viscous flow conditions through capillary tubes. The double viscous flow inlet systems have two sample connections and two variable sample containers with fine adjustment of the storage pressure, measured with a micromanometer. Such arrangements allow quick comparison of two samples or a sample with a standard.

In gas stream *monitoring* instruments designed to sample continuously from atmospheric or higher pressure, it is customary to employ a nonfractionating viscous flow pressure divider to drop the sample pressure to about 1 torr pressure. Typical dimensions for the capillary tube are 1 m length and 0.1 mm inside diameter. To provide for continuous sample flow, i.e., to reduce "dead space," the downstream end of the viscous leak is connected to a mechanical pump. While most of the gas sample passing through the capillary is pumped away, a small portion is continuously introduced into the ion source through a conventional molecular inlet. Two important considerations must be made in the interpretation of the spectra. First, due to the time constant of the viscous leak, changes in the composition of the gas stream will appear at the molecular leak only a few seconds later. Second, the scanning of the required peaks may take several seconds during which changes in the composition of the stream may occur.

V. INTRODUCTION OF LIQUIDS

Volatile liquids can often be simply introduced by the *freezeout* technique. Here a small amount of sample is frozen out with Dry Ice or liquid nitrogen in the bottom of a small tube connected to the sample manifold. First the refrigerant is applied, then the tube is evacuated, and finally the sample is warmed to room temperature (or higher) and allowed to evaporate into the sample reservoir. When liquid mixtures are analyzed by any of the techniques to be described, relative volatilities must be carefully considered to avoid errors in quantitative analysis.

The use of *mercury-covered sintered disks* for the introduction of volatile liquids was first suggested by Taylor and Young (8). The method is good for liquids with vapor pressure >0.1 torr. The sample is transferred in a micropipet which is submerged into the mercury and touches the disk.

The liquid quickly evaporates through the disk, while for mercury the porous disk is leak-tight. As the liquid is evaporating, mercury enters the capillary. The quantity of sample introduced is obtained in weight by multiplying the volume of the pipet by the density of the liquid. Capillary sizes range from 0.003 to 0.0003 cc. The time required for passage through the disk increases from a few seconds to several minutes as the molecular weight of the sample increases. Various mercury-sealed orifice systems have been developed to reduce sample introduction time. In one design (9) a small orifice is closed by a well-fitting plug, with mercury again providing the seal. When the plug is removed, mercury slowly flows into a collection tube. The sample is quickly introduced by means of a short pipet ("dipper") which is submerged into the mercury so that its lower tip touches the orifice. A commercial version is shown in Figure 7-2.

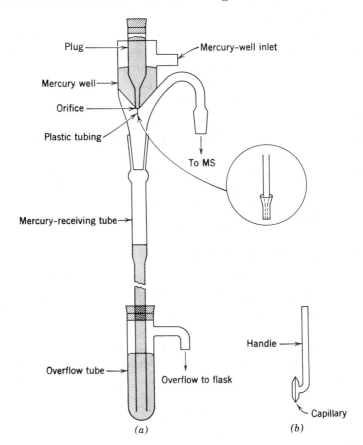

Figure 7-2. Mercury-sealed orifice sample introduction system for liquids. (Courtesy of Consolidated Electrodynamics Corp.).

For less volatile liquids and solids heated inlets are necessary. Both single and dual inlet systems are featured as accessories with commercial instruments in the forms of "150°C inlets" and "350°C inlets." These systems can operate at any desired temperature between ambient and the temperature indicated, and are completely bakeable including connecting lines, all valves, and gauges. Both all-glass and all-metal versions are available. The systems operated at 350°C are often made of glass to eliminate catalytic decomposition of samples. All-glass systems are also easier to clean. With all types of heated inlets one must be careful also to heat the connecting line between the leak and the ion source. The temperature of this line is usually kept about 100°C below that of the reservoir.

Since the vapor pressure of mercury is about 1 mm at 125°C, it cannot be used in heated inlet systems. Introduction of liquid samples at temperatures up to 300°C is most conveniently done in the popular gallium inlet systems. Several versions have been described (10–12). Gallium is a low-melting metal (m.p. 30°C, b.p. 1450°C which can be used with sinters in essentially the same way as described for mercury. The sample is placed in a pipet (capacity 0.1–1.0 µl) which is touched to the sintered surface under warm gallium and the sample evaporates into the reservoir. A commercial version is shown in Figure 7-3. Several cups are provided in parallel because there is a tendency for the sinters to become blocked with impurities. Also the surface of the gallium becomes oxidized and the oxide tends to stick to the dipper pipet. The two sample reservoirs permit the

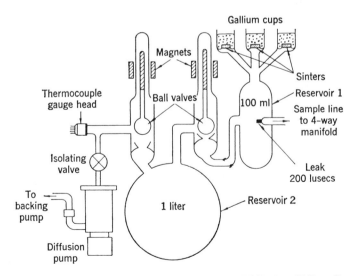

Figure 7-3. Gallium inlet system. (Courtesy of AEI/Picker X-Ray Corp.)

introduction of small samples at the same pressure as larger ones. Gallium inlets are simple and fast and the only disadvantage, apart from surface oxidation, is that reaction sometimes occurs between gallium and the sample.

Vapors of liquids may conveniently be introduced through serum caps by heatable *hypodermic* needles. This technique is becoming popular due to the availability of leak-tight and heat-stable cap materials.

VI. INTRODUCTION OF SOLIDS

Volatility is, of course, the major consideration in the introduction of solids. However, thermal sensitivity may prevent the use of heat to achieve the necessary vapor pressure and also often the sample quantity is very small. Waxes, tars, and many organic solids have vapor pressures of 0.5–0.01 torr at 300°C and thus can be analyzed in heated inlet systems. Low-melting solids may be introduced through heated syringes.

In Caldecourt's technique (13), a heatable stainless steel inlet block has a tapered hole covered by a somewhat oversized Teflon plug. The solid (or liquid) sample to be introduced is weighed into a Teflon cup of the same outside diameter as the plug and a duplicate plug is used to cap it. Next, sample cup and cap are placed in a split ring in a recess at the top of the inlet port, and the original plug plus the cup with the sample is pushed into the heated reservoir with a pusher rod. At the same time the hole is closed by the new Teflon plug. The small amount of air which enters does not usually interfere with organic analysis.

A. Direct Admission

Recently, direct admission techniques have been developed for solids with low vapor pressure and for small samples. Most commercial mass spectrometers are now available with some sort of direct inlet system. There are two basic types: in *direct evaporation* the sample is evaporated into the ion source from the outside, while in *direct insertion* the sample is inserted in solid form into the ionization chamber at the end of a suitable probe and evaporation takes place inside the source. Here the temperature required to obtain a pressure of 10^{-5} to 10^{-6} torr is 100°C or more lower than that needed for the 10^{-1} to 10^{-2} torr in a batch-type reservoir system.

An all-glass direct evaporation system is shown in Figure 7-4. The sample is loaded in solid form on the tip of a fine glass tube which is then broken off and dropped into the bulb or, more conveniently, it is loaded in a dis-

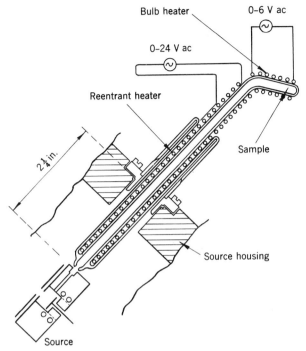

Figure 7-4. Sample introduction by direct evaporation from a reentrant glass inlet tube. (Courtesy of AEI/Picker X-Ray Corp.)

persion or solution from which the solvent can be evaporated. There are separate controls for the temperature of the sample bulb and the reentrant tube, providing flexibility in temperature programming. Sample analysis is performed by repeated scanning of the spectrum.

A direct sample insertion probe is shown in Figure 7-5; similar systems are featured by other manufacturers. A sample as little as 10^{-9} to 10^{-10} g may be adequate for matrix identification; however, much more is needed for extensive studies. The probe is introduced into the source through a vacuum lock (see later) and the sample is evaporated directly into the ion source. Spectra may be taken in an isothermal mode, i.e., at a steady state after thermal equilibrium has been established, or the probe may be temperature programmed manually or automatically. The maximum available temperature is usually 350°C.

As discussed in Section 10-III, in order to accurately measure the mass of an ion with a high-resolution mass spectrometer, it is necessary to introduce, along with the sample, a substance which gives ions at known masses as a reference. Perfluorokerosene is a substance often used for this

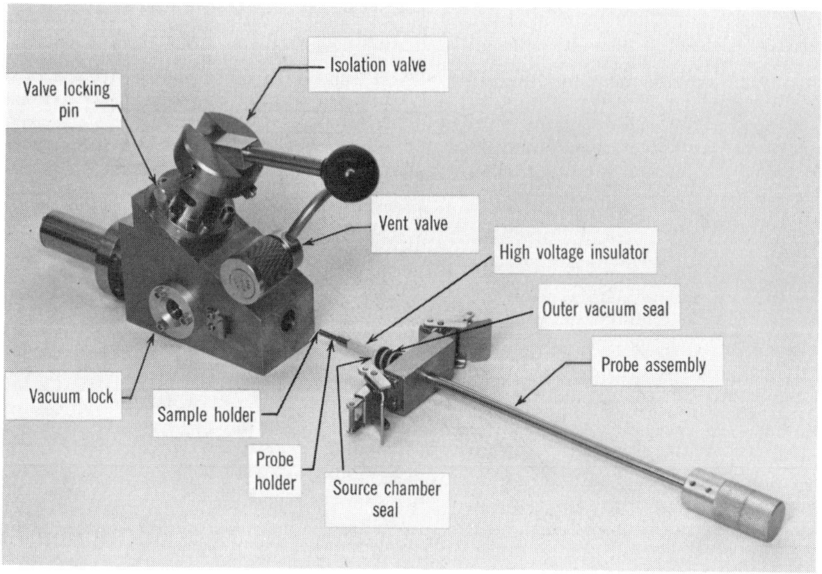

Figure 7-5. Vacuum lock assembly and solid direct insertion probe. (Courtesy of Consolidated Electrodynamics Corp.)

purpose; it is normally introduced through a heated inlet system. Shadoff and Westover recently described (14) a hollow direct probe which permits the *controlled* introduction of this or other types of mass calibrant through the existing direct introduction system of the mass spectrometer.

B. Knudsen Cell

A molecular beam which is a representative sample of a vapor in thermodynamic equilibrium with a condensed phase can be obtained in the Knudsen effusion cell (15). A Knudsen cell is a small crucible fitted with a tight cover which has a small hole or slit. The area of the orifice must be small compared to the area of the vaporizing sample, and the diameter of the orifice must be small (by a factor of 10) compared to the mean free path inside the cell. When a solid sample is vaporized at high temperature inside the crucible, a thermodynamic equilibrium is established between the vapor phase and the condensed sample—as long as there is no chemical reaction between sample and container material. The small quantity of vapor leaving the cell under such conditions (molecular effusion) is truly representative of the sample to be studied. The molecular beam is introduced, after passing a number of defining slits, into a more or less conventional electron bombardment source where the emerging neutral atoms and/or molecules are ionized.

A Knudsen cell ion source configuration is shown in Figure 7-6. There are many problems in designing a cell that is suitable for high-temperature studies. Construction materials must be carefully selected, with consideration for both the requirements of the high-temperature operation and the nature of the sample material: molybdenum, tungsten, and tantalum are the most frequently used materials. To compensate for shifts in the cell

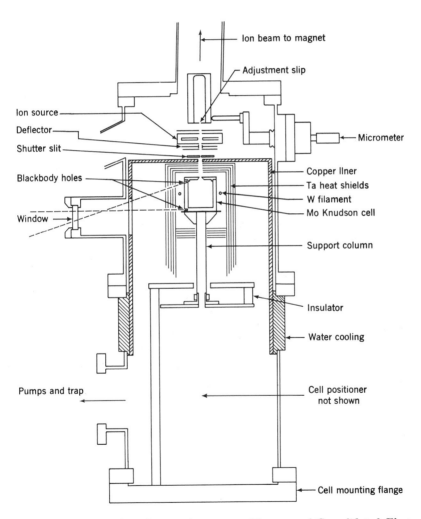

Figure 7-6. Knudsen cell and ionization source. (Courtesy of Consolidated Electrodynamics Corp.)

position during the heating cycle, mechanical alignment with the source is made with the aid of an externally adjustable micrometer screw.

Heating up to 1400°K is accomplished by radiation from a tungsten filament which surrounds the cell. Temperatures as high as 2500°K can be obtained by electron bombardment from the filament; as much as 1000 W of power can be concentrated on the crucible. Temperature is usually maintained within about 1% by carefully controlling such parameters as power, potential, etc. Heat loss is reduced and temperature uniformity is increased by carefully positioned tantalum heat shields. These are highly polished and are perforated on the side to facilitate temperature measurements by means of an optical pyrometer under blackbody conditions. Below 1800°K, temperature can be measured with a Pt/Pt/Rh thermocouple located at the base of the cell.

The cell and the ion source are separated by a shutter-slit assembly, the functions of which are to shut the molecular beam from the ion source completely or to permit the entrance of the beam in steps to determine its profile. A deflector between the collimating slits and the ion source removes positively charged particles from the beam. The number of charged particles leaving the Knudsen cell is a function of many parameters, including temperature and ionization potential of the material under study. Knudsen cells usually have their own separate pumping systems.

Knudsen cells are extensively used in modern high-temperature chemistry to study vaporization processes of refractory materials, group III–V compounds, etc. Vapor pressures, heats of vaporization, dissociation energies, etc. can be determined by this technique. The main advantages of the mass spectrometric detection technique as compared to other methods are high sensitivity, wide range of pressure (10^{-5} to 10^{-12} torr) that can be measured, uniform sensitivity for all components present, and an ability to identify the various species present in the vapor phase. The thermodynamic considerations underlying high-temperature chemistry and applications are outlined in Section 12-II.

Temperatures higher than 2500°K cannot be reached by heated crucible type arrangements because of mechanical instability due to thermal weakening, radiation losses, temperature control problems, etc. Temperatures up to 4000°K can be attained by focusing the radiant energy of a high intensity carbon arc onto spots 1 cm in diameter. Another possibility is heating by laser beams.

VII. VACUUM LOCKS

To overcome the necessity of breaking vacuum when changing solid samples, many types of vacuum locks have been developed (16–19), and most commercial instruments are now equipped with vacuum lock as standard or optional accessories. The sophistication of these vacuum locks vary considerably. Some are very simple and serve but a single purpose, others are multipurpose constructions permitting the switching between several types of inlets. Important considerations in selecting a commercial model or designing a new lock include: What kind, if any, gasket material is used, how precise is the external control in positioning the sample inside the ion chamber, how much additional volume is added, and how long it takes to perform a complete sample changing cycle. A source pressure of 10^{-6} torr should certainly be maintained with a well-designed vacuum lock.

Figures 7-4 and 7-5 show the use of vacuum locks in direct evaporation and direct insertion inlets. Figure 7-7 shows the operation of another type of commercially available vacuum lock (19). Ion source chamber and lock chamber are separated by a hinged flap which is pressed against a Teflon ring by a rod which also carries the interchangeable ion source inserts. The first step in changing samples consists of venting the lock chamber, inserting the new sample, and reevacuating the lock chamber. Next, the sample holder rod is retracted by a microswitch-controlled motor and the lock flap is opened. The rod is turned 180° by the motor and introduced into the ion source region. With further rotations and back-and-forth manipulations of the rod, the old sample casket is removed and the new one is placed. Finally, the rod is withdrawn and the flap is closed, thus completing the cycle. Samples may be changed in less than one minute, and

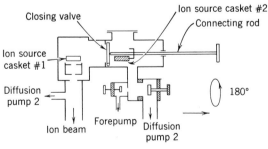

Figure 7-7.   Operation of a vacuum lock. (After 19.)

the pressure in the source region is restored to the 10^{-7} torr level in a few minutes.

References

*1. Inghram, M. G., and R. J. Hayden, *A Handbook on Mass Spectroscopy* (Nuclear Science Series, Report Number 14), National Academy of Sciences— National Research Council, Washington, D.C., 1954.

*2. Honig, R. E., *J. Appl. Phys.*, **16**, 646 (1945).

*3. Zemany, P. D., *J. Appl. Phys.*, **23**, 924 (1952).

4. Marks, R., *Rev. Sci. Instr.*, **28**, 381 (1957).

5. Mattraw, H. C., R. E. Patterson, and C. F. Pachucki, *Appl. Spectry.*, **8**, 117 (1954).

6. Nester, R. G., *Rev. Sci. Instr.*, **27**, 874 (1956).

7. Harrison, E. R., *J. Sci. Instr.*, **30**, 170 (1953).

8. Taylor, R. C., and W. S. Young, *Ind. Eng. Chem., Anal. Ed.*, **17**, 811 (1945).

9. Purdy, K. M., and R. J. Harris, *Anal. Chem.*, **22**, 1337 (1950).

10. O'Neil, M. J., and T. P. Wier, *Anal. Chem.*, **23**, 830 (1951).

11. Lumpkin, H. E., and B. H. Johnson, *Anal. Chem.*, **26**, 1719 (1954).

12. Beynon, J. H., and G. R. Nicholson, *J. Sci. Instr.*, **33**, 376 (1956).

13. Caldecourt, V. J., *Anal. Chem.*, **27**, 1670 (1955).

14. Shadoff, L. A., and L. B. Westover, *Anal. Chem.*, **38**, 1048 (1966).

15. Chupka, W. A., and M. G. Inghram, *J. Phys. Chem.*, **59**, 100 (1955); see also references in Section 12-II.

16. Munro, R., and R. G. Ridley, *Rev. Sci. Instr.*, **36**, 1538 (1965).

17. Christie, W. H., and A. E. Cameron, *Rev. Sci. Instr.*, **37**, 336 (1966).

18. Junk, G. A., and H. J. Svec, *Anal. Chem.*, **37**, 1629 (1965).

19. Brunnèe, C., *Z. Instrumentenkunde*, **68**, 97 (1960).

Chapter Eight

Commercial Instruments

I. AVAILABILITY AND PRICE

Commercial mass spectrometers have become a multimillion dollar business during the past 5 years. The total market in the USA is estimated to be 10–15 million dollars annually. This sum is about evenly divided among high, intermediate, and low resolution instruments, the latter group including leak detectors. A large number of instruments have also been sold in Europe and Japan during the past few years.

Mass spectrometer manufacturers and representatives are listed in Table 8-1. For some of these companies mass spectrometry is major business, for others it is only a sideline; this is indicated to a certain degree by the specialties listed. Residual gas analyzers are available from most major mass spectrometer manufacturers and also from a number of firms producing high vacuum equipment. There are a few additional manufacturers in the USA, France, Italy, and Japan offering specialized and custom-made machines but their market is limited. Every important type of mass spectrometer is manufactured in the Soviet Union by the State Combined Design Bureau (GSKB), but relatively little is known about their features and performance (1,2).

Many of the companies listed in Table 8-1 entered the field of mass spectrometry during the last few years. The resulting competition is encouraging the manufacturers to spend considerable efforts on research and development. This, of course, is most welcome to the customers.

Purchase price depends much more on the performance offered than upon the principle of mass separation employed. Resolution usually determines the "size" and also the price of an instrument. Residual gas analyzers with resolution up to 80 cost $3000–5000, while resolution up to 150 cost $5000–7000. Quadrupole and monopole residual analyzers with resolution as high as 500 sell for between $8000 and $13,000. Several magnetic and quadrupole type residual analyzers are also available in full mass spectrometer versions, i.e., with vacuum and inlet systems; these instruments

Table 8-1 Instrument Manufacturers and Representatives

Company[a]	Address	Specialty
Aero Vac Corp.	Troy, N.Y.	Single-focusing magnetic; magnetic residual
AEI: Associated Electrical Industries, Ltd. Marketed in USA by Picker X-Ray Corp.	Manchester, England, 1275 Mamaroneck Ave., White Plains, N.Y.	Full line single/double focusing magnetic
AVCO Corp., Electronics Division	10700 E. Independence St., Tulsa, Oklahoma	Custom-made 90° single-focusing magnetic with variety of sources
Bendix: The Bendix Corp., Cincinnati Division	3130 Wasson Road, Cincinnati, Ohio	TOF instruments
CEC: Consolidated Electrodynamics Corp. A subsidiary of Bell & Howell Corp.	1500 S. Shamrock Ave., Monrovia, Calif.	Full line single/double focusing magnetic; magnetic residual
EAI: Electronic Associates, Inc.	4151 Middlefield Rd., Palo Alto, Calif.	Quadrupole
Finnigan Instruments Corp.	2625 Hanover St., Stanford Industrial Park, Palo Alto, California	Quadrupole
GCA Corp., Technology Division	Bedford, Mass.	Double-focusing ion microprobe
GE: General Electric Co. Vacuum Products Division	Schenectady, N.Y.	Monopole residual
		Magnetic residual

Granville-Phillips Co.	5675 East Arapahoe Ave., Boulder, Colo.	Quadrupole residual
Hitachi: Hitachi Ltd. Marketed in USA by Perkin-Elmer Corp.	Tokyo, Japan Main Ave., Norwalk, Conn.	Full line single/double focusing magnetic
JEOL: Japan Electron Optics Laboratory Co.	477 Riverside Ave., Medford, Mass.	Full line single/double focusing magnetic
LKB Instruments, Inc.	1221 Parklawn Drive, Rockville, Maryland	60° single-focusing magnetic for gas chromatography work
NUCLIDE: Nuclide Analysis Associates	642 E. College Ave., State College, Pa.	Full line single/double focusing magnetic; custom-made instruments
The Perkin-Elmer Corp.	Main Ave., Norwalk, Conn.	Nier–Johnson intermediate-resolution double-focusing
Process & Instruments Co.	15 Stone Ave., Brooklyn, N.Y.	Custom-made 60° deflection magnetic
Sloan Instruments Corp.	P.O. Box 4608, Santa Barbara, Calif.	Omegatrons
Thomson-Houston Corp.	173 Bd Haussmann, Paris, France	Full line single-focusing magnetic
Ultek Corp., subsidiary of The Perkin-Elmer Corp.	Box 10920, Palo Alto, Calif.	Quadrupole residual
Vacuum Electronics Corp.	Terminal Drive, Plainview, N.Y.	Magnetic deflection residual
Varian Associates	611 Hansen Way, Palo Alto, Calif.	Cycloidal, intermediate-resolution; syrotron; quadrupole residual
Varian/MAT, formerly Fried. Krupp Mess und Analysen Technik	Wolterhauser Str. 442, Bremen, Germany	Full line single/double focusing magnetic; quadrupole residual

[a] Abbreviations preceding full names are used in text to identify manufacturers

cost in the order of $25,000. General-purpose medium resolution (up to 3000) mass spectrometers and time-of-flight instruments (resolution up to 700) are in the $30,000–50,000 range, not counting special accessories. High-resolution double-focusing instruments, including spark versions, can be purchased for about $100,000. Most instruments are featured with a wide range of desirable optional accessories so that the final bill may be as much as one-third higher than the basic price. In addition, data reduction and computer equipment may cost almost as much as the instrument itself. The cost of a well-equipped mass spectrometer laboratory for the computerized study of high-resolution organic spectra can easily run up to a quarter of a million dollars.

II. SELECTION OF INSTRUMENTS

Modern commercial mass spectrometers are designed in modules. Once the basic analyzer system has been selected, there are a number of ion sources and detectors available to suit particular problems. Turning to a multiplier detector from a conventional amplifier often involves nothing more than turning a switch. Important factors to be considered when selecting an instrument are summarized in Table 8-2.

Table 8-2   Considerations in Instrument Selection

- Resolution needed (unit-usable)
- Mass measurement capability
- Mass range
- Coverage (broad-specific)
- Sensitivity (detection limit)
- Available inlet systems
- Speed of scanning and response
- Vacuum
- Reputation and experience of manufacturer

Resolution. Maximum required resolution is the most important consideration in selecting an instrument. As a rule, resolution cannot significantly be increased without extensive modifications. Maximum resolution in magnetic instruments is achieved using the narrowest possible slits (plus maximum accelerating voltage). Commercial instruments often feature variable resolution by providing adjustable source and collector slits. Continuous variation is accomplished by micrometer screws; sometimes slits have to be replaced by others of different size. Narrow slits, of course, result in sensitivity loss, so a compromise must be made.

Numerical resolution values that follow are based on the "10% valley" definition. It is recalled that according to this definition two adjacent peaks of equal height are said to be resolved when the valley between them is equal to, or less than, 10% of the height of either peak (Fig. 1-3). For comparison purposes, a resolution figure based on 5% cross-talk is approximately twice that based on the 10% valley definition.

Resolving power of 50–80 is sufficient for most residual gas analyzers. In these analyzers it is, perhaps, the most important to have an ability to distinguish between carbon dioxide (at m/e 44) and the residual hydrocarbon peak at m/e 43. A number of hydrocarbons have relatively large pattern contribution to the 43 peak, making this peak important for their detection.

There are a number of small-radius magnetic deflection instruments (cycloidal, 180°, and sector field) featuring resolution of 150–500. Such resolution is adequate for a number of problems including the analysis of inorganic gases, analysis of gases in metals, respiratory gas analysis, many problems involving isotope dilution, and gas chromatographic eluent identification of many types of organic compounds. This resolution range includes also the time-of-flight instruments and quadrupole/monopole mass spectrometers. The primary advantage of these machines is speed. Many reaction mechanism studies, analysis of discharges, flames, etc. do not require resolution higher than 500.

Conventional "medium resolution" single-focusing mass spectrometers achieve unit resolution in the 700–3000 range using analyzers with a radius of 15–30 cm (6–12 in.). This is sufficient for a large variety of analytical problems and several new instruments have been introduced recently in the medium or intermediate resolution range. In organic chemistry applications these instruments are perfectly adequate for the identification of gas chromatography effluents of a large variety of organic compounds and are routinely being used for this purpose by many laboratories in petroleum chemistry and other chemical industries. This resolution range is also adequate for Knudsen cell studies, for most work with thermal sources, and for appearance potential measurements.

When setting resolution requirements for the analysis of solids by the spark source technique (Mattauch-Herzog double-focusing spectrograph), one must consider that the elements have many isotopes and also that many multiply charged ions are produced with relatively high intensity as compared to the faint lines of traces one is searching for. Selection of resolution can be made with reference to Fig. 8-1 (3) which shows the percentage of the elements resolved from their nearest neighbor as a function of the available resolving power. In preparing this table, isotope abundances were considered in such a way that abundances were sacrificed by a factor of not more than 3, if by so doing a more remote nearest neighbor was ob‐ tained. This cannot, of course, be done with monoisotopic elements (some 27% of all elements). It is seen from the figure that a resolution of about 2500 permits about 35–40% of the elements to be resolved from their nearest neighbors. Separation of additional elements would require a large, and very expensive, increase in resolving power. Another problem is the resolution of impurity lines from background lines. The 2500–3000 resolu‐ tion usually available in commercial spark instruments is adequate to resolve peaks in the mass 28 area, as shown in the following tabulation:

Doublet	ΔM, amu	$M/\Delta M$
$^{56}\mathrm{Fe}^{++}-^{28}\mathrm{Si}^{+}$	0.0095	2950
$^{28}\mathrm{Si}^{+}-^{12}\mathrm{C}^{16}\mathrm{O}^{+}$	0.0180	1550
$^{12}\mathrm{C}^{16}\mathrm{O}^{+}-^{14}\mathrm{N}_2^{+}$	0.0112	2500
$^{14}\mathrm{N}_2^{+}-^{12}\mathrm{C}_2^{1}\mathrm{H}_4^{+}$	0.0252	1100

An enlargement of a photographically recorded mass spectrum showing the triplet at m/e 28 at a resolution of about 2500 is shown in Figure 8-2a.

In organic chemical applications, resolution requirements fall broadly into two classes. To differentiate between adjacent peaks around the

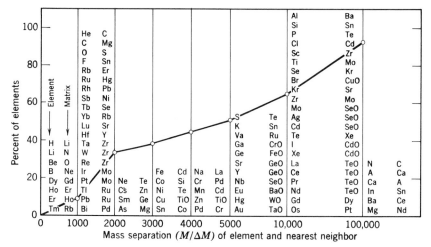

Figure 8-1.   Resolution requirements in spark source analysis (3).

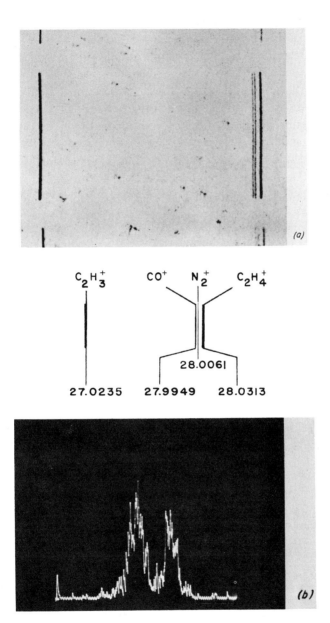

Figure 8-2. (a) Enlargement of photographically recorded mass spectrum showing the triplet at nominal mass 28 at a resolution of 2500. (The ends of lines at the top and bottom of the figure show how multiple exposures may be made on a photoplate by vertically moving the entire plate assembly.) (b) Oscillographic recording of the $C_6H_6^+$–$C_6H_4D^+$ doublet at nominal resolving power of 50,000. (Courtesy of Consolidated Electrodynamics Corp.)

molecular ion requires resolution of 600–1000. To separate doublets involving hydrogen, carbon, nitrogen, and oxygen requires considerably higher resolution. Table 8-3 lists a number of mass doublets encountered in organic work and also shows the resolving power necessary for separation. For 10 millimass units at mass 100 a resolution of about 10,000 is required. The last column in the table gives the resolution for each particular doublet to be fully resolved at mass 100. The resolution required at any other mass number can be obtained by multiplying by the factor $M/100$, where M is the mass of the doublet concerned (4). The required resolving power of

Table 8-3   Some Common Doublets Encountered in High Resolution Mass Spectra (after ref. 4)

m/e	Doublet	$\Delta M \times 10^3$	$100/\Delta M$
2	H_2-D	1.548	64,600
12	$H_{12}-{}^{12}C$	93.900	1,065
13	${}^{12}CH-{}^{13}C$	4.467	22,385
	$\frac{1}{2}{}^{12}C_2D-{}^{13}C$	3.693	27,080
14	${}^{12}CD-{}^{14}N$	11.028	9,070
	${}^{13}CH-{}^{14}N$	8.109	12,330
	${}^{12}CH_2-{}^{14}N$	12.576	7,950
	${}^{13}CH-\frac{1}{2}{}^{12}C^{16}O$	13.726	7,285
15	${}^{14}NH-{}^{15}N$	10.789	9,270
	${}^{12}CH_3-{}^{15}N$	23.365	4,280
	${}^{12}HD-{}^{15}N$	21.817	4,585
16	${}^{12}CH_4-{}^{16}O$	36.386	2,750
	${}^{14}NH_2-{}^{16}O$	23.810	4,200
	${}^{13}CH_3-{}^{16}O$	31.919	3,130
	${}^{15}NH-{}^{16}O$	13.021	7,680
18	${}^{15}NH_3-{}^{18}O$	24.424	4,095
	${}^{17}OH-{}^{18}O$	7.795	12,830
	${}^{14}NH_2D-{}^{18}O$	33.665	2,970
28	${}^{14}N_2-{}^{12}C^{16}O$	11.234	8,900
	${}^{13}C^{12}CH_3-{}^{14}N_2$	20.685	4,835
	${}^{12}C_2H_2D-{}^{14}N_2$	23.604	4,240
	${}^{12}C_2H_4-{}^{28}Si$	54.375	1,840
32	${}^{16}O_2-{}^{32}S$	17.755	5,630
	${}^{12}CH_8-{}^{32}S$	90.527	1,105
	${}^{12}CH_8-{}^{14}N^{18}O$	60.365	1,655
	${}^{13}C^{12}CH_7-{}^{14}N^{18}O$	55.898	1,790
44	${}^{12}C_3H_8-{}^{14}N_2{}^{16}O$	61.538	1,625
	${}^{14}N_2{}^{16}O-{}^{12}C^{32}S$	28.989	3,450
	${}^{12}C_3H_8-{}^{28}Si^{16}O$	90.761	1,100

some common doublets as a function of mass is shown graphically in Figure 8-3. It is concluded that the minimum resolution requirement for high-resolution organic work is about 15,000. Resolution up to 35,000 is featured on many commercial instruments. Figure 8-2*b* shows the C_6H_6–C_6H_4D doublet at m/e 78.

Ultrahigh resolution refers to resolution around 500,000 and over. Such resolution is required only in extremely accurate mass determinations in basic physics research and in the structural study of certain unknown natural products. Although JEOL offers a commercial instrument with ultrahigh resolution, this is a highly specialized field and most work is done in university laboratories.

Mass Measurement Capability. The second feature of mass spectrometers to be considered is mass measurement capability. In low and intermediate resolution instruments one usually attempts only to determine unit masses, normally by means of comparisons with a calibration compound. Fractional mass units are frequently encountered and can readily be determined within a few tenths of a mass unit. For the determination of empirical formulas by mass spectrometry, high resolution must be combined with an ability to measure masses to an accuracy of at least ± 10 ppm. Methods for measur-

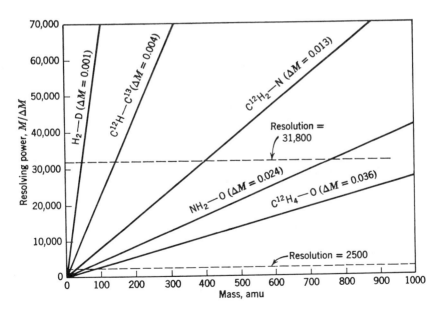

Figure 8-3. Required resolving power of some common doublets in organic analysis. (Courtesy of Consolidated Electrodynamics Corp.)

ing mass with such accuracy and the use of the technique for empirical formula determination is discussed in Section 10-IV.

Mass Range. It is obvious that the mass range of an instrument must be such that the highest mass anticipated is covered. In residual gas analysis and inorganic work the upper limit needed is about 300. In organic applications the upper limit is set by sample volatility, as molecules over mass 1000 are usually very difficult to introduce into the mass spectrometer. The mass range of commercial instruments is usually from 1 to 2000, though masses as high as 3500 can be measured (Fig. 10-10) in certain instruments at low acceleration voltages (1–2 kV). Electrical scanning is usually limited to a mass range of a factor of 10 so that separate magnetic field settings are needed to cover the entire mass range. With magnetic scanning the entire mass range may be scanned in a single sweep. Whenever stepwise scanning is necessary, sufficient overlapping must be provided.

Coverage. The mass spectrometric technique and consequently the selection of source is greatly influenced by the kind of information desired in the investigation. By coverage is meant whether the information sought is broad, covering for example the whole periodic table (e.g., spark technique) or very specific (e.g., thermal ionization).

Sensitivity. Although the overall sensitivity of a mass spectrometer is the result of such parameters as source efficiency, transmission, etc., it is the detector system where one has the most flexibility. In modern multipurpose single-focusing instruments, collector systems have two modes of operation: a standard mode, using a dc electrometer or vibrating reed electrometer, or a high sensitivity mode where the collector plate is pulled out of its position exposing the conversion dynode of a multiplier. In some double-focusing instruments a combination electrometer/electron multiplier and photographic detection is often featured as optional. This provides much flexibility in sensitivity and recording speed.

In connection with sensitivity, it is recalled that resolution can be traded for sensitivity provided that there is resolution to spare.

Available Inlet Systems. Vapor pressure of the sample is the most important consideration in the selection of the proper inlet system. The sample inlet system is usually the most divergent section of commercial mass spectrometers and generally two or three different types are needed to meet requirements posed by the vapor pressure of the samples. Inlet systems were discussed in detail in the previous chapter.

Speed of Scanning and Response. Speed of scanning the mass spectrum, together with the speed of response of the detector system, have recently achieved prominence among performance characteristics. Speed is very important in the analysis of gas chromatographic effluents where scan time is normally selected in such a way that a complete spectrum can be obtained in a time interval less than the half-width of the chromatographic peak. High speed scanning is needed in studies involving transient species, fast reactions, free radicals, etc. In these studies dynamic instruments are often used; their fast scanning capability has been discussed earlier in the description of their operation.

In modern magnetic instruments, variable scan speeds are standard features. The range covers scan speeds from a few tenths of a second to tens of seconds per mass decade. Faster scanning and/or response is usually paid for by a decrease in sensitivity and accuracy.

Voltage versus magnetic scanning is another question often considered in instrument selection. We have already touched the problem several times in Chapters 2 and 3. Voltage scanning is generally believed to be more reproducible than magnetic scanning. Magnetic scanning involves a variation of the magnet current to change the field and when the current returns to the starting point the field strength may not be the same due to hysteresis. When the magnet current is increased linearly the peak spacing is nonlinear (tied to the square of the magnetic field strength) and, as with voltage scanning, the peak spacing decreases at higher masses. When linear magnetic scanning is employed the peaks are equally spaced and peak counting is easier. A disadvantage is a broadening of the peak shape as the mass scale is scanned. Linear magnetic scanning is accomplished by using two Hall crystals in series to produce a voltage proportional to the square of the magnetic field and thus proportional to the mass.

The rather involved problem of mass discrimination with voltage and magnetic scanning cannot be discussed here, and the reader is referred to advanced texts on ion optics. All types of mass spectrometers exhibit mass discrimination, and the degree of discrimination depends on instrument design, operating conditions, and the masses of the ions involved. Mass discrimination with voltage scanning originates partially from the fact that the ions in the source do not possess zero initial energies, and partially from changes in the ion source optics as the accelerating field is changed. To keep discrimination to a minimum, the highest possible acceleration voltages should be used.

It is stressed that the ultimate resolution of a mass spectrometer does not depend on scanning speed, and is also independent of the detector and recording system. However, when the speed of scanning is increased to the point where the detector and/or recorder cannot faithfully follow the rising and falling ion current signal, peak "clipping" results. Observed resolution will decrease, and also quantitative measurements will become erroneous when amplifier and/or recorder response speed is inadequate. For example, the amplifier of the instrument may have a pass band of 1 kHz. This, combined with a 1 kHz response galvanometer recorder, is adequate for rather fast recording at low or intermediate resolution. When high resolution, e.g., 10,000 is to be achieved together with fast scanning, the response time of the amplifier has to be much higher, e.g., 10 kHz. Of course, the recorder must also have fast response, and either a photographic oscillograph or magnetic tape recording becomes necessary.

Vacuum. To state that vacuum in a mass spectrometer should be as good as possible is commonplace. Vacuum conditions acquire, however, special significance in trace analytical applications. Scattering in the analyzer, for example, significantly reduces the ultimate sensitivity of photoplates due to diffuse darkening. Another example is the analysis of ions from neutral species in a sputtering source where poor vacuum can considerably falsify results. Separate pumping systems for the source, analyzer, and detector sections are now common practice in modern instruments; vacuum of at least 10^{-8} torr is usually featured. Special attention must be paid to baking facilities, gaskets employed, and the location of pressure gauges.

Reputation of Manufacturer. Reputation of the manufacturer is included in the list of considerations in instrument selection for several reasons. The analytical precision achievable with a given kind of instrument depends to a great extent upon the design of electronic components, the quality of components used, and the care with which the instrument is assembled. Maintainance problems, spare parts, repair service, technical assistance, are well-understood key words underlying the importance of this phase of the instrument selection.

Manufacturers attempt to make their instruments as versatile and flexible as possible and there is a high rate of change from year to year. Detailed performance specifications together with information on the latest models is readily available from manufacturers upon request (see Table 8-1 for addresses). In what follows the more significant features of commercial instruments are summarized. The information presented is based on manufacturers' published specifications, advertisements, and private communications. While basic instrument design is usually kept unchanged for a particular type of mass spectrometer until some major advancement occurs, performance specifications are often changed on account of minor modifications.

Presently there are at least 75 different models of mass spectrometers commercially available. Tables of numerical characteristics would be outdated in a short time and the listings of instruments no longer in existence are quite irritating; therefore, no such tables are presented in this text. Instead, representative models in the various categories are cited and basic performance data are given for similar types of instruments. Reference to a make or model should not be interpreted as endorsement; rather it reflects only the author's familiarity with it. Instruments are discussed under headings of low (<150), intermediate (up to 3000), and high resolution, and further subdivision is based upon either geometry or field of application.

III. LOW-RESOLUTION INSTRUMENTS

Low-resolution (<150) mass spectrometers are mainly utilized as residual gas analyzers. There are at least a dozen commercial types available and just about every type of mass analyzer has been utilized for residual gas analysis including 60°, 90°, 180° magnetic, cycloidal, quadrupole, rf, omegatron, and TOF instruments. The use of residual analyzers in high vacuum technology and the problems associated with residual gas analysis are discussed in Section 13-IV. The performance factors and operating features of at least 20 commercial residual gas analyzers have been compared by Hultzman (5). Among the performance factors, sensitivity, minimum detectable partial pressure, maximum operating total pressure, mass range, resolution, and scanning time must be considered. The main operating features include reliability, flexibility, convenience, and construction.

Among the magnetic type low resolution full scale instruments, quite popular are the *cycloidal* analyzers. CEC's Model 21-620A and Model 21-130 have first-order focusing for both velocity and direction, resulting in true trapezoidal peak shapes. Both the electron impact source and the dc amplifier collector system are combined in a single unit totally immersed in the magnetic field (permanent magnet, 4500 gauss). A schematic of this instrument is shown in Figure 3-10. Mass range is m/e 2–230 with unit resolution around 200. Both batch and continuous inlets (for process monitoring) are available. Readout options are direct-writing oscillograph, strip chart recorder, or digital output. These instruments are popular for the analysis of trace inorganic gases, gases in metals, low mass organic samples, and low mass chromatographic effluent identification.

AEI's Model MS-10 instrument is similar in performance and applications to the cycloidal instruments. It is a 180° magnetic analyzer with a 5 cm radius. Its resolution is about 100 with a sensitivity of 5×10^{-5} A/torr for nitrogen. AeroVac's Model 686 mass spectrometer is designed for trace inorganic gas analysis. Its analyzer utilizes two stages of 60° deflection 5 cm radius sectors in tandem, employing a magnetic field of 5000 G. With unit resolution of 250 and mass range 1–500 amu the instrument is optimized for high sensitivity. Abundance sensitivity of 0.02 ppm and partial pressure sensitivity of 10^{-13} torr (for nitrogen) is achieved with the multiplier detector. A special suppressor in the collector assembly eliminates collection of ions scattered at energies less than that appropriate to the mass to which the analyzer is tuned. High vacuum combined with baking results in very low background which is a major requirement in trace component analysis.

Nuclide's Model 4.5-60-RSS mass spectrometer (Fig. 8-4) is the duplicate of an instrument designed by Reynolds (6) for the analysis of *ultrasmall* quantities of noble gases. It is a 11.4 cm (4.5-in.) radius, 60° sector machine with an all glass envelope, bakeable to 450°C. It has an electron impact source, a small total volume, and normally uses a Faraday cup collector connected to a vibrating reed electrometer. Usable resolution is about 135 (xenon). Analyzer pressure is as low as 10^{-10} torr and for many applications getter-ion pumping is recommended. It is used primarily for "static" analysis, i.e., there is no pumping during analysis. Samples as small as 10^{-7} cc (STP) can be analyzed routinely, while the least abundant isotope of xenon can be detected in a sample of 10^{-11} cc (STP) when multiplier detection is employed. This instrument is primarily used in geochemical studies (Section 13-II).

Varian's Model M-3 (Fig. 8-5) and AEI's Model MS-4 are 180° mass spectrometers primarily designed for medical use to study *respiratory gases*. As shown in the figure, four masses can be monitored simultaneously. Resolution with normal slit sizes is about 120. Response time is so fast (100 msec level) that single breath analysis can be carried out.

IV. INTERMEDIATE RESOLUTION INSTRUMENTS

One of the best known and most widely used intermediate resolution mass spectrometers in the USA is the CEC Model 21-103 instrument (Fig. 8-6). Introduced some 20 years ago and updated several times, the 21-103

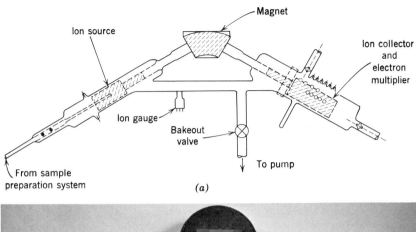

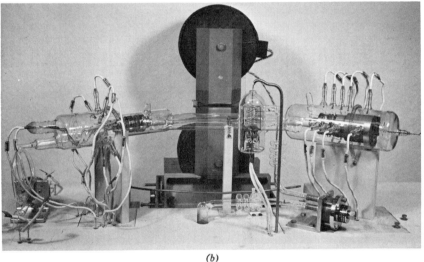

Figure 8-4.   Reynolds-type mass spectrometer for the analysis of ultrasmall gas samples. (*a*) Schematic drawing; (*b*) photograph. (Courtesy of Nuclide Corp.)

has become a kind of industrial standard, particularly in the petroleum industry. Most of the API spectra (Section 9-I) were obtained with this instrument. It is a 5-in. radius (12.7 cm), 180° deflection, direction-focusing instrument. Unit resolution of about 350 (2% valley definition) is achieved with a 0.007 in. resolving slit width; usable mass separation can be obtained up to mass 700. Threshold sensitivity is 1/1000 mole %. In analytical applications, components at the 1–5% level can be determined with a maximum deviation of 5–1%, while components at the 0.05% level are measured with a maximum deviation of 100%. The ion source ("Isatron") is of the electron impact type with adjustable ionization current (10–100

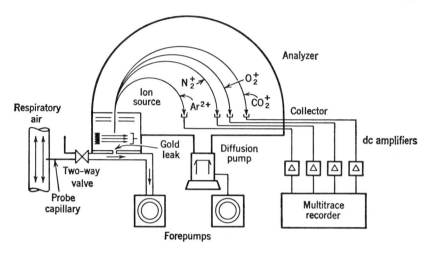

Figure 8-5. Mass spectrometer for breath analysis. (Courtesy of Varian/MAT).

mA) and ionization potential (50–100 V, 70 V normal). Readout is on a five-galvanometer oscillograph. The 21-103 mass spectrometer has been distinguished for its long-term stability and good analytical accuracy. Although it has recently been replaced with a much improved new instrument (Model 21-104), and several other instruments of equal or better specifications have been introduced by other manufacturers, the 21-103 instruments are still being used in many laboratories and their use will certainly continue for a number of years.

Just about every major manufacturer of mass spectrometers offers a *"general purpose"* single-focusing mass spectrometer in the intermediate resolution range. Most of these instruments have been introduced during the past 5 years and this is reflected in their versatility. They are usually designed around a basic analyzer system consisting of a 60°, 90°, or 180° magnetic analyzer of 5–12 in. radius to which a variety of interchangeable ion sources and detectors can be attached. Electron impact and thermal sources are most frequently employed, but others such as Knudsen cells and Fox sources are also available. All these instruments employ electrical detection, and combination electrometer and multiplier detectors are becoming common. The resolution of these instruments is in the 1000–4000 range. Of course, higher resolution is paid for by reduced sensitivity, part of which, however, may be gained back by multiplier detection.

Table 8-4 summarizes the geometrical features of several representative multipurpose single-focusing mass spectrometers currently available. The resolution of these instruments may be estimated from their sizes;

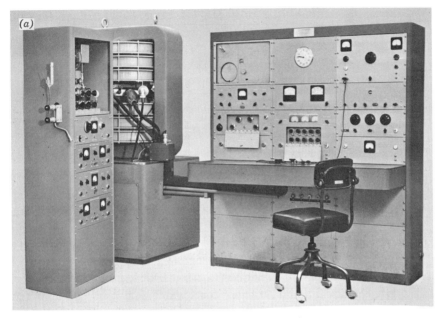

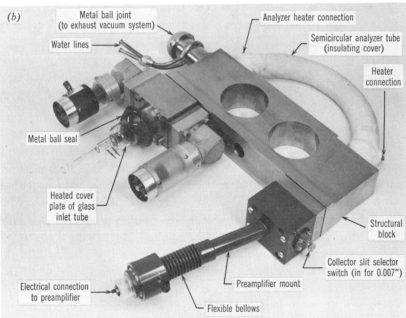

Figure 8-6. (a) CEC Model 21-103C mass spectrometer; 180° magnetic deflection, single-focusing. (b) "Isatron" ion source. (Courtesy of Consolidated Electrodynamics Corp.)

Table 8-4   Geometry of Single-Focusing Instruments

			Radius	
Company	Model	Deflection	cm	in.
AEI	MS-12	90°	30	12
CEC	21-104	180°	13	5
Hitachi	RMU-6E	90°	20	8
JEOL	JMS-05SA	90°	30	12
LKB	9000	60°	20	8
Nuclide	12-60	60°	30	12
Nuclide	6-90	90°	15	6
Varian MAT	CH 5	60°	23	9

resolution can be varied with adjustable slits. The number and variety of optional accessories is indeed impressive. Vacuum locks, heated inlet systems, direct sample introductions, mass markers, peak selectors, digitizers, etc. can assist in almost every conceivable analytical problem. The Varian CH 5 is available with a novel dual electron bombardment–field emission source (Fig. 4-10). The AEI, Hitachi, and Nuclide instruments may be equipped with a double-focusing attachment (i.e., an electrostatic sector) by which resolution is extended up to as high as 15,000. The Varian and Hitachi instruments are shown in Figures 8-7 and 8-8.

Two more magnetic deflection instruments should be mentioned in connection with the multipurpose designs; both have been introduced rather recently. Varian's Model M-66 mass spectrometer is of *cycloidal* design. It has a resolution of 2000 and a mass range of 10–2000. Due to the linear magnetic scanning, peaks are equally spaced and mass measurement accuracy of 10 ppm is claimed. Another new instrument is Perkin-Elmer's Model 270-DF mass spectrometer. It is a small double-focusing instrument of the *Nier-Johnson geometry* with a 90° electrostatic sector (12.7 cm radius) followed by a 60° magnetic sector (10.16 cm radius). Unit mass resolution of 850 is obtained with magnetic field scanning. Mass range is up to 3000 with reduced acceleration voltage. The instrument is primarily intended, similarly to Varian's cycloidal machine, for the analysis of gas chromatography effluents.

There are at least four commercial magnetic deflection instruments designed for *isotopic abundance measurements* in solids; all are based on a design published by Nier in 1947 (7). These instruments are uses in the nuclear industry for isotope ratio and isotope dilution analyses, in age

Figure 8-7.   Varian/MAT CH 5 mass spectrometer system. (*a*) Overall picture; (*b*) closeup of sample inlet system (oven-type) and ion source. (Courtesy of Varian/MAT.)

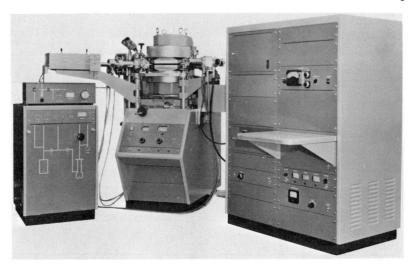

Figure 8-8. Hitachi Model RMU-6E mass spectrometer system. (Courtesy of the Perkin-Elmer Corp.)

determinations, nuclear cross section measurements, etc. All four machines are 12-in. radius instruments. Nuclide's Model 12-90-SU and AEI's Model MS-5 employ 90° deflection, while CEC's Model 21-703 and Nuclide's Model 12-60-SU use 60° sector fields. They all have comparable features and specifications. Resolution is 600–1000. Triple-filament thermionic sources (single-filament for Model 12-60-SU) are used together with fast vacuum locks for routine analyses. A fraction of the total beam is collected on one collector in the double collector assembly, while the other collector serves for single masses; each is connected to a vibrating reed electrometer. For ion currents smaller than 10^{-15} A, the single collector is pulled back to expose a multistage electron multiplier.

Two basic models of tandem mass spectrometers (Section 3-II-A) are offered by Nuclide. The two 90° sector analyzers of 12 or 15 in. radii can be arranged either in a C configuration (zero dispersion) or in an S shape (double dispersion). A surface ionization source is used and either a Faraday cage or an electron multiplier type detector is employed. At the focal point of the first-stage analyzer, an adjustable defining slit controls the ion beam entering the second-stage analyzer. Two identical electromagnets are used and the relative field strengths are adjustable from a common control. According to performance specifications, resolution is 1000 (15 kV acceleration) and abundance sensitivity is 10^6 for uranium isotopes. Stability of the accelerating voltage is better than 1 part in 20,000 and that of the magnet control is better than 1 part in 50,000. Ion detection

Figure 8-9. Nuclide Model TZD tandem mass spectrometer. (Courtesy of Nuclide Corp.)

sensitivity is determined by the type of detector used. It is 10^{-15} A with the vibrating reed electrometer and 10^{-20} A with the secondary electron multiplier. Available accessories include NMR equipment for magnetic field strength control, isolation valves, vacuum locks, and various pumping arrangements. Nuclide's Model TZD tandem spectrometer is shown in Figure 8-9.

Many instruments are available for the precision measurement of isotope abundances in gases, e.g., hydrogen, nitrogen, argon, oxygen. The mass range requirement is often only 5–150, and a modest resolution of 300 is normally adequate. Abundance sensitivity (Section 3-II-A) as high as 1 part in 10^4 can be achieved at mass 44. Conventional Nier-type electron impact source and double collectors are standard components. The less abundant species are normally measured with a vibrating reed electrometer, the more abundant ones with a multiplier. Several types of "balance panels" are available for the final ratio measurement. AEI's Model MS-3, Varian's Model M-86, and Nuclide's Model 6-60-RMS are representative instruments in this category.

For *high temperature* studies Knudsen-cell sources are available for most general-purpose instruments, and also for TOF spectrometers, as optional accessories. Knudsen cells may, of course, also be attached to the large double-focusing instruments and options are indeed offered by several manufacturers. CEC's Model 21-703B and Nuclide's Model 12-90-HT and 12-60-HT are specifically designed for physical and chemical studies at temperatures as high as 2500°C. The basic Knudsen cell design (Section 7-VI-B) is essentially the same in all commercial versions, but there are significant differences in such constructional details as heat shielding and electrical connections. It is not easy to measure performance in these instruments. Sensitivity may best be defined in terms of signal-to-noise ratio; for example, with silver in the cell a typical value for the Ag^+ ion/noise ratio is 10 at 900°K.

At the present time there are two commercial instruments available for *ion bombardment* studies. With considerable current interest in this field, new instruments are likely to appear soon. Nuclide's 6-60-DB mass spectrometer is a 6 in., 60° magnetic sector instrument equipped with a special source for ion bombardment (Fig. 4-9). Part of the ion source is of the oscillating-electron type supplying a beam of inert ions which is directed onto the sample target. The second part of the source is a conventional electron impact source for the ionization of sputtered neutrals. Both Faraday cup and multiplier sources are offered. Optional features include a synchronous source–detector system (8) and equipment for laser beam sputtering. The availability of an ion-microprobe mass spectrometer for solids analysis has recently been announced by the GCA Corporation. The secondary ions formed in the ion bombardment source are focused electrostatically into a double-focusing analyzer which in addition to energy focusing, also provides two-dimensional directional focusing. Thus, a point rather than a line image is formed, permitting the use of electron multiplier detection. Resolution is adjustable from 250 to 5000 by adjusting slit widths from 0.1 to 0.005 in. Mass range is up to 1000. Sensitivity is claimed to be in the ppb range for many elements, and areas as small as 0.1 mm² can be sampled.

Turning now to dynamic instruments, it is first noted that *time-of-flight* mass spectrometers are available only from the Bendix Corporation. There are two basic models, Model 3012 (Fig. 8-10) and Model 3015. The former has a spectral frequency of 10 kc/sec and ion energy of 2.8 kV; the latter has spectral frequency from 10 to 100 kc/sec and ion energy from 2 kV to 6 kV. The Model 3012 is a general-purpose instrument while the Model 3015 is primarily intended for fast reaction studies. Resolution is in the range of 200–600 depending upon the mode of operation. Ion path length is 180 cm. Mass range covered is 1–3000 with oscilloscope recording and 1–900 with analog (electric) recording. The Model 12-107 and Model 14-107 versions are based on earlier designs and offer more limited performance at considerably lower cost. An important feature of TOF instruments is the versatility of the output systems (Section 5-I-B-6 and Fig. 5-10). Simul-

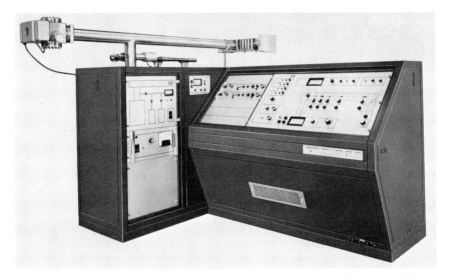

Figure 8-10.   Bendix Model 3012 time-of-flight mass spectrometer. (Courtesy of Bendix Corp.)

taneous recording of up to six ion species or simultaneous scanning of six spectral ranges is possible at various speeds ranging from 2 sec to 50 min for the entire mass range. The magnetic electron multiplier provides high sensitivity, while the omnipresent oscilloscope allows observation of the entire spectrum at all times. A number of inlet systems are available, including heated molecular leaks, vacuum locks, direct inlet systems, Knudsen cell, manifold for gas chromatographic eluent analysis, nude-source for residual analysis, and ion-molecule source. A laser-microprobe source for minute spot analysis, surface contaminant identification, and vaporization with minimum thermal degradation has recently been developed. Many examples of the use of TOF instruments are given in Part II.

Mass spectrometers with quadrupole and monopole analyzers have recently been offered by EAI, Finnigan, and GE. Both instruments are outgrowths of instrumentation developed for residual gas analysis. Resolution up to 500 is available and the fast scanning capability of both the quadrupole and monopole systems can be utilized in many analytical problems; multiplier detection permits high sensitivity. The sample inlets and the general vacuum systems of these instruments are quite similar to those used with conventional mass spectrometers. Present efforts in instrumentation development are concentrated around the combination with gas chromatographs through a proper interface (Section 11-V-C).

It is expected that quadrupole and monopole analyzers which proved quite successful in residual gas analysis will also find widespread use as full-scale mass spectrometers in the future.

V. HIGH–RESOLUTION INSTRUMENTS

High-resolution double-focusing mass spectrometers have become available only during the last few years. Consequently, they are the most versatile of all commercial instruments, and just about every conceivable accessory is available. Three types of high-resolution instruments are offered: intermediate resolution sector instrument plus electrostatic analyzer, Nier-Johnson geometry, and Mattauch-Herzog geometry. At least six kinds of sources are advertised as options: electron impact, spark, thermal, field ionization, Knudsen cell, and Fox-type monoenergetic. Of these only the first two are currently in extensive use. The gas source is utilized in organic structure determinations and gas chromatographic effluent identifications, the spark source for the trace analysis of solids. (It is recalled that the spark source requires the Mattauch-Herzog geometry.) The other source types, which do not necessarily require double focusing, will eventually also become of widespread use as laboratories with new instruments expand their range of operations.

As mentioned previously, single-focusing sector machines may be converted into double-focusing instruments with high resolution by the *addition of an electrostatic analyzer*. This approach has been pioneered by Hitachi. The addition of a 45° deflection, 25 cm radius electrostatic sector (with associated vacuum equipment) converts the Model RMU-6E instrument into a double-focusing system with resolution of the order of 15,000 (Fig. 8-8). Resolution is controlled by adjustable slits. Nuclide offers a cylindrical electrostatic analyzer (36.6 cm radius) for the conversion of its Model 12-90G spectrometer.

At the present time there is only one *Nier-Johnson type* double-focusing high-resolution mass spectrometer on the market. AEI's Model MS-9 is shown in Figure 8-11. This design achieves first-order double focusing and second-order angular focusing (Section 3-II-B-4). Both the electric and magnetic sectors have 90° deflection angle, and the radii are 15 and 12 in., respectively. Resolving power, based on the 10% valley definition, is about 20,000, but usable resolution as high as 50,000 has been achieved. Rapid changing from high to low resolution is provided by adjustable slits. Both acceleration voltage and magnetic field are variable (1–8 kV and 400–12,000 G). The normal mass range of 2–800 can be extended to 6400 at 1

Figure 8-11. AEI Model MS-9 double-focusing high-resolution mass spectrometer; Nier-Johnson geometry. (Courtesy of AEI/Picker X-Ray Corp.)

kV acceleration. Precise mass measurement, a major feature of the instrument, is accomplished by the peak-matching technique (Section 10-IV-A): the unknown peak and a reference peak are alternately displayed once every second on a long persistence oscilloscope screen, and the acceleration voltage necessary to bring the peaks into coincidence are measured. The accuracy of mass measurement is 5 ppm for mass differences up to 2% and 10 ppm for mass differences to 10%. Fast scanning and collector systems for both electrometer and multiplier detection are standard features. An ion current monitor between the analyzers interrupts about 50% of the total ion current. To obtain element maps (Section 11-IV), data handling accessories of varying sophistication are available. In addition to the conventional sample inlet system for gas introduction at both room temperature and 350°C, insertion probes are also available for direct evaporation of nonvolatile and thermally unstable materials into the ion source.

The AEI Model MS-702, Varian Model SM-1, CEC Model 21-110C, JEOL Model JMS-01S series, and Nuclide Model Graf-3 are modular, multipurpose instruments designed for "materials research." They represent the utmost in systems design. Although spark and gas source versions are offered separately, each of these instruments can be used with both sources with the appropriate additional circuitries. Indeed, an almost "total" mass spectrometry capability may be built around these instruments.

All these machines follow the basic Mattauch-Herzog design (Section 3-II-B-3) using a 30° electrostatic field followed by a 90° magnetic field (Fig. 3-13). The radii of the electrostatic and magnetic analyzers is different for the different makes. The radius of the electrostatic analyzer is 15–25 in; the maximum radius in the magnetic analyzer is 8–12 in., and the minimum radius is 1–2 in. In some models a spherical condenser is used as an electrostatic analyzer to focus the beam in both directions normal to the ion velocities; this should result, at least in principle, in improved focusing and decreased mass discrimination. Resolution is 1500–3000 with a spark source, and as high as 35,000 with a gas source. Mass range is $1:35M$ in a single magnetic setting with photographic detection. The gas source versions have both Faraday cage–multiplier combination *and* photoplate detection. The electrical detector is located at the point where maximum dispersion between adjacent masses occurs (e.g., at the 12 in. radius on the CEC Model 21-110C). Merits and demerits of electrical versus photographic detection are discussed in Section 11-V; for spark work, photographic detection is universally used. Mass scale can be extended to well beyond 3000 with proper ion acceleration and magnetic field setting.

The spark source is usually of the rf spark type (Section 4-IV), although vacuum vibrator types are also offered as an option. Solid samples (0.1 × 0.1 × 0.5 in.) are mounted in sample vises which can be positioned from outside the source housing. Sparks are produced by a pulsed power supply (1Mc/sec) in the 10–100 kV range. Sparking voltage, pulse length, and repetition rate are adjustable for selecting optimum operating conditions. In the gas source versions, source design is essentially the same as in single-focusing instruments. Also, all the accessories, including heated inlets, direct solid inlets, and gas chromatography interfaces are readily available. Also, thermal, Knudsen, and Fox sources are featured as options.

Acceleration of the ion beam from the source into the electrostatic analyzer is accomplished by a 2–20 kV variable supply. For the spark source only a few per cent stability is required, but for the gas source voltage stability of at least 5×10^{-1} must be provided for the duration of the analysis. Condenser voltage is normally variable in the 0.5–3.0 kV range, and must be regulated to 1 part in 100,000 or better for over the total exposure interval. The value of the condenser voltage is often kept one-tenth of the ion acceleration voltage; some instruments feature a direct coupling to maintain the ratio within a few per cent.

A beam monitor collector is located at the entrance of the magnetic sector and intercepts a fixed portion of the total ion beam. It is recalled that no mass separation has yet occurred at this point. Both instantaneous and integrated ion current measurements are provided by a dc integrating amplifier. The former assists in adjusting for optimum ion transmission; the latter serves as a measure of exposure when photographic detection is used.

The magnetic sector consists of an electromagnet and its power supply and the detection system. Magnetic field strength can be varied from 0 to 20,000 G both continuous and stepwise. Regulation of 20 ppm must be provided for sharp lines.

The vacuum system features differential pumping between the various sectors and baking is normally available. Pressure must be at least 10^{-8} torr. Plate-locks and prepumping systems prevent the need to vent the analyzer when inserting plates. Since photoplates are movable in the focal plane, as many as 15 exposures may be made on a single plate.

In evaluating instrument performance, spark and gas machines should be considered separately. For spark work, sensitivity is the single most important performance parameter. Overall sensitivity is usually in the 0.01 ppm level. It is important to know how long an exposure is needed to reach

Figure 8-12. AEI Model MS-702 double-focusing mass spectrometer, spark source version; Mattauch-Herzog geometry. (Courtesy of AEI/Picker X-Ray Corp.)

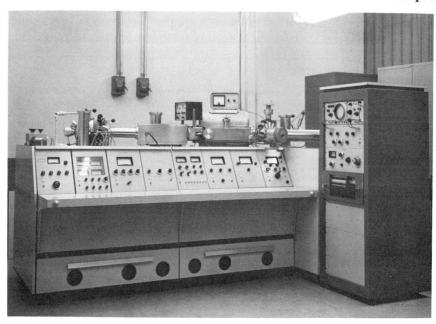

Figure 8-13. CEC Model 21-110B double-focusing high-resolution mass spectrometer; gas source version; Mattauch-Herzog geometry. (Courtesy of Consolidated Electrodynamics Corp.)

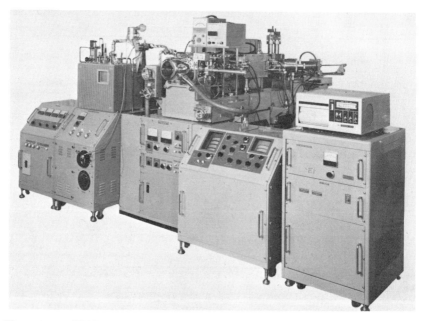

Figure 8-14. JEOL Model JMS-01SG double-focusing high-resolution mass spectrometer; spark source version; Mattauch-Herzog geometry. (Courtesy of Japan Electron Optics Laboratory Corp.)

such sensitivity. In organic work, resolution and accuracy of mass measurement are the important performance characteristics. Detailed discussion of the analytical techniques recently developed with these instruments is presented in the appropriate sections of Part II. Figures 8-12, 8-13, and 8-14 show commercial instruments of the Mattauch-Herzog geometry.

References

1. Pavlenko, V. A., A. E. Rafalson, and A. M. Shereshevskii, *Instruments and Exptl. Techniques* (Transl. of *Pribory i Technika Experimenta*), **3**, 319 (1958).
2. Shumulovskii, N. N., and R. I. Stakhovskii, *Mass-Spektral'nye Metody* (in Russian), Energiya, Moscow, 1966.
3. Ahearn, A. J., *11th Ann. Conf. Mass Spectrometry*, ASTM Committee E-14, San Francisco, Calif., 1963, *Proc.*, p. 223.
*4. Saunders, R. A., and A. E. Williams, in *Mass Spectrometry of Organic Ions*, F. W. McLafferty, Ed., Academic Press, New York, 1963, p. 360.
5. Hultzman, W. W., NASA Tech. Publications TM-X-1281, Aug. 1966.
*6. Reynolds, J. H., *Rev. Sci. Instr.*, **27**, 928 (1956).
*7. Nier, A. O., *Rev. Sci. Instr.*, **18**, 398 (1947).
8. Smith, A. J., L. A. Cambey, and D. J. Marshall, *J. Appl. Phys.*, **34**, 2489 (1963).

Part II

APPLIED MASS SPECTROMETRY

Chapter Nine

Types of Ions in Mass Spectra

I. CRACKING PATTERNS

The array of peaks in the complete spectrum of a pure substance is referred to as a "cracking pattern." Chapter 1 briefly dealt with the fact that when the energy of the bombarding electrons is just sufficient to cause ionization, i.e., 5–15 eV for elements and 8–12 eV for most organic compounds, atomic or molecular ions begin to form. It was also mentioned that for good ionization efficiency, electron impact sources are customarily operated in the 70–80 eV electron energy range. Although in this energy range a wide variety of ionization and fragmentation processes may take place in addition to parent ion formation, cracking patterns are relatively insensitive to small changes in electron energy. This insensitivity, together with the good reproducibility of the patterns, is utilized in many analytical applications.

The fragmentation processes caused by the collision of energetic electrons with polyatomic molecules are considered to be a series of competing and consecutive unimolecular reactions, similar to the rate processes characterizing ordinary chemical reactions. The energy content of the impinging electrons is adequate to break every chemical bond in the molecule. The fact that certain groups remain associated after ionization of the parent molecule is explained by a theory which assumes that the excited particles produced upon impact do not dissociate immediately but have a finite, though small, lifetime, during which the excess energy is distributed equally among all bonds, and subsequent dissociation occurs in such a manner that the fragments have the smallest possible energy. It is the objective of the theory of mass spectra to explain why a given cracking pattern is observed, and then to calculate the patterns from basic physico-chemical properties. The theory of mass spectra is briefly discussed at the end of this chapter.

A practical approach to understanding the nature of mass spectra is to obtain the cracking patterns of a large number of compounds, catalog them in an orderly manner, and try to establish correlations between observed spectra and molecular structure. This empirical technique has proved exceptionally useful. Thousands of compounds have been analyzed, and many features of the spectra of various types of compounds have been explained either from direct clues or with the aid of rationalization, employing such concepts of modern physical organic chemistry as hyperconjugation, resonance, etc. Correlations between spectrum and structure are discussed in Section 10-III-D.

In Tables 9-1 and 9-2 mass spectral data are given for *n*-butane and isobutane. These tables are reproduced from a compilation published by the American Petroleum Institute (API Research Project 44). The reader should become familiar with these forms since they are widely used in everday practice; other available compilations are listed in Chapter 14 on Information and Data. An actual spectrum of *n*-butane is shown in Figure 9-1, while line diagrams for the butanes are depicted in Figure 9-2.

When a mass spectrum is taken, ion currents are normally measured in arbitrary units of peak height (chart divisions) rather than in units of current. This is practical in routine work, and the sizes of peaks are referred to as "peak intensities" or "peak heights." (For the evaluation of basic instrument sensitivity, the knowledge of actual ion currents is required.) Peak heights are usually *normalized* with respect to the largest peak in the spectrum (*base* peak),

$$\text{Pattern coefficient} = \frac{\text{Peak height (div)}}{\text{Peak height (div) of base peak}} \times 100 \quad (9\text{-}1)$$

These normalized values are alternatively called "pattern coefficients," relative abundances, or relative intensities. In a complete cracking pattern all peaks, with intensities to the 0.01 level, must be listed (Tables 9-1 and 9-2, and left ordinate of Fig. 9-2). When the cracking pattern of a pure compound is determined for reference, care must be taken to remove from the pattern contributions from instrument background and possible impurities.

The peak corresponding to the molecular ion is called the *parent* peak, or "*P* peak," or "*M* peak." Frequently the parent peak is much smaller than the base peak, and occasionally it does not appear at all (Section 10-III-C). Peaks may, of course, be normalized with respect to any peak in the spectrum, e.g., the parent peak, if such tabulation appears advantageous in calculations or interpretation.

Table 9-1

MASS SPECTRAL DATA

American Petroleum Institute Research Project 44

Contributed by the National Bureau of Standards, Mass Spectrometry Laboratory, Washington, D. C.

n–Butane (gas)   Serial No. 4   October 31, 1947

Mass-Charge Ratio (m/e)	Type of Peak	Relative Intensities 50 volts	70 volts	Mass-Charge Ratio (m/e)	Type of Peak	Relative Intensities 50 volts	70 volts	Mass-Charge Ratio (m/e)	Type of Peak	Relative Intensities 50 volts	70 volts
1		.69	1.11	48		.02	.06				
2		.07	.10	49		.18	.40				
				50		.94	1.29				
12		.05	.13	51		.95	1.05				
13		.13	.26	52		.27	.26				
14		.70	.96	53		.85	.74				
15		4.83	5.30	54		.21	.19				
16	1,r	.11	.12	55		.97	.93				
				56		.75	.72				
19	d	.02	.04	57		2.46	2.42				
19.5	d	--	.01	58	p	12.6	12.3				
20	d	.02	.02	59	i	.54	.54				
24		.03	.03								
24.1	m	.01	.01								
25		.20	.46								
25.1		.12	.11								
25.5	d	.16	.36								
26		5.36	6.17								
26.5	d	.04	.08								
27		37.9	37.1								
27.5	d	.05	.05								
28		33.3	32.6								
29		44.8	44.2								
30	1	.98	.98								
30.4	m	.13	.14								
31.9	m	.20	.20								
35.1	m	.01	.02								
36		.02	.08								
37		.68	1.01								
37.1	m	.06	.06								
38		1.66	1.89								
39		13.0	12.5								
39.2	m	.47	.46								
40		1.71	1.63								
41		28.4	27.8								
42		12.4	12.2								
43		100.	100.								
44	1	3.36	3.33					Sensitivity for base peak *in divisions per micron*			
45	1	.04	.05					43		37.5	39.6

ADDITIONAL INFORMATION

METASTABLE ION TRANSITIONS

39.2 (43+) → (41+) + 2
37.1 (41+) → (39+) + 2
35.1 (39+) → (37+) + 2
31.9 (58+) → (43+) + 15
30.4 (58+) → (42+) + 16
25.1 (29+) → (27+) + 2
24.1 (28+) → (26+) + 2

		Sensitivity for n-Butane
		43
		Relative Intensities for n-Butane
		15
		27
		29
		43
		58

SYMBOLS: p=parent peak   r=rearrangement   d=doubly-charged ion
               i=isotope peak   m=metastable ion
                                 (diffuse peak)

COMPOUND	MASS SPECTROMETER
Name: n–Butane	Model: Consolidated, # 21-102

Electron current (catcher):   9.0   *microamperes*

Molecular Weight	Molecular Formula	Semi-structural Formula
58.12	C_4H_{10}	$CH_3(CH_2)_2CH_3$

Ion accelerating voltages:	(m/e)	volts	(m/e)	volts
	2	1185		
	28	1330		

Source: Phillips Petroleum Company

Purity 99.78 ± 0.08 *mole percent*

Temperature of ionization chamber: 245   °C

Basis of pressure measurement: Manometer

LABORATORY: Mass Spectrometry Laboratory, National Bureau of Standards

Date of measurement   March 27, 1947

Table 9-2

MASS SPECTRAL DATA

American Petroleum Institute Research Project 44

Contributed by the National Bureau of Standards, Mass Spectrometry Laboratory, Washington, D. C.

2-Methylpropane (Isobutane) (gas)   Serial No. 5   October 31, 1947

Mass-Charge Ratio (m/e)	Type of Peak	Relative Intensities 50 volts	70 volts	Mass-Charge Ratio (m/e)	Type of Peak	Relative Intensities 50 volts	70 volts	Mass-Charge Ratio (m/e)	Type of Peak	Relative Intensities 50 volts	70 volts
1		.60	.98	48		--	.04				
2		.05	.06	49		.10	.27				
12		.06	.14	50		.62	.89				
13		.16	.32	51		.67	.74				
14		.84	1.18	52		.16	.15				
15		5.55	6.41	53		.54	.50				
16	1,r	.15	.18	54		.08	.07				
19	d	.01	.05	55		.40	.42				
19.5	d	--	.02	56		.37	.34				
20	d	.02	.04	57		3.01	3.00				
20.5	d	--	.01	58	p	2.75	2.73				
24		--	.02	59	1	.11	.11				
25		.07	.22								
25.1	m	.02	.04								
25.5	d	.06	.21								
26		1.95	2.36								
26.5	d	.01	.02								
27		27.9	27.8								
28		2.51	2.62								
29	1,r	6.37	6.16								
30	1	.14	.13								
35.1	m	.01	.02								
36		.02	.10								
37		.94	1.41								
37.1	m	.07	.08								
38		2.27	2.77								
39		16.6	16.5								
39.2	m	.70	.72								
40		2.46	2.37								
41		37.7	38.1								
42		33.9	33.5								
43		100.	100.								
44	1	3.32	3.33								
45	1	.03	.03								

Sensitivity for base peak
in divisions per micron

43   44.6   46.4

ADDITIONAL INFORMATION

Sensitivity for n-Butane

43   37.5   39.6

METASTABLE ION TRANSITIONS

39.2 (43+) → (41+) + 2
37.1 (41+) → (39+) + 2
35.1 (39+) → (37+) + 2
25.1 (29+) → (27+) + 2

Relative Intensities for n-Butane

15	4.83	5.30
27	37.9	37.1
29	44.8	44.2
43	100.	100.
58	12.6	12.3

SYMBOLS: p=parent peak   r=rearrangement   d=doubly-charged ion
           i=isotope peak   m=metastable ion
                                (diffuse peak)

COMPOUND	MASS SPECTROMETER

Name: Iso-Butane (2-Methylpropane)

Model: Consolidated, #21-102

Electron current (catcher):   9.0 *microamperes*

Molecular Weight	Molecular Formula	Semi-structural Formula
58.12	C_4H_{10}	$(CH_3)_3CH$

Ion accelerating voltages:

(m/e)	volts	(m/e)	volts
2	1185		
28	1330		

Source: Phillips Petroleum Company

Purity
99.88 ± 0.06
mole percent

Temperature of ionization chamber: 245 °C

Basis of pressure measurement: Manometer

LABORATORY:

Mass Spectrometry Laboratory, National Bureau of Standards

Date of measurement

March 27, 1947

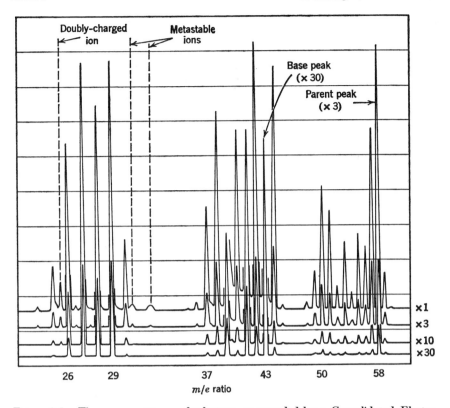

Figure 9-1.   The mass spectrum of *n*-butane, as recorded by a Consolidated Electrodynamics Co. Model 21-103 spectrometer. Simultaneous tracings by galvanometers of different sensitivities provide a dynamic range up to 30,000 in this instance. (After ref. 2 in Chapter 10.)

It is important to realize that cracking pattern values have no absolute physical significance. Although patterns are remarkably constant for a given instrument as long as experimental conditions are unchanged, and patterns taken on similar instruments are readily comparable, different types of mass spectrometers may, and indeed do, yield significantly different patterns.

Cracking patterns are sometimes given as percentages of the *total* ion current. These "per cent Σ_m" values are obtained by summing all peak heights or normalized values from a selected mass, m, to the molecular ion peak and then calculating the per cent contribution of the various peaks to this total. Such values are shown in the right ordinate of Figure 9-2. When the only large peak in the spectrum is the base peak, its $\%\Sigma_m$ value is large; when several large peaks are present, the $\%\Sigma_m$ value of each is small. This technique is useful in studying the significance of individual peaks in a fragmentation process.

There are about 50 peaks listed in the spectra of n-butane and isobutane. In inorganic compounds the number of peaks is considerably smaller (Table 9-3), while complex organic molecules may exhibit hundreds of peaks. Some peaks can be identified with ease: the m/e 43 peak obviously results from the removal of a CH_3 group from the butane molecule, and it is clear that the m/e 20 peak in the spectrum of argon refers to Ar^{2+} ions. The peak at m/e 39.2 in the n-butane spectrum, however, is not so simple to explain. It is the objective of this chapter to describe the various types of peaks that appear in mass spectra.

Table 9-3   Cracking Patterns of Inorganic Gases[a]

m/e	N_2	O_2	CO	CO_2	H_2O	Ar
12		0.12	3.18	5.40		
13		0.03	0.08			0.01
14	6.85		1.02			
16		8.96	0.98	7.85	3.01	
17					27.09	
18					100.00	
19						
20						19.10
22				2.00		
27	0.96					
28	100.00	3.40	100.00	6.60		
29	0.89		1.06	0.07		
30			0.25			
32		100.00				
33		0.10				
34		0.40				
36						0.36
38						0.08
40						100.00
44				100.00		
45				1.14		
46				0.41		

[a] Patterns obtained on CEC Model 21-130 mass spectrometer at 80 μA ionization current.

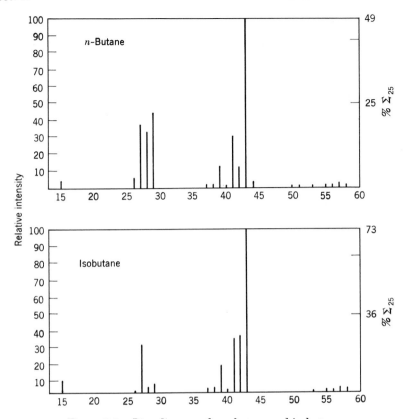

Figure 9-2.   Line diagrams for *n*-butane and isobutane.

II. TOTAL IONIZATION

Total ionization is defined as the sum of the abundances of all the peaks in the spectrum multiplied by the sensitivity, measured in ion current per unit of pressure, of the base peak in the spectrum. This is an important concept, both theoretically and in practice. Total ionization is normally expressed on a relative scale, using *n*-butane or *n*-hexadecane as standard. In the API and similar catalogs, sensitivities for both the compound in question and *n*-butane are given, so that total ionization can be computed using the cracking pattern. Since instrumental effects are largely canceled out by taking the ratio, total ionization may be considered as a measure of the *relative tendencies* of compounds toward ionization. On this basis, spectra obtained on different instruments may be compared. On the other hand, differences in mass discrimination among instruments can markedly change the value found for total ionization.

In an extensive study of a large number of hydrocarbons (from C_1 to C_{10}), Mohler et al. (1) established that relative total ionization within a given compound class increases both with molecular weight and with the degree of unsaturation; for structural isomers, however, the values remained constant.

In their often quoted study of ionization cross sections, Ötvös and Stevenson (2) made two basic postulates: (a) atomic ionization cross sections of the elements are proportional to the number of outer (valence) electrons, the mean radii of the electrons (estimated using hydrogen wave functions) serving as weighting factors; (b) relative total ionization cross sections are constitutive molecular properties, and molecular ionization cross sections can be obtained by summing the cross sections of the constituent atoms. The total ionization, T, of hydrocarbons, for example, is given by

$$T = \text{const.} \; (n_H Q_H{}^i + n_C Q_C{}^i) \tag{9-2}$$

where n_H and n_C are the number of H and C atoms in the molecule, and the Q^i's refer to respective ionization cross sections. Numerical values are given in reference 2. The principle of additivity is well established for hydrocarbons and some other compounds, but its general validity, particularly for inorganic compounds, is still questioned.

Recently, analytical methods based on total ionization measurement have been developed for compound type analysis of organic mixtures (Section 10-II-D). Many commercial instruments feature a monitor electrode which intercepts a certain percentage (e.g., 50%) of the total ion current before mass dispersion takes place. The ion current thus obtained is proportional to the total number of ions leaving the ion source. Since the densities of many hydrocarbons are essentially the same, total ion intensity may be used as a measure of sample quantity; it is believed to be more accurate than measuring liquid sample volume (3).

III. MOLECULAR IONS

Molecular or parent ions are formed from neutral molecules by removing one electron from the parent compound,

$$M + e^- \rightarrow M^+ + 2e^- \tag{9-3}$$

Both *n*-butane and isobutane have their parent peaks at m/e 58 (Tables 9-1 and 9-2) but their relative intensities with respect to the base peak at

m/e 43 are significantly different. Since branching favors fragmentation, the P peak of n-butane is more intense than that of isobutane.

Molecular ions are perhaps the most important of all types of ions in a mass spectrum because they reveal the molecular weight of the compound investigated. (The mass of the electron is negligible compared to the total mass.) In high resolution mass spectrometry a single mass determination may be the only measurement needed for compound identification (see Section 10-III-C). The first step in evaluating the spectrum of an unknown compound is to locate the parent peak at the high mass end. Parent peaks appear in the spectra of most compounds, although peak intensity may be very low compared to the base peak. There are compounds, such as CCl_4, for which the parent peak is too weak to be recognized. Methods for parent peak identification are discussed in Section 10-II-C.

At the customary 70 eV electron energy level, molecular ions are formed with excess energy (e.g., vibrational), and decomposition (fragmentation) may also occur during the approximately 10^{-5} sec interval which is required for the removal of ions from the ionization region and for full acceleration. The stability of molecular ions, and consequently their relative abundance in the cracking pattern, mainly depends on chemical structure and size. The presence of π-electron systems, for example, increases stability, while certain functional groups, such as OH, tend to weaken bonds and stabilize fragments. In agreement with these considerations, aromatic and alicyclic compounds show a high abundance of molecular ions, while ethers, branched hydrocarbons, and alcohols are at the low end of the list. The probability of parent ion decomposition, W_z, is given by Pahl (4) as

$$W_z = \frac{\Sigma I_f}{\Sigma I_p + \Sigma I_f}$$

where ΣI_p is the sum of the intensities of the undecomposed parent ions (including isotopic ions), and ΣI_f refers to the total intensities of all other singly charged ions. From this equation the stability of the parent ions, W_p, is obtained as

$$W_p = 1 - W_z = \frac{\Sigma I_p}{\Sigma I_p + \Sigma I_r} \qquad (9\text{-}4)$$

From the numerical values given (4), it can be seen that the stability of molecule ions increases from alkanes through alkenes to alkines for compounds with less than five carbon atoms. Fine details of comparative stabilities were obtained by refining the above considerations to include the number of bonds within the molecules.

The abundance of molecular ions relative to that of other ions in the spectrum can be significantly increased by reducing the energy of the bombarding electrons to the 10 eV level. This technique, known as *low-voltage mass spectrometry*, has recently become a powerful tool in organic analysis (Section 10-I-E).

IV. FRAGMENT IONS

The formation of molecule ions may be considered as ionization without fragmentation. At 70 eV electron energy many kinds of fragment ions may form. The fragmentation process can usually be understood by chemical logic, though at times rather involved explanations are necessary. The relative probability of the formation of a particular fragment depends heavily on molecular construction which determines how the acquired energy, distributed throughout the molecule, is going to be utilized. However, relative abundances of fragment ions are constant in a spectrum as long as experimental conditions are unchanged. The spectrum thus becomes a "fingerprint" of the molecule, and even complex mixtures can be analyzed on the basis of cracking patterns (Section 10-I).

The cracking patterns of Tables 9-1 and 9-2 indicate that the most probable process following electron impact in both *n*-butane and isobutane is the rupture of a carbon–carbon bond with the removal of a CH_3 group. Other peaks appearing as a result of C—C bond ruptures are at m/e 15 and 29. From the $C_3H_7^+$ ion several fragments form by rupturing carbon–hydrogen bonds: $C_3H_6^+$, $C_3H_5^+$ ... C_3^+. Similarly, CH_3^+, $C_2H_5^+$, and $C_4H_{10}^+$ fragments serve as starting points for carbon–hydrogen bond ruptures. Such cleavage processes are rather straightforward, and often even the magnitudes of peaks can be estimated. One would expect, for example, that the m/e 43 peak should be stronger in isobutane than in *n*-butane, since there are three possible ways to split off a CH_3 group in the former, while there are only two possibilities in the latter.

The discussion in the previous paragraph is certainly oversimplified. For example, the $C_3H_5^+$ ion is derived, in part, from $C_4H_9^+$ by loss of CH_4. In general, the most abundant fragment ion peaks of a particular compound correspond to the products of the most favorable decomposition reaction pathways, leading to the most stable final products, both ionic and neutral. Reactivities within the decomposing ions often parallel reactivities known from chemical reactions in solution, and this fact has great significance to the chemist trying to reconstruct the underlying chemistry in fragment ion formation. Although certain fragment peaks, such as metastable peaks and rearrangement peaks, are discussed separately later in this chapter, it is stressed that there is an essential unity of the reactions responsible for the formation of all types of fragment ions. For details the reader is referred to texts on organic mass spectrometry.

V. MULTIPLY CHARGED IONS

Although most ions in mass spectra are singly charged ($e = 1$ in m/e), multiply charged ions do occur and have been known since the early days of mass spectrometry. Doubly charged ions appear at $\frac{1}{2}$ mass, triply charged ions at $\frac{1}{3}$ mass, etc. Removal of the second, third, etc. electron from an atom or molecule requires, of course, more energy than in single ionization. Appendix II shows second, third, etc. ionization potentials for the elements, and Figure 9-3 shows ionization efficiency curves for mercury. It is seen that the threshold energy for triply charged ions is greater than 70 eV, and the intensity of multiply charged ions decreases rapidly with increasing charge. In electron impact sources, ions with charge greater than 2 are formed only in negligible quantities; argon is perhaps the only gas which has routinely been observed to appear in triply charged form (at m/e 13.3) in certain mass spectrometers. In thermal ionization and

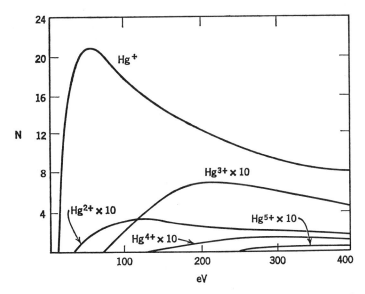

Figure 9-3. Ionization efficiency of mercury as a function of the energy of the bombard - ing electrons for the multiply charged ions formed. (After W. Bleakney, *Phys. Rev.,* **35,** 139 (1930).)

field emission sources multiply charged ions have little importance, but in the discharge-type sources several degrees of ionization is anticipated.

The formation of a doubly charged ion of an element or compound can be described by

$$M + e^- \rightarrow M^{2+} + 3e^- \tag{9-5}$$

There are many polyatomic organic compounds that yield doubly charged ions in electron impact sources, both without and with concurrent fragmentation. The abundance of such ions, however, is always low, normally below 1% of the base peak. In the spectrum of n-butane (Table 9-1), peaks at m/e 19.0, 19.5, 20.0, 25.5, 26.5, and 27.5 are doubly charged ions, all with relative intensities below 1%. Doubly charged ions appear at half-integer mass when the mass of the singly charged ion is an odd value, such as in compounds with an odd number of nitrogen atoms. Doubly charged ions from even-mass parent ions are often indistinguishable from singly charged fragment ions and high resolution is required for separation. A small peak, due to the ^{13}C isotope and appearing $\frac{1}{2}$ mass unit higher than the composite peak, may be used to evaluate the doubly charged peak contribution. When the ionization energy is reduced below about 30 eV doubly charged peaks disappear. Doubly charged ions occur in higher abundance (as high as 20%) in aromatic molecules and molecules containing conjugated systems, due to the π-electrons. This fact can be utilized in compound characterization and structure determination: multiply charged ions indicate relatively high stability and are also useful in the recognition of the molecule ion.

In the various discharge sources multiply charged ions of the elements are produced abundantly, and ions with up to five or six charges appear. The intensity of such ions in spark sources decreases by a factor of about 5 with each degree of ionization. In low-voltage sources, on the other hand, doubly charged ions may be more intense than singly charged ions. For high-precision mass measurement the presence of multiply charged ions on the photoplates is advantageous because of the many doublets formed (Section 10-III-A). In the trace analysis of solids, multiply charged ions are undesirable.

Multiply charged ions and neutral residual molecules may collide at various locations within the mass spectrometer, giving rise to *charge transfer* processes. In the general case (5)

$$X^{n+} + Y \rightarrow X^{m+} + Y^{(n-m)+} \tag{9-6}$$

where n and m represent charges. In spark-source instruments, charge-exchange collisions may take place at various points inside the analyzer system. The result of charge transfer in the electrostatic analyzer is a continuum extending downwards from the $(n/m^2)M$ line, where M is the isotopic mass of X. When the location of collisions is the area between the electrostatic and magnetic analyzers sharp lines will appear at a mass of $(n/m^2)M$. Finally, collisions within the magnetic analyzer will give rise to a continuum extending upwards from the $(1/m)M$ line. Charge-exchange lines are a definite disadvantage in analytical work since their presence reduces the sensitivity: faint impurity lines are difficult to detect inside a continuum. In direct current sources more multiply charged ion form than in radiofrequency sources, and thus charge-transfer lines are more intense. Improved vacuum is the obvious route to the elimination of charge-transfer lines.

VI. ISOTOPE IONS

Since the majority of the elements have stable isotopes (all naturally occurring isotopes and their relative abundances are listed in Appendix I), and since mass spectrometers separate ions according to their masses, it is obvious that isotope peaks appear in practically all spectra. In the spectrum of nitrogen (Table 9-3), the m/e 29 and 30 peaks represent $(^{14}N^{15}N)^+$ and $(^{15}N^{15}N)^+$ ions, respectively. Peaks at m/e 14 and 15 consist of doubly charged ions $(^{14}N^{14}N)^{2+}$ and $(^{15}N^{15}N)^{2+}$ and also of $^{14}N^+$ and $^{15}N^+$ ions. The ion $(^{14}N^{15}N)^{2+}$ appears at the fractional mass of 14.5. Similarly, in the spectrum of carbon dioxide, the peak at m/e 46 is made up by $^{12}C^{16}O^{18}O^+$ and $^{13}C^{16}O^{17}O^+$, and other isotope peaks can also readily be identified. In the cracking patterns of hydrocarbons isotope peaks appear due to the presence of ^{13}C and heavy hydrogen (D) isotopes. The m/e 59 peak, for example (n- and isobutane), is composed of ions of $^{12}C_3^{13}CH_{10}^+$, $^{12}C_4H_9D^+$, $^{12}C_2^{13}C_2H_9^+$, etc.

Since peak heights in the spectra are proportional to the number of ions at each mass, relative abundances can be calculated from measured peak intensities. Applications of isotope abundance determinations are discussed in Section 13-I. Conversely, if the abundances are known, the heights of the peaks to be expected from isotopes can be predicted. The number of possible isotope peaks increases rapidly with the number of elements in the compound (there are 30 combinations for CH_4), and also with the number of isotopes within each element, but peak intensities are often too small for detection. Deuterium contribution, for example, is usually negligible when the compound has only a few hydrogen atoms.

The presence of isotope peaks results in the appearance of peaks 1, 2, 3, 4, etc. units higher than the "monoisotopic" peak, i.e., the peak containing the most abundant isotopes of the elements forming the compound in question. The abundance of the $P + 1$, $P + 2$, etc. peaks depends on the number of atoms present in the molecule and on the relative abundance of the isotopes in these elements. The following list summarizes those heavy isotopes and their per cent relative abundance with respect to that of the lowest mass, which contribute significantly to organic spectra:

$P + 1$	^{13}C	1.11%
	^{2}H	0.015%
	^{15}N	0.37%
	^{33}S	0.75%
$P + 2$	^{18}O	0.20%
	^{34}S	4.22%
	^{37}Cl	24.5%
	^{81}Br	49.5%

The contribution of isotopes may be estimated by employing the binomial expansion

$$(a + b)^n \tag{9-7}$$

where a and b are the natural abundances of the light and heavy isotopes, respectively, and n is the number of atoms of the element present in the molecule. For a compound containing two carbon atoms, $a = 98.892$, $b = 1.108$, and $n = 2$. After expansion and division by 100, one obtains three terms with numerical values 97.79, 2.19, and 0.01. The first term is the relative abundance of the ion containing no ^{13}C atom, and if this was a molecular ion the peak is referred to as the P peak. The second term gives the abundance of the species containing one ^{13}C atom, and it is seen that the $P + 1$ contribution of heavy carbon is 1.1% of the P peak for each carbon atom present. The $P + 2$ term is significant only when the number of carbon atoms present in the molecule is large. The $(P + 1)/P$ abundance ratio for compounds containing carbon, hydrogen, nitrogen, and oxygen can be calculated (6) but the equations are rather complex and also there is an inaccuracy due to changes in the natural abundances of isotopes depending on the origin of the molecule (Section 13-I).

In the routine analysis of hydrocarbons (Section 10-I) the presence of isotope peaks is a hindrance. Isotope peaks can be readily computed with the aid of "isotope distribution coefficients," and one can obtain a mono-isotopic spectrum, i.e., one that contains only ^{12}C and ^{1}H contributions. A number of such tabulations are available for hydrocarbons, and also for oxygenated and sulfur compounds.

In qualitative organic analysis, isotope peaks can be readily utilized for identification. For three chlorine or bromine atoms in a molecule, $n = 3$ and the expansion contains 4 terms corresponding to the four possibilities of combining the two isotopes, e.g., $^{35}Cl_3$, $^{35}Cl_2{}^{37}Cl$, etc. For chlorine the ratio a/b is equal to $3:1$ because the natural abundance of ^{35}Cl is 75.5 and that of ^{37}Cl is 24.5 ($a = 3$, $b = 1$). Substituting into the expanded form of equation 9-7 yields the relative abundances of the four peaks as $27:27:9:1$. For bromine $a = 50.5$, $b = 49.5$, so that one would expect a $1:3:3:1$ ratio for the four peaks. When both chlorine and bromine appear in the molecule, the general expansion becomes

$$(a + b)^n (c + d)^m \qquad (9\text{-}8)$$

where a, b, c, d are abundances, while m and n refer to the number of each species present. Figure 9-4 depicts some characteristic multiplets of chlorine and bromine atoms in a molecule. The usefulness of such patterns in compound identification is obvious. It is noted here that both fluorine and iodine are monoisotopic. Their presence is often revealed by peaks appearing at unusual masses.

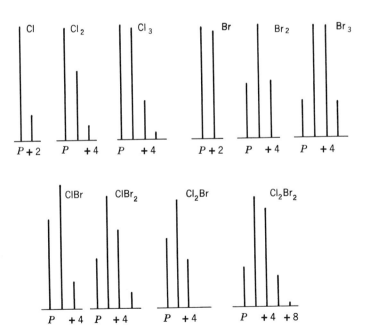

Figure 9-4. Characteristic multiplets of peaks, spaced two mass units apart due to the isotopes of chlorine and bromine. *P* refers to the parent ion containing only light isotopes.

In spark-source mass spectrography, isotope peaks often abound since all isotopes of the matrix appear both in the singly and multiply charged forms. Even when available resolution is adequate for separating isotope lines from impurities, the presence of the latter reduces detection sensitivity. On the other hand, isotope lines may be helpful in the identification of certain impurities, and are also employed for photographic plate calibration using the internal standard method (Section 11-III-B). Measurement of isotope ratios is the basis of the isotope dilution technique (Section 11-III-A), and isotopically labeled molecules are invaluable in the study of chemical and biological reaction mechanisms (Section 12-III-A).

VII. REARRANGEMENT IONS

Rearrangement ions are formed as a result of intramolecular atomic reorganizations accompanying decomposition. Their origin cannot be explained by a simple (single or multiple) cleavage mechanism in the parent ion. One of the first rearrangement ions studied, some 25 years ago (7), was the $C_2H_5^+$ ion in the spectrum of isobutane. The m/e 29 peak in n-butane is simply understood, in fact it is expected to be a strong peak (Table 9-1), but in isobutane its appearance can only be explained by assuming isomerization followed by dissociation,

$$i\text{-}C_4H_{10} \rightarrow n\text{-}C_4H_{10} \rightarrow C_2H_5^+ + C_2H_5 + e^- \tag{9-9}$$

or possibly by considering $C_4H_9^+$ as the immediate precursor of the $C_2H_5^+$ ion with the second step proceeding via a methylated cyclopropane intermediate.

Many other rearrangement ions have subsequently been reported, first in saturated hydrocarbons, later in compounds with heteroatoms. These peaks are often intense, and even form the base peak. Since rearrangement peaks are difficult to analyze it is understandable that organic chemists for a long time doubted that the study of fragmentation in mass spectrometers could ever become a significant tool for structure determination. In recent years, successful attempts (8,9) to systematize available information have resulted in a better understanding of many seemingly chaotic rearrangement processes.

One should consider a rearrangement process just another possible path competing with "normal" fragmentation. On this basis, the probability of occurrence of a particular type of rearrangement is determined by the stability of the product ions and product neutral fragments, together with the ease of formation of the transition state of the reaction, as compared to the energy requirements of the competing normal fragmentation processes. Rearrangement processes often involve six-member transition states which are energetically favorable because a new bond is formed for each one broken. The stereochemistry must also be favorable for transfers to occur.

The most common type of rearrangement involves the intramolecular migration of hydrogen atoms in molecules containing heteroatoms (10). The general scheme of these "McLafferty-type" rearrangements is illustrated in reaction 9-10. The atoms A, B, C, D, and E, and the group G

$$\text{(9-10)}$$

can vary widely as long as the conditions of the multiple bond between D and E, and the availability of a γ-hydrogen are fulfilled. Compounds exhibiting McLafferty-type rearrangement include ketones, aldehydes, amides, substituted aromatic systems, etc. In alkyl ketones, for example, the rearrangement proceeds as shown in reaction 9-11. When one side chain of the ketone is at least three carbon atoms long, the proton on the γ carbon adjacent to the carbonyl oxygen migrates to the oxygen, and the β and γ carbons are eliminated as a neutral olefin molecule. The rearrangement proceeds through a sterically favorable six-membered cyclic transition state, and the driving force is the stability of the end products. The protonated carbonyl fragment retains the positive charge and is probably resonance stabilized. The stable neutral molecule cannot, of course, be seen in the mass spectrometer. The mechanism of the McLafferty-type rearrangement has been thoroughly investigated using stable isotope labeling techniques (Section 12-III-A).

$$\text{(9-11)}$$

parent radical ion   fragment radical ion   stable neutral substituted ethylene

It is noted here that in current usage single-headed arrows ("fish-hook") are employed to show the transfer of one electron, while double-headed arrows indicate the movement of two electrons in the fission or formation of a bond. The sign $\overset{+}{\cdot}$ indicates localized positive charge for odd-electron ions, + denotes localized charge for even-electron ions. It must be kept in mind, however, that reaction mechanisms do not necessarily reveal information on the actual electron motion inside the molecule.

Many skeletal rearrangement processes encountered in mass spectrometry are accompanied by the ejection of a stable neutral molecule from a nonterminal position in the original molecule. A classic example (11) is the elimination of carbon monoxide from anthraquinone ($m/e = 208$) resulting in the appearance in the spectrum of fluorenone ($m/e = 180$) and diphenylene ($m/e = 152$) peaks, as shown in reaction 9-12,

(9-12)

These peaks are rather intense, corresponding to 78 and 51% of the parent peak, respectively. Species corresponding to M − CO have been observed in several other types of compounds, e. g., ethers, phenols, and sulfoxides.

Other neutral entities which may be ejected include carbon dioxide, sulfur dioxide, and formaldehyde. Several examples are discussed in reference 9.

There is much current interest in the study of rearrangement processes. The motivation is provided by the presently popular "mechanistic" approach to the elucidation of organic structures. This approach is based on the premise that the processes leading to the appearance of rearrangement and metastable ions in the electron impact induced fragmentation of organic compounds can be explained by the rational application of everyday concepts of organic chemistry. The availability of high resolution instruments, and the development of the "element-mapping" technique (Section 10-III-C) has given additional impetus to rearrangement studies.

VIII. METASTABLE IONS

The majority of ions formed by electron impact are either stable with insufficient excitation energy to decompose before collision, or unstable with adequate energy to decompose before leaving the ionization region. Ions with a half-life of the order of 10^{-6} sec are sufficiently long-lived to be accelerated out of the ionization chamber, but decompose in transit. These ions are called metastable ions, the process is a metastable transition, and the peaks resulting from these transitions are universally termed metastable peaks. In the fragmentation process the metastable parent ion, m_1, decomposes into a stable fragment ion, m_2, and a neutral particle of mass m_3,

$$m_1^+ \rightarrow m_2^+ + m_3 \tag{9-13}$$

The neutral fragment is normally a stable molecule or radical, e.g., H_2, CH_3, CO, C_3H_7.

Metastable peaks have four apparent characteristics: (*1*) They appear at both integral and non-integral masses; (*2*) their shape is broad and diffuse, often extending over several mass units; (*3*) abundances are usually low, about 0.1–1.0% of the base peak or even lower; and (*4*) the relative intensity of metastable peaks with respect to that of fragment ions can be varied by changing the exit slit width (change in resolution), or by altering the ion repeller voltage (change in time spent in the source). Metastable ions in the spectrum of *n*-butane are shown in Figure 9-1, and are designated by *m* in cracking pattern tables (Tables 9-1 and 9-2).

Although decomposition may take place at any place between source and collector, the probability for collection is the greatest for those metastables which form in the vicinity of the entrance slit of the magnetic analyzer. Assuming the dissociation of m_1^+ to take place at the boundary between accelerator and analyzer, the total kinetic energy of the parent ion is going to be distributed between the newly formed ion m_2^+ and the neutral particle *m*. The m_2^+ ions enter the magnetic field with kinetic energy

$$T_{m_2} = eV(m_2/m_1) \tag{9-14}$$

Since the kinetic energy of these m_2^+ ions is lower than that of those which formed in the ion source and received full acceleration, they will be deflected to a greater extent in the magnetic field, and will be recorded at a lower apparent mass than m_2^+. The apparent mass is obtained by substituting equation 9-14 into the basic mass spectrometer equation (eq. 2-19), and m^*, the apparent mass of the dissociated ions is obtained as (12)

$$m^* = m_2^2/m_1 \qquad (9\text{-}15)$$

In the case of n-butane, the metastable peak at mass 39.2 ($= 41^2/43$) is the result of an ion of mass 43 dissociating into an ion of mass 41 and a neutral particle of mass 2,

$$\begin{array}{ccc} C_3H_7^+ \rightarrow & C_3H_5^+ & + \; H_2 \\ (43)^+ & (41)^+ & 2 \end{array}$$

For the metastable peak at m/e 31.9 there are two possible transitions, based on placing the pairs 43–37 and 58–43 into equation 9-15. The process actually takes place according to

$$\begin{array}{ccc} C_4H_{10}^+ \rightarrow & C_3H_7^+ & + \; CH_3 \\ (58)^+ & (43)^+ & 15 \end{array}$$

since the formation of a neutral particle of mass 6 required by the $(43)^+ \rightarrow (37)^+$ process is highly unlikely. Several metastable transitions are summarized in Tables 9-1 and 9-2.

In practice m^* is measured and transitions must be found by the trial-and-error method, inserting m_1 and m_2 values into the equation and then looking for "logical" processes. It can be shown (13) that log m_2 is the mean of the logs of m_1 and m^*, and nomograms can easily be constructed. For compounds of high molecular weight, the number of metastable peaks may be large, and to determine which particular transition actually produces the observed metastable ions, all possible transitions must be calculated for process 9-13 using equation 9-15. A computer program for such calculations has recently been described by Rhodes et al. (14). The true process is selected by eliminating theoretically unreasonable transitions.

Figure 9-5 shows, for several types of analyzers, the regions where metastable transitions can occur which give rise to collectable ion currents. A detailed study of metastable ion transitions in sector instruments is given by Hipple et al. (12), while 180° instruments are considered by Coggeshall (15). Barber and Elliott (16) showed that in double-focusing instruments the optimum resolution for metastable studies is about 1000; at higher resolution, peak intensities decrease rapidly. These authors also discuss the advantages of the Nier-Johnson geometry in investigating metastable ions.

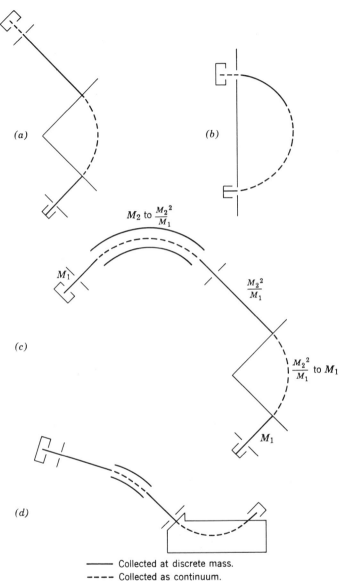

M_2 to $\dfrac{M_2{}^2}{M_1}$

M_1

$\dfrac{M_2{}^2}{M_1}$

$\dfrac{M_2{}^2}{M_1}$ to M_1

M_1

——— Collected at discrete mass.
- - - - Collected as continuum.

Figure 9-5. Formation of metastables in (*a*) 90° sector, (*b*) 180°, (*c*) Nier-Johnson double-focusing, (*d*) Mattauch-Herzog double-focusing instruments (16).

In conventional quantitative analysis the presence of metastable peaks is undesirable and they are normally suppressed (Section 5-I-A). In qualitative organic analysis metastable peaks have considerable practical importance. The appearance of metastables is generally considered as proof of a *one-step* decomposition process: the parent and daughter peaks must have originated from the same molecule and the existence of particular neutral group entities may be quite helpful in reconstructing the original molecule. In dihydrorobustic acid ($C_{22}H_{22}O_6$), for example, a metastable peak at mass 144.6 revealed (17) that a $C_9H_8O_2$ unit exists as a structural unit in the unbroken molecule. Metastable transitions of primary fragments into secondary ones provide useful information about consecutive stages of decomposition. Such information may serve as supporting evidence for a proposed breakdown mechanism.

If a metastable transition of the type shown in equation 9-13 takes place in a time-of-flight mass spectrometer, all particles have the same velocity and arrive at the collector at the same time. Separation may be obtained by introducing a potential barrier at some point along the flight. Neutral fragments are undeflected, while both parent and daughter ions will be

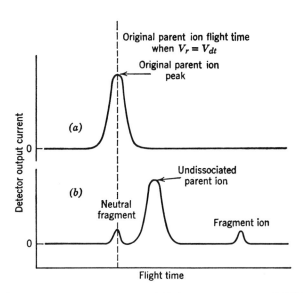

Figure 9-6. Hypothetical mass spectrum of AB^+ ion dissociation in TOF drift tube: (*a*) with no potential barrier applied, there is only one peak corresponding to the characteristic flight time of AB^+. (*b*) With a flat-top potential barrier applied, there are three separate peaks, corresponding to the neutral fragments (at the original flight time of AB^+), the undissociated AB^+, and the fragment ions, respectively (18). The terms V_r and V_{dt} refer to the voltages applied to the retarding gird and the drift tube, respectively.

decelerated in proportion to their respective kinetic energies as shown in Figure 9-6. A considerable amount of knowledge about the nature of *neutral* fragments has been accumulated from such studies of metastable transitions, particularly as a result of the efforts of Hunt and his co-workers (18).

In cycloidal instruments the m/e value of metastable ions may be greater than that of either the parent or daughter ion, and the apparent mass varies according to the point on the trajectory of the parent ion at which fragmentation occurs, and also with injection voltage.

IX. ION–MOLECULE REACTIONS

In ion–molecule reactions a chemical reaction takes place in the source region between ions formed by electron impact (primary ions) and neutral species. At the low pressure that ordinarily prevails in ion sources, ion-molecule collisions are relatively rare and the resulting ion intensities are low, since only the un-ionized portion of the sample under investigation is available for reactive collisions. For the study of ion–molecule collisions, high source pressure (10^{-4} to 10^{-1} torr or even higher) must be employed, requiring the deliberate introduction of a neutral gas to provide optimum conditions for collisions to take place. The ionization chamber must be as gastight as possible and efficient differential pumping must be provided in the analyzer region.

While most ions in a mass spectrometer are formed in unimolecular reactions, ion–molecule reactions are *binary* processes, and consequently peak intensities are proportional to the *square* of the sample concentration, i.e., sample pressure. This provides a practical means to identify ion-molecule reaction peaks: when the sample pressure is increased, these peaks increase relatively more than other peaks in the spectrum. Ion-molecule reactions are very rapid, because the ions exert a strong polarizing force on the neutral molecules. This makes the probability of collisions orders of magnitude larger than in ordinary chemical reactions.

The only kind of ion–molecule reaction that is significant in ordinary mass spectrometry practice is the *abstraction* of a hydrogen from a neutral molecule by the molecular ion. The resulting peak is 1 mass unit higher than the molecular weight ($P + 1$ peak),

$$X + YH \rightarrow XH^+ + Y \tag{9-16}$$

A classical example is the reaction

$$H_2^+ + H_2 \rightarrow H_3^+ + H$$

which has been known since the early days of mass spectrometry. Another example of the radical or atom abstraction type of ion–molecule reaction is the much investigated reaction between the parent ion of methane and neutral methane molecules to form the methanium ion,

$$\text{CH}_4^+ + \text{CH}_4 \rightarrow \text{CH}_5^+ + \text{CH}_3 \tag{9-17}$$

An interesting, although somewhat ambiguous, hydrogen transfer reaction is found in the mass spectrum of iso-butane. The m/e 57 peak (Table 9-2) is in excess of the normal isotope peak and increases with the second power of pressure. The ion–molecule process is most likely,

$$\text{i-C}_4\text{H}_8^+ + \text{i-C}_4\text{H}_8 \rightarrow \text{C}_4\text{H}_9^+ + \text{C}_4\text{H}_7$$

The $P + 1$ peak arising in ion–molecule reactions is superimposed upon the $P + 1$ peak originating from heavy isotope contribution, and high resolution is required for their separation. There are some organic compounds, e.g., ethers, esters, amines, nitriles, which yield rather unstable molecule ions (i.e., low intensity or no parent peak) but whose protonated molecules are quite stable. The $M + 1$ peaks of such molecules are very helpful in the determination of the molecular weight. One must, however, clearly establish that the peak in question indeed originated in an ion–molecule reaction. In addition to the pressure dependence test (plot of abundance vs. pressure), such peaks may be recognized by changing the ionization conditions that influence the ion residence time in the source (e.g., repeller voltage). Most ion–molecule peaks originate inside the ion source and are therefore sharp, but those originating from collisions along the ion path toward the collector are somewhat diffuse, as if they were metastables.

Direct investigation of reactions between ions and molecules in the gas phase has become rather popular during the last decade. The objective is to obtain rate constants with the aid of which the kinetics of reactions may be understood, and reaction mechanisms may be proposed for radiation processes in conventional gas-phase (or even liquid) systems.

The main requirement of the ion sources designed for such studies is that secondary ion formation must be enhanced and provisions must be made to eliminate (or at least reduce) the collection of primary ions. A simple source, designed by Cermàk and Herman (19), is shown in Figure 9-7. The voltage difference between the filament and the chamber is too low for the electrons to acquire adequate energy for ionization, and only after the additional acceleration between chamber and trap are they capable of ionization. Positive ions can thus be formed only inside the trap

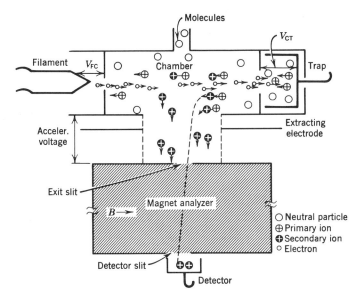

Figure 9-7. Čermák-Herman type source for ion–molecule reactions (19). Secondary ions are created in charge-transfer collisions with monoatomic primary ions. Voltages in ion source of conventional mass spectrometer are adjusted so that primary ions are created only in the region of the trap and charge-transfer reactions occur in the ion source chamber.

chamber, and when they move out into the main ionization region secondary ion formation begins in collisions with the neutral molecules present. The secondary ions, which are formed mainly in *charge transfer*, or dissociative charge transfer reaction, are drawn out from the analyzer, and the high velocities of the primary ions in the direction parallel to the slit prevent them from leaving the chamber. The result is an almost pure secondary ion spectrum.

To determine the rate constant of interest, one must know the ratio of secondary to primary ion currents, the concentration of the neutral molecules (pressure in source), and the average residence time of the ions in the reaction zone, i.e., in the source. The first two parameters can be measured, while the residence time is estimated from source geometry and the known potential gradients. Reaction rates are obtained in units of cm^3 $mole^{-1}$ sec^{-1}, and are usually of the order of 10^{-9}; cross sections are of the order of 10^{-16} cm^2 $molecule^{-1}$. Reaction 9-17 has been thoroughly investigated, employing both pressure and potential gradient (conventional mass spectrometer), and time (TOF instrument) as the experimental variable. The bimolecular reaction rate was found to be 1×10^{-9} cm³ mole⁻¹ sec⁻¹ over a pressure range of 10^{-2} to 10^{-4} torr (20).

A recently developed branch of gaseous ion–molecule chemistry, *chemical ionization* mass spectrometry, is defined by Field (21) as a form of mass spectrometry wherein the ionization of the substance under investigation is effected by reactions between the molecules of the substance and a set of ions which serve as ionizing reactants. The reactant ions are formed in a combined electron impact and ion–molecule ionization process. Reactant ions have been obtained exclusively from methane (though other gases may also be used), and the reactions are carried out at a source pressure of 1–2 torr. Chemical ionization mass spectra, which consist of relatively small numbers of ion types, differ from both electron impact and field ionization spectra, and provide certain structural information not available with the other modes of ionization. In addition to molecular structure determinations, chemical ionization is also expected to be utilized in compound-type analysis of organic mixtures.

X. NEGATIVE IONS

Although many negative ions were observed and identified by J. J. Thomson in his parabola mass spectrograph, the study of negative ions has received attention only in recent years. The main reason for this is that the efficiency of ion formation of negative ions is lower, by about a factor of 1000, than that of positive ions. Otherwise, most mass spectrometers can be employed for negative ion studies by suitably reversing the fields. A mass spectrometer which simultaneously collects both negative and positive ions has been described by Flesch and Svec (22). Secondary electron multipliers are generally employed to increase sensitivity. The negative ion acceleration potential is maintained about 2 kV higher than that of the first dynode to prevent ion repulsion by the high negative potential on the conversion dynode. Filaments must be made of corrosion-resistant materials (e.g., thoria-iridium), since most negative ion producing compounds are highly reactive. Differential pumping is needed to maintain high source pressure.

Negative ions may be formed from neutral molecules upon electron impact by three mechanisms:

(*a*) Resonance electron capture,

$$\text{AB} + e^- \rightarrow \text{AB}^- \tag{9-18}$$

(*b*) Dissociative resonance capture,

$$\text{AB} + e^- \rightarrow \text{A} + \text{B}^- \tag{9-19}$$

(*c*) Ion pair production,

$$AB + e^- \rightarrow A^+ + B^- + e^- \tag{9-20}$$

In the first two processes there are no product electrons to carry excess kinetic energy, and ion formation occurs only over a narrow range of electron energy (0–10 eV). The ionization efficiency curve of the ion pair production mechanism is similar to that shown in Figure 1-4.

Negative ion spectra are normally much simpler than those of positive ions. An important exception is perchlorylfluoride (23), ClO_3F, where a wide variety of ions are formed by attachment; ions of F^-, Cl^-, ClO^-, ClO_2^-, ClO_3^-, and ClO_2F^- appear in great abundance. The fact that halogenated compounds almost always yield negative halogen ions is explained by considering that electron affinities of halogen atoms are higher than those of other elements. Another interesting compound yielding negative ions is SF_6 (24,25). Among hydrocarbons, only methane, the C_2 hydrocarbons, and *n*-butane have been studied. In Table 9-4 the negative ion mass spectrum of *n*-butane is summarized for 90 eV ionizing electrons.

Table 9-4   **Negative Ion Mass Spectrum of *n*-Butane (26)**

       (90 eV ionizing electrons)

m/e	Negative ion	Relative abundance
12	C	2.9
13	CH	13.3
14	CH_2	7.0
15	CH_3	2.2
24	C_2	29.5
25	C_2H	100
26	C_2H_2	5.6
27	C_2H_3	0.2
36	C_3	2.6
37	C_3H	2.2
38	C_3H_2	0.2
39	C_3H_3	0.2
48	C_4	1.4
49	C_4H	2.9
50	C_4H_2	0.2
51	C_4H_3	0.2

Ratio $C_3H_7^+/C_2H^- \simeq 33,000$

The ratio $C_3H_7^+/C_2H^-$ is about 3×10^4, and this ratio between the most abundant positive and negative ions is also found for the other hydrocarbons studied (26). The abundance of parent ions in negative ion spectra is usually low compared to that of fragment ions.

Negative ions have not yet been used in chemical analysis. Possible advantages are seen in electronegative compound analysis; also the relatively low negative mass background may find analytical applications. In structure studies negative ion spectroscopy has been useful for highly fluorinated and polyhalogenated compounds which normally do not yield intense positive molecular ions. Other fields of possible applications include catalytic studies, the chemistry of the atmosphere, and identification of the products of a unimolecular decomposition.

XI. THEORY OF MASS SPECTRA

A. Franck-Condon Principle

Figure 9-8 shows the variation of potential energy with internuclear distance of a diatomic molecule. The shape of such *potential energy curves* results from the unharmonic nature of the nuclear vibrations. A simple parabola (dashed line) would result if nuclear oscillations were simple harmonic. The horizontal lines indicate various vibrational levels; the zero-point energy, E_0 is the difference between the bottom of the curve and

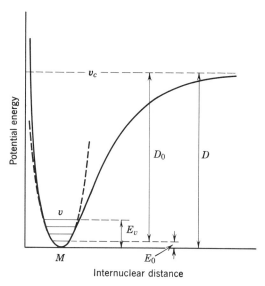

Figure 9-8.   Potential energy curve.

the first horizontal line ($v = 0$). At higher vibrational levels the anharmonicity of the oscillations increases, and at v_c dissociation occurs, i.e., the distance between nuclei becomes infinite. The heat of dissociation at $0°K$ is D_0, while D ($= D_0 + E_0$) refers to the bottom of the potential energy curve.

A potential energy curve of the type shown in Figure 9-8 can be constructed for every electronic state in the molecule. Figure 9-9 shows the lowest (A) and one of the higher (B) electronic states within a diatomic molecule. The minima of the curves, i.e., the nuclear separation at equilibrium, are different, and so are the heats of dissociation. The vertical distance between the two minima represents the energy difference (E_e) between the two electronic states. A transition from a given vibrational level in a lower state into any level in the upper state can occur when radiation is absorbed; the reverse process is accompanied by radiation emission.

An electron with 10 eV energy travels, according to equation 2-3, with a velocity of about 2×10^6 m/sec. Taking the average molecular diameter as 1×10^{-10} m (10 Å), the passage of an electron "through" the molecule takes about 2×10^{-16} sec. The frequencies of vibrations of bonds, on the other hand, are of the order of 10^{14} sec^{-1} in most molecules, showing that

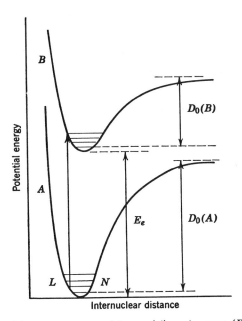

Figure 9-9. Potential energy curves of lower (A) and upper (B) electronic states, diatomic molecule.

the impacting electrons pass the molecule in a fraction of the vibrational period. Consequently, no changes occur during the course of an electronic transition in the positions and velocities of the nuclei or, in other words, the *nuclear configuration* of the system does not change during the transition. This is the essence of the Franck-Condon principle. Another aspect of this principle relates to the *probability* of the transitions. Transitions from one state to another are most probable when the vibrational kinetic energy is zero. This becomes clear when it is considered that the nuclei spend their longest time at the extreme positions on a given level* (*L* and *N* in Fig. 9-9). Transitions are represented by vertical lines on the potential energy diagram and the most probable transitions are shown to commence from the extreme positions on a given level.

The dissociative ionization of the molecule XY can be studied with the aid of Figure 9-10 (27) which shows potential energy curves for the ground state (*1*) and ionic states (*2* and *3*). The vertical dashed lines indicate the limits of the Franck-Condon region and it is immediately seen that the XY$^+$ ion has very little probability of being formed in its lowest vibrational level since the minimum on curve *2* falls outside the effective Franck-Condon region. Since a portion of this region crosses curve *2* *above* the dissociation limit, transitions to this state produce both stable XY$^+$ ions, which are vibrationally excited, and fragment ions of X$^+$. Transition into the repulsive state (state *3* in Fig. 9-10) results in dissociation into fragments which have considerable kinetic energy. Curves *4* and *5* indicate the predicted energy distribution for the dissociation products of states *2* and *3*, respectively. (These are drawn by reflecting the square of the eigenfunction of the lowest vibrational level of state *1* onto the curves for states *2* and *3*.) The horizontal dashed line represents the final energy level of a transition. In Figure 9-10 the symbols E_k and E_e represent the kinetic and excitational energy, respectively, of the particle indicated within the parentheses. The dissociation energy of XY is $D(XY)$, while I and A refer to ionization and appearance potentials, respectively (Section 12-I).

B. Quasi-Equilibrium Theory

The objectives of the theory of mass spectra are to establish correlations between fragmentation patterns and the physical and chemical structure of the molecules and to understand and explain the mechanism of ion for-

---

* The square of the vibrational eigenfunction is greatest near the extreme positions of the nuclei.

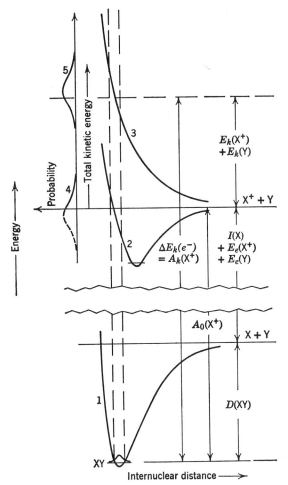

Figure 9-10. Potential energy curves involved in the dissociative ionization of a diatomic molecule by electron impact (27). For description see text.

mation. It is therefore a branch of chemical kinetics. The subject is difficult because a thorough knowledge of statistical mechanics is required. The increasing interest in the theoretical aspects of mass spectrometry is evidenced by the rapidly growing number of publications on the subject. Only a very brief and superficial discussion is given here; the interested reader is referred to a review article by Lester (28) and a more detailed treatment by Rosenstock and Krauss (29).

The so-called *quasi-equilibrium* theory of mass spectra, originally developed by Eyring and his co-workers (30), is based on the following assumptions:

1. The first step in ion formation is the transfer of energy (normally 50–100 V electrons) from the bombarding electron to the neutral molecule, resulting in the formation of a molecular ion. Ionization is a "vertical" Franck-Condon transition, and there is no change in the position and kinetic energy of the nuclei. The removal of a valence shell electron, however, brings the parent molecule ion into an electronically excited state.

2. Since the parent molecule ion, which contains an odd electron, is of low symmetry, and the many low-lying electronic states form a continuum, radiationless transitions can transfer electronic energy into vibrational energy. The excess energy is not retained at the bond from which the electron was removed in ionization, but rather it is rapidly distributed over all internal degrees of freedom. Fragmentation of the parent ions occurs when sufficient vibrational energy (i.e., equal to the dissociation energy) is concentrated in a particular bond. There is thus a *time delay* of as much as 10^{-5} sec between the formation of the parent ion and its dissociation. This explains why fragmentation, and the resulting cracking patterns, are so sensitive to chemical structure that no two compounds have exactly the same mass spectrum.

3. The fragment ions that are observed in the mass spectra are formed in a series of competing and consecutive unimolecular reactions which are similar to the rate processes of conventional chemical kinetics and for which rate constants can be calculated on the basis of the absolute rate theory.

4. The fragment ions thus formed may again have a sufficient amount of excitation energy (originating from the parent molecular ion) to undergo further decomposition in the same manner.

The decomposition of the molecular ion is considered to be a quasi-equilibrium unimolecular reaction of an isolated system. The term unimolecular refers to the *molecularity* of the reaction, i.e., to the number of molecules involved in the rate-determining step. The system is treated as an *isolated* one because of high vacuum conditions. Reference to quasi-equilibrium is based on the assumption that transitions between the accessible states is rapid enough not to appreciably disturb the equilibrium of distribution.

A chemical reaction, according to the absolute reaction rate theory, proceeds as follows: the reactants from the initial state first pass over the top of a potential barrier (transition state) and than descend into the final product state. In the transition state an activation complex is believed to be formed, one which is in equilibrium with the reactants. The rate of this reaction is given by the number of activated complexes crossing over the saddle point per unit time.

In ordinary chemical kinetics, molecules are expected to continuously energize and deenergize in collisions and the distribution of the reactants and activated complexes among their accessible states is described in terms involving the temperature. The equilibrium constant is obtained by using partition functions. In mass spectrometer ion sources there are no molecular collisions, and theoretical considerations lead to the use of a "density of states" function, $\rho(E)$, instead of partition functions. The expression $\rho(E)\delta E$ gives the number of quantum states with energy in the E to $E + \delta E$ interval. The decomposition rate constant, k, is related to the excitation energy, E, by

$$k = \int_0^{E-\epsilon_0} \frac{1}{h} \frac{\rho^{\ddagger}(E,\epsilon_0,\epsilon_t)}{\rho(E)} \, dE \qquad (9\text{-}21)$$

where the distribution of the states corresponding to the activated complex through which the decomposition proceeds is represented by $\rho\ddagger(E,\epsilon_0,\epsilon_t) \, dE$. This density function includes those energy states with total energy between E and $E + dE$, with potential energy ϵ_0 and kinetic energy (along the path of decomposition)ϵ_t. The internal energy of the transition complex is given by $E - \epsilon_0$.

Since very little is known about the density functions of the above equation, several assumptions must be made for practical calculations. When the molecular ion is considered as a system consisting of N loosely coupled harmonic oscillators with frequencies ν_i, one obtains

$$k = \left(\frac{E - \epsilon_0}{E}\right)^{N-1} \frac{\prod_{i=1}^{N} \nu_i}{\prod_{j=1}^{N} \nu_j^{\ddagger}} \qquad (9\text{-}22)$$

where $\nu_j{}^{\ddagger}$ refers to the activated complex. The first step in applying the theory to a practical case is to list all the possible competing modes of decomposition, and then to calculate reaction rates choosing a set of "reasonable" parameters. For propane, four decomposition schemes were suggested,

$$\begin{array}{l}
C_3H_8{}^+ \xrightarrow{k_1} n\text{-}C_3H_7{}^+ \longrightarrow n\text{-}C_3H_5{}^+ \longrightarrow C_3H_3{}^+ \longrightarrow C_3H^+ \\
\quad\;\; \xrightarrow{k_2} C_2H_5{}^+ \longrightarrow C_2H_3{}^+ \\
\quad\;\; \xrightarrow{k_3} C_2H_4{}^+ \longrightarrow C_2H_2{}^+ \\
\quad\;\; \xrightarrow{k_4} C_3H_6{}^+ \longrightarrow C_3H_4{}^+ \longrightarrow C_3H_2{}^+
\end{array}$$

The rate constants are calculated from the activation energies of the various postulated processes which, in turn, may be evaluated from appearance and ionization potential measurements. The final results are expressed in the form of cracking patterns which may then be directly compared with experimental observations. For propane good agreement is obtained between calculated and measured values. As seen from the data in Table 9-5 (31) three different distribution functions for the excitation energies have been employed in these calculations. Many other compounds, including alkanes and esters, have been investigated using this theory. Results are usually in semiquantitative agreement with experiments.

Table 9-5 Calculated and Experimental Mass Spectra of Propane (31)[a]
(Three distribution functions used)

Distribution function	C_3H_8	C_3H_7	C_2H_5	C_2H_4	C_2H_3	C_2H_2
1	0.102	0.064	0.249	0.161	0.272	0.031
2	0.114	0.071	0.279	0.178	0.220	0.027
3	0.112	0.074	0.300	0.185	0.194	0.024
Experimental	0.090	0.067	0.310	0.183	0.122	0.027

[a] Total ionization normalized to unity.

Recent work in theoretical mass spectrometry is mainly concerned with the improvement of the rate expressions and with the methods to acquire knowledge about such parameters used in the calculations as activation energy, frequency factors, energy distribution factors, etc. The basic assumptions of the statistical theory are also being questioned and re-examined, particularly the one on random energy distribution in the parent ions.

References

1. Mohler, F. L., P. Brandt, and V. H. Dibeler, *J. Res. Natl. Bur. Std.*, **60**, 615 (1958).
*2. Ötvös, J. W., and D. P. Stevenson, *J. Am. Chem. Soc.*, **78**, 546 (1956).
3. Crable, G. F., and N. D. Coggeshall, *Anal. Chem.*, **30**, 310 (1958).
4. Pahl, M. Z., *Z. Naturforsch.*, **9 b**, 188 and 418 (1954).
5. Craig, R. D., G. A. Errock, and J. D. Waldron in *Advances in Mass Spectrometry*, Vol. II, J. D. Waldron, Ed., Pergamon Press, New York, 1959, p. 136.
*6. Beynon, J. H., *Mass Spectrometry and its Applications to Organic Chemistry*, Elsevier, New York, 1960, p. 294.

7. Stevenson, D. P., and J. A. Hipple, *J. Am. Chem. Soc.*, **64**, 1588 (1942).

8. McLafferty, F. W., *Anal. Chem.*, **31**, 82 (1959); also in *Mass Spectrometry of Organic Ions*, F. W. McLafferty, Ed., Academic Press, New York, 1963, pp. 309–342.

* 9. Brown, P., and C. Djerassi, *Angew. Chem., Intern. Ed.*, **6**, 477 (1967).

*10. McLafferty, F. W., *Anal. Chem.*, **34**, 2, 16, 26 (1962).

11. Beynon, J. H., G. R. Lester, and A. E. Williams, *J. Phys. Chem.*, **63**, 1861 (1959).

12. Hipple, J. A., R. E. Fox, and E. U. Condon, *Phys. Rev.*, **69**, 347 (1946).

*13. Beynon, J. H., *Mass Spectrometry and Its Applications to Organic Chemistry*, Elsevier, New York, 1960, p. 546.

*14. Rhodes, R. E., M. Barber, and R. L. Anderson, *Anal. Chem.*, **38**, 48 (1966).

15. Coggeshall, N. D., *J. Chem. Phys.*, **36**, 1640 (1962).

16. Barber, M., and R. M. Elliott, *12th Ann. Conf. Mass Spectrometry, ASTM, Committee E-14, Montreal, Canada, 1964, Proc.*, p. 150.

17. Johnson, A., A. Pelter, and M. Barber, *Tetrahedron Letters*, **1964** [20], 1267.

18. Hunt, W. W., R. E. Huffman, and K. E. McGee, *Rev. Sci. Instr.*, **35**, 82 (1964); also W. W. Hunt, R. E. Huffman, J. Saari, G. Wassel, J. F. Betts, E. H. Paufve, W. Wyess, and R. A. Flugge, *Rev. Sci. Instr.*, **35**, 88 (1964).

19. Cermàk, V., and Z. Herman, *Nucleonics*, **19**, 106 (1961).

20. Hand, C. W., and H. von Weyssenhoff, *Can. J. Chem.*, **42**, 195 (1964).

*21. Field, F. H., *Accounts Chem. Res.*, **1**, 42 (1968).

22. Flesch, G. D., and H. J. Svec, *Rev. Sci. Instr.*, **34**, 897 (1963).

23. Dibeler, V. H., R. M. Reese, and D. E. Mann, *J. Chem. Phys.*, **27**, 176 (1957).

24. Ahearn, A. J., and N. B. Hannay, *J. Chem. Phys.*, **21**, 119 (1953).

25. Hoene, J. von, and W. M. Hickam, *J. Chem. Phys.*, **32**, 876 (1960).

26. Melton, C. E., and P. S. Rudolph, *J. Chem. Phys.*, **31**, 1485 (1959); also C. E. Melton in *Mass Spectrometry of Organic Ions*, F. W. McLafferty, Ed., Academic Press, New York, 1963, pp. 163–205.

*27. H. D. Hagstrum, *Rev. Mod. Phys.*, **23**, 185 (1951).

*28. Lester, G. R., *Brit. J. Appl. Phys.*, **14**, 414 (1963).

29. Rosenstock, H. M., and M. Krauss, in *Mass Spectrometry of Organic Ions*, F. W. McLafferty, Ed., Academic Press, New York, 1963, pp. 1–64; also in *Advances in Mass Spectrometry*, R. M. Elliott, Ed., Pergamon Press, Oxford England, 1963, pp. 251–284.

30. Rosenstock, H. M., M. B. Wallenstein, A. L. Wahrhaftin, and H. Eyring, *Proc. Natl. Acad. Sci. U.S.*, **38**, 667 (1952).

31. Field, F. H., and J. L. Franklin, *Electron Impact Phenomena*, Academic Press, New York, 1957, p. 78

Chapter Ten

Analytical Techniques: Organic

I. QUANTITATIVE ANALYSIS

A. Component Analysis

The widespread use of mass spectrometry in quantitative as well as qualitative analysis is a direct consequence of the availability of highly stable and reliable commercial instruments. The basic principles of quantitative analysis were developed by Washburn and his co-workers (1), who investigated the analytical conditions required and worked out methods for calculations of complex mixtures. There are four assumptions in quantitative mass spectral analysis:

1. Each chemical compound has its own characteristic mass spectrum or, in other words, cracking patterns are different. Most often there are peaks at different masses in the spectra of the various compounds, but the difference in cracking patterns may only be restricted to different ion abundances of the important peaks, e.g., *n*-butane and isobutane. There are cases where the patterns are not sufficiently dissimilar to permit analysis in a mixture, e.g., *cis*-2-butene and *trans*-2-butene.

2. Both cracking patterns and sensitivities are constant, as long as experimental conditions are unchanged. The sensitivity in quantitative analysis is usually expressed in units of chart divisions of the base peak (proportional to the ion current) divided by sample pressure in the sample reservoir,

$$\text{Sensitivity} = \frac{\text{Peak height (divisions) of base peak}}{\substack{\text{Sample pressure } (\mu) \text{ in reservoir at the} \\ \text{beginning of the analysis}}} \qquad (10\text{-}1)$$

Since cracking patterns are constant, the sensitivity of any particular peak may be calculated from the knowledge of the base peak sensitivity and the cracking pattern. Such sensitivity values are characteristic only of the particular instrument operated under given conditions, e.g., at 20 μA ionization current. For comparative purposes, sensitivity for a reference compound, such as n-butane, is normally also listed in compilations. Instead of sensitivities, sometimes "pressure factors," i.e., reciprocal of sensitivity (μ/div), are employed.

3. Each compound behaves in the mass spectrometer as if it were present alone. In the spectrum of a mixture, therefore, peaks caused by every component are linearly additive.

4. The measured intensities of the ion beams of the various components are proportional to the partial pressures in the reservoir of the respective components.

These requirements are fulfilled remarkably well in commercial mass spectrometers. Deviations from ideal behavior result in a decrease of analytical accuracy. The validity of the above assumptions is discussed in connection with accuracy later in this section.

Component analysis in mixtures is very simple when there are *unicomponent* peaks present in the mixture spectrum. A unicomponent peak is one that results from only one component of the mixture. In a mixture of methane, nitrogen, and argon, for example, peaks at m/e 16, 28, and 40 are all unicomponent peaks. The first step in any calculation is to read the peak heights, make the necessary zero level corrections (Fig. 10-1), and

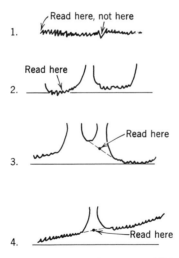

Figure 10-1. Direction for zero level reading (Consolidated Electrodynamics Corp., Computing Manual). (*1*) Read general average in center of noise level. (*2*) When zero level is higher on right than on left, read zero at left. (*3*) When zero level is lower on right than on left, project zero line across base of peak and read under maximum point of peak. (*4*) If zero is drifting consistently up or down, project across base and read under maximum point.

multiply by the attenuation factor employed on the strip chart recorder or galvanometer.

The partial pressure (μ) of a component is obtained from a unicomponent peak by dividing the peak height by the sensitivity of the component (obtained in a calibration run),

$$\text{Partial pressure} = \frac{\text{Peak height (div)}}{\text{Sensitivity (div}/\mu)} \tag{10-2}$$

Sensitivity normally refers to base peak sensitivity. The partial pressures of all components are calculated in a similar manner. The calculated total pressure should agree within a few per cent with the original sample pressure measured in the sample reservoir at the beginning of the analysis. The mole fraction of every component is next calculated,

$$\text{Mole \% of component} = \frac{\text{Partial pressure of component}}{\text{Total pressure}} \times 100 \tag{10-3}$$

Since the partial pressures are calculated in a constant volume (the reservoir) percentage values are in mole per cent. The sum of the mole percentages should be 100.

The completeness of the analysis should be checked by subtracting from every peak in the mixture spectrum the contributions made by every component. In the above example of the mixture of methane, nitrogen, and argon, this is a very simple task, since almost all peaks are unicomponent peaks (m/e 14 peak has contributions from both nitrogen and methane). The residual peaks (Δ peaks) after subtraction must be small, generally less than 1% of the original peak height. It often happens that the mixture spectrum contains both unicomponent and polycomponent peaks. Calculations can be made considerably easier if all unicomponent peaks are first removed. If the above mixture also contained oxygen, the m/e 16 peak of methane could be obtained after "peeling off" the oxygen contribution. In this case, it would be more accurate to use the m/e 15 peak in the calculations.

B. Methods for Calculation

Most frequently it is not possible to subtract successively all components on the basis of unicomponent peaks. When the components in the mixture spectrum overlap, simultaneous linear equations must be set up and solved. In these equations the unknowns are the peak heights contributed by the individual components, the constant terms are the measured mixture peaks, and the coefficients are the cracking pattern coefficients, or fractional pattern coefficients, i.e., $1/100$ of the coefficients with the base peak equal to 1.00.

In a mixture of n-butane and isobutane (Tables 9-1 and 9-2), for example,

$$0.1230x_1 + 0.0273x_2 = \text{peak at } m/e\ 58$$
$$x_1 + x_2 = \text{peak at } m/e\ 43$$

where x_1 and x_2 are the contributed peak heights to the m/e 43 peak by n-butane and isobutane, respectively. Once the unknowns are determined, the contributions to the 58 peak can be calculated utilizing the cracking patterns, and from there on calculation of the partial pressures and mole percentages are straightforward.

In general, there must be as many equations as there are unknowns, and every possible contribution to the mixture peak must be included. When there are unicomponent peaks present, their contributions can be determined on the basis of their cracking patterns and subtracted from the mixture peak. In a mixture consisting of n components of partial pressure $p_1, p_2, \ldots p_n$, the measured ion beam intensities (peak heights) are H_1, H_2, $\ldots H_m$, where m is the number of peaks in the spectrum. Each of the H values represents the sum of the contributions to that peak by the various components. In the general discussion it is assumed that every component of the mixture contributes to every peak. The principle of linear additivity of peaks is expressed by m equations, containing n unknown partial pressures,

$$
\begin{aligned}
h_{11}p_1 + h_{12}p_2 + \ldots + h_{1n}p_n &= H_1 \\
h_{21}p_1 + h_{22}p_2 + \ldots + h_{2n}p_n &= H_2 \\
&\;\;\vdots \\
h_{m1}p_1 + h_{m2}p_2 + \ldots + h_{mn}p_n &= H_m
\end{aligned}
\tag{10-4}
$$

where h_{mn} is the ion current at mass m due to component n. Pattern coefficients (relative to base peak) are normally used for h_{mn}, and since the H values are in units of peak heights, the unknowns are also obtained in peak heights which, in turn, must be converted to partial pressures (divide by sensitivity), and finally to mole per cent (divide by total pressure). Equations 10-4 provide a set of m equations in n unknowns where, in general, $m > n$.

In setting up the analysis, the first step is to identify all components present. In routine laboratories the composition is generally known and only possible impurity peaks must be checked prior to calculations. Qualitative analysis is discussed later in this chapter. The selection of the "best" peaks for the mixture analysis requires judgment and experience, and for common mixtures standard procedures, usually computerized, are developed by individual laboratories. In the ideal case, only monoisotopic peaks are used. Failing this, the objective is to select peaks with as little contribution from other components as possible. In addition to convenience in calculations, accuracy is also affected by this choice. In the case of a mixture of n-butane and isobutane, for example, the 58 peak provides good "leverage" with the 43 base peak, while the use of the 39 or 50 peaks would be a poor choice. Also, pattern variability and possible impurity interferences must also be considered: the m/e 12 peak in the butane spectrum is more variable than the 43, and the 28 peak may have air contribution.

A very instructive example of mixture analysis, given by Melpolder and Brown (2), is reproduced in Tables 10-1 and 10-2. The mixture is a hydrocarbon sample containing C_4 paraffins, monoolefins, and propane. The 44 peak of propane is essentially a monoisotopic peak. Both butanes have the same pattern coefficients at m/e 44 since this peak originates from the base peak at 43 due to heavy carbon and hydrogen isotopes. The propane calibration at mass 44 is precorrected for its contribution to the 43 peak. The propane peak at m/e 44 is then obtained as 84.6 div minus 1570 × 0.0329

Table 10-1   Calibration Spectra for Analysis of C_3–C_4 Hydrocarbon Mixture (2)

m/e	Propane	n-Butane	Isobutane	Isobutene	1-Butene	2-Butene (av. *cis* and *trans*)
39	0.749	0.171	0.225	0.519	0.405	0.386
40	0.105	0.025	0.034	0.119	0.071	0.068
41	0.494	0.308	0.420	1.00	1.00	1.00
42	0.226	0.126	0.330	0.036	0.035	0.032
43	0.884	1.00	1.00	0.002	0.001	0.001
44	1.00[a]	0.0329	0.0329	0	0	0
55		0.011	0.005	0.164	0.182	0.223
56		0.008	0.004	0.429	0.372	0.463
57		0.025	0.030	0.018	0.017	0.022
58		0.126	0.026	—	—	—
S[b]	14.7	67.3	76.3	61.1	58.6	53.1

[a] Monoisotopic peak.

[b] Sensitivity of base peak (underlined)

Table 10-2   Calculation of a C_3–C_4 Hydrocarbon Mixture—Contributions of Components (2)

m/e	Mixture spec- trum	Pro- pane	n- Butane	Iso- butane	Iso- butene	1- Butene	2- Butene	Residual spectrum
39	736	24.6	155.6	141.7	173.3	155.9	81.4	+3.5
40	128	3.4	22.7	21.4	39.7	27.3	14.3	−0.8
41	1491	16.2	280.2	264.6	334.0	385.0	211.0	
42	364	7.4	114.6	207.9	12.0	13.5	6.8	+1.8
43	1570	29.1	909.7	629.9	0.7	0.4	0.2	
44	84.6	32.9[a]	29.9	20.7	0	0	0	
55	185		10.0	3.1	54.8	70.1	47.0	
56	394		7.3	2.5	143.3	143.2	97.7	
57	58.4		22.7	18.9	6.0	6.5	4.6	−0.3
58	131		114.6	16.4				
p_n		2.24	13.52	8.26	5.47	6.57	3.97	Σ40.03
Mole %		5.6	33.8	20.6	13.7	16.4	9.9	Σ100.0

[a] Monoisotopic peak.

(= 32.9) div. The third column of Table 10-2 shows the "peeling down" of the propane contributions on the basis of its cracking pattern.

In the next step, contributions of n-butane and isobutane are calculated using two simultaneous equations based on peaks at masses 43 and 58. There is no contribution by the other components. The equations are

$$1.00 \ \ n\text{-butane} + 1.00 \ \ \text{isobutane} = 1540.9 \quad (43 \ \text{peak})$$
$$0.126 \ n\text{-butane} + 0.026 \ \text{isobutane} = 131 \quad (58 \ \text{peak})$$

The 43 peak is obtained as 1540.9 div after subtracting 29.1 div (propane) from the mixture spectrum. Thus, one obtains 909.7 div for n-butane, and 629.9 div for isobutane. Columns 4 and 5 show the determination of the butenes: three simultaneous equations are needed,

Isobutene		1-Butene		2-Butene			
1.00	+	1.00	+	1.00	=	930.0	(41 peak)
0.164	+	0.182	+	0.223	=	171.9	(55 peak)
0.429	+	0.372	+	0.463	=	384.2	(56 peak)

The solution of these equations yields the mass 41 values for the butenes. The overall accuracy of this peeling down process can be checked by observing the residual peaks; all values are less than 1% of the original peak height. Finally, the partial pressures and mole per cent values are calculated in a routine manner.

When there are a large number of equations to be solved, the methods of matrix algebra must be used. First, a suitable choice of equations is needed to reduce the number of equations to the number of unknowns ($m = n$ in eq. 10-4). When only a limited number of samples are to be analyzed the Crout method (3) is fast and simple and requires only a desk calculator. For a large number of analyses it is worthwhile to use a reciprocal or inverse matrix technique (4). Laboratories with large sample loads now employ digital computers, and with the almost complete automation of data handling, thousands of man-hours can be saved (Section 5-1-D-4). A review of the digitization of mass spectral data is given by Dudenbostel and Klaas (5). Although multicomponent mixture analysis in the oil industry has been largely taken over by gas–liquid chromatography, there are still many problems requiring the knowledge of the handling of mixture spectra.

The calculation technique based on equation 10-4 requires that the cracking pattern and sensitivity data be available for all components, and frequent recalibration must be performed to maintain accuracy. In the "peak ratio" method of analysis (6) of two-component mixtures, the cracking pattern of a known mixture is compared with an unknown mixture. There is no need to measure absolute pressures and the achievable accuracy depends upon that of the synthetic blend. Since micromanometers are now standard equipment, this technique is employed only in special problems.

C. Precision and Accuracy

The basic assumptions of mixture analysis, discussed at the beginning of this chapter, are not strictly correct; several deviations are considered in the following. In routine applications, particularly for multicomponent systems requiring complex calculations, continuous recalibration is not practical. The time stability of cracking patterns is influenced by such factors as temperature and surface condition of the filament, temperature changes in the ion source, and spurious potential changes due to the deposition of insulating layers within the source. Cracking patterns are ratios with respect to the base peak, and some ratios fluctuate more than others. For example, isotope peak ratios are usually quite stable, while those of fragments where many hydrogens are stripped off often exhibit larger variability. Short-term (day to day) variations in a good analytical instrument are normally less than 2%, but weekly variations may be as high as 3–5%. Sensitivity values also exhibit variations due to changing or fluctuating source conditions (e.g., repeller voltage, etc.). Short-term variations are usually within 3–4%, and a large change in sensitivity usually indicates instrument malfunction. Sensitivity changes normally influence all peaks to about the same degree, so that periodic checking of a standard compound (n-butane, argon) is adequate to detect troubles.

The principle of linear superposition may, on occasion, be disobeyed. When the introduction of a gas influences the sensitivity or the cracking pattern of another gas, the phenomenon is called interference. Filament conditioning (Section 4-II), and preconcentration techniques (Section 11-II-A) may be used to reduce interference.

The precision and accuracy of the analyses depends strongly upon the nature of the samples. Many reports have been published on repetitive and comparative masurements. Table 10-3 is a summary of a classical study made by Washburn (7), who evaluated the results of some 92 analyses of a synthetic hydrocarbon mixture on 24 instruments. For each instrument the mixture was independently synthesized and then run three or more times. The average error is about 0.2 mole % for all components with the exception of the butenes for which it is 0.2–0.8 mole %. Total butanes can be determined with an average error of 0.27%.

Table 10-3   Accuracy of Mass Spectrometer Analyses as Determined from a Large Number of Analyses of a Synthetic C_1–C_4 Paraffin Olefin Mixture (7)

Components	Composition, mole %	Average error, mole %	90% of errors less than
Methane	15	±0.14	±0.4
Ethane	20	±0.22	±0.4
Propane	20	±0.21	±0.5
Propylene	10	±0.17	±0.5
Isobutane	10	±0.19	±0.5
n-Butane	8	±0.18	±0.4
Isobutene	7	±0.33	±0.8
Butene-1	5	±0.83	±1.4
Butene-2	5	±0.77	±1.2
Total butenes	17	±0.27	±0.6

Accuracy decreases when certain components, e.g., oxygenated compounds, become adsorbed on the walls, or when there is background interference. The availability of double inlet systems (Section 7-IV) permits increased accuracy by use of a comparative method in which carefully prepared mixtures are analyzed prior to and following the analysis of the unknown. This technique is particularly useful for small impurity determination.

D. Compound Type Analysis

The determination of compound types rather than specific compounds in complex hydrocarbon mixtures was first suggested by Brown (8). He found that the spectra of such principal compound types as paraffins, cycloparaffins, cycloolefins, acetylenes, and aromatics possessed characteristic ions, and types could be quantitatively characterized by using the sums of specific mass peaks for each type. Peaks at m/e 43, 57, 71, 85, and 99, for example, are most abundant in paraffins, while other peaks occur mostly in other compound types. Each group of peaks used to characterize a compound type is assigned a so-called *sigma* value: $\Sigma 43$ refers to paraffins, $\Sigma 77$ to alkylbenzenes, etc. A typical calibration matrix showing the dissimilarity among the different types of compounds is shown in Table 10-4. Cycloparaffin and monoolefin contributions are determined in separate runs on the original and on acid-treated (olefin-free) samples.

Table 10-4   A Typical Calibration Matrix in Gasoline Analysis (2)

	$\Sigma 43$	$\Sigma 41$	$\Sigma 67$	$\Sigma 77$	Sensitivity, div/μ
Paraffins	*1.00*	0.266	0.005	0.002	180
Cycloparaffins	0.15	*1.00*	0.153	0.012	121
Monoolefins	0.37	*1.00*	0.058	0.017	108
Cycloolefins, etc.	0.13	0.75	*1.00*	0.08	133
Alkylbenzenes	0.004	0.012	0	*1.00*	156

The technique of compound type analysis is illustrated by a brief description of a method recently approved by the ASTM (9) for the analysis of hydrocarbon types in propylene polymers. Experimental procedure is simple: at a sample pressure of 15–30 μ in a heated inlet system (125–325°C), a standard magnetic field scan is performed over the mass range 40–294 at 70 eV ionization voltage and 10–70 μA ionizing current. Hydrocarbon types are determined on the basis of summing the peak heights of characteristic mass numbers. The average number of carbon atoms per molecule is obtained from spectral data, and final results, in per cent by liquid volume, are calculated using calibration data. Calculations are somewhat complex, but once a proper scheme is set up digital computers may be used. Table 10-5 shows precision data obtained in a cooperative test of a number of laboratories. Here repeatability refers to variability associated with a single operator working on a particular instrument, while reproducibility is a measure of variability involving different laboratories.

Table 10-5   Mass Spectrometer Analysis of Tetramer Propylene Polymers; Results of Cooperative Test (9)

	Sample 1			Sample 2		
	Volume % (Mean)	Sr[a]	Sx[b]	Volume % (Mean)	Sr[a]	Sx[b]
C_nH_{2n+2}	0.3	0.0	1.0	1.3	0.8	2.0
C_nH_{2n}	96.4	0.2	1.7	85.3	0.6	2.5
C_nH_{2n-2}	2.8	0.1	1.1	11.6	0.3	1.4
C_nH_{2n-4}	0.02	0.0	0.04	0.8	0.1	0.2
C_nH_{2n-6}	0.5	0.3	0.3	1.0	0.1	0.7
C_nH_{2n-8}	0.0	0.0	0.0	0.0		
C_9	0.06	0.01	0.2	0.3	0.03	0.7
C_{10}	0.5	0.2	0.6	1.6	0.2	0.7
C_{11}	9.6	0.3	0.4	14.5	0.3	1.3
C_{12}	81.4	0.05	1.9	48.3	0.5	1.9
C_{13}	4.2	0.08	0.2	10.4	0.2	0.6
C_{14}	0.6	0.02	0.4	8.9	0.2	0.7
C_{15}	0.04	0.01	0.07	1.3	0.1	0.5
Number of laboratories	14	9	14	14	9	14
Number of analyses	15	18	15	15	18	15

[a] Sr = Estimate of the repeatability standard deviation.
[b] Sx = Estimate of the reproducibility standard deviation.

The application of total ionization measurements (Section 9-II) in compound type analysis (10) permits calibrations that can be employed more universally. Applications of high resolution mass spectrometry are discussed in Section 10-III-D.

E. Low-Voltage Mass Spectrometry

When the energy of the bombarding electrons is reduced, the amount of fragmentation decreases. This can be carried to the point that the available electron energy is just adequate to ionize the molecule (i.e., slightly above ionization potential), but not sufficient to cause fragmentation (i.e., below the lowest appearance potential). The technique was first utilized for compound type analysis by Field and Hastings (11), who found that at 6.9 V voltage between filament and ionization chamber, olefinic and aromatic hydrocarbons yielded measurable parent ions, while aliphatic paraffins did not show at all and cycloparaffins appeared only at very much reduced intensities.

The mass spectra of a mixture of 4-n-propylpyridine and 3-methylindole (12) are shown in Figure 10-2 at various electron energy levels. It is seen that at the 9 V level only the parent peaks appear, and a mixture analysis would be relatively simple. Ion intensities are, of course, also significantly reduced in low-voltage mass spectrometry. However, the spectra are simple and without interference, and also sensitivity may be increased by using higher ionization currents.

Correlation between molecular structures and low-voltage ion current intensities have been established for several classes of compounds by Crable and co-workers (10). The availability of high resolution instruments which make possible parent peak identification for several compound types has only added to the attractiveness of this technique in the analysis of complex mixtures. A new set of low-voltage calibration data for aromatic hydrocarbons has been published by Lumpkin and Aczel (13), who also studied the correlations between ion current magnitude and molecular structure at given ionizing energy levels.

F. Quantitative Analysis by Field Ionization

It was pointed out in the discussion of the field ionization source (Section 4-VI) that field ion spectra are much less complex than electron impact spectra. This is illustrated in Figure 10-3, which shows both types of spectra for n-heptane. The field ionization spectrum consists essentially of the parent peak, and the most intense fragment peak ($C_2H_5^+$) has an intensity of less than 1% (14). The simplicity of the spectra, together with good reproducibility (particularly with wire-type emitters) have suggested applications similar to those of low-voltage mass spectrometry. The difference between the two techniques lies in the fact that field ionization is a step function of the applied field strength, while ion intensity significantly decreases with electron energy. The most important consequence of this is that the probability of ionization is essentially the same for most organic compounds in field ionization.

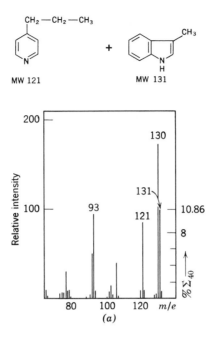

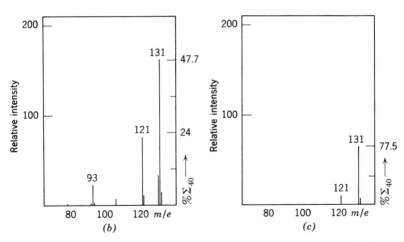

Figure 10-2. Mass spectra of a mixture of 4-*n*-propylpyridine and 3-methyl-indole (12). (*a*) with 70-V electrons; (*b*) with 12-V electrons (uncorrected), and (*c*) with 9-V electrons. For all three ordinates, *m/e* 131 in (*a*) = 100.

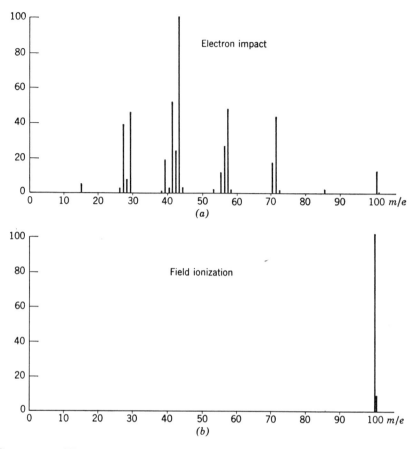

Figure 10-3. Electron impact and field ion mass spectra of *n*-heptane obtained by means of the combined ion source (one-wire field ion emitter) (14).

Field ionization analysis of hydrocarbon mixtures has been shown (14) to be a highly useful addition to gas–liquid chromatography, particularly in cases where the large number of isomeric components makes gas chromatographic identification difficult and time-consuming. In gasoline samples, where gas chromatography would show hundreds of isomeric compounds, field ionization reveals only a relatively small number of peaks since all isomeric compounds of a specific mass number appear as one peak.

Figure 10-4 shows the field ionization of an artifically made seven-component hydrocarbon mixture (14). Once a calibration is made with a known mixture, quantitative analysis can be made with essentially the same accuracy as provided by gas chromatography. Accuracy is, of course, much poorer when one deals with mixtures where no calibration is possible

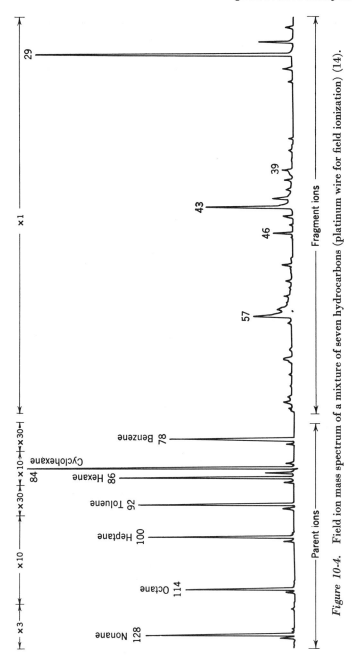

Figure 10-4. Field ion mass spectrum of a mixture of seven hydrocarbons (platinum wire for field ionization) (14).

(or available). Extension of the field ionization technique to analysis of heavy petroleum fractions, e.g., paraffin waxes in the 300–500° boiling point range, requires high resolution instruments. The basic technique and need for proper calibration are essentially the same as with low resolution instruments (15).

II. QUALITATIVE ANALYSIS

A. Identification Techniques

An organic substance is fully identified only when both its molecular formula and its structural formula are known. Mass spectrometric analysis provides information in both respects. Determination of the mass of the molecular ion leads to the molecular formula, and mass and abundance data on the fragment ions, combined with a good knowledge of the principles of fragmentation, permit the solution of the jigsaw puzzle of the molecular structure. For final proof, the spectrum of the unknown must be compared to the spectrum of one whose identify is known. If such a compound is not available it must be synthesized.

In addition to mass spectrometry, there are four other techniques available for the organic chemist in organic structure studies. The *infrared* spectrum originates in the interatomic vibrations of the molecule and is therefore characteristic of the whole molecule. A peak-by-peak correlation of an unknown with a reference is good evidence for identity. In addition, individual bands are associated with particular functional groups present in the molecule regardless of structure. Thus useful structural information may be obtained by study of characteristic group frequencies. The most interesting area of the infrared spectrum for the organic chemist is the 2.5–15 μ range (4000–660 cm⁻¹). *Ultraviolet* spectra originate from the excitation of the electronic energy levels associated with certain resonating atomic groupings (chromophores) in the molecule. An ultraviolet spectrum, similarly to infrared, consists of a plot of the wavelength (in mμ) or frequency (wave numbers) versus the absorption intensity (transmittance or absorbance). The ultraviolet portion of the electromagnetic spectrum extends from 100 to 380 mμ. While all organic compounds show characteristic absorption in the infrared, only certain types of compounds exhibit ultraviolet absorption. This selectivity, combined with high detection sensitivity, is very useful in structure determinations.

Nuclear magnetic resonance spectroscopy is based on the absorption of electromagnetic radiation in the radiofrequency region by the organic compound in the presence of a strong applied magnetic field. The frequencies at which nuclei in various molecules come to resonance are slightly different depending upon magnetic shielding effects of the electrons within the molecules and these "chemical shifts" can be used to characterize certain functional groups. In addition, the "spin coupling" of certain magnetic nuclei, separated only by a few chemical bonds, results in a hyperfine structure from which stereo chemical information may be obtained. Most studies in NMR spectroscopy have been limited to proton spectra, although ^{19}F, ^{13}C, ^{31}P, ^{11}B, and ^{27}Al may also be employed. *Gas–liquid chromatography* does not provide structural information, and its chief use is certainly in quantitative analysis. The combination of chromatographic separation with mass spectrometric analysis, however, has recently become an extremely powerful technique in structural elucidation technique. GC-MS combination methods are discussed later in this chapter.

Infrared, ultraviolet, nuclear magnetic resonance, and gas chromatography–mass spectrometry techniques should be used in a complementary fashion, since each furnishes vital information about the structure of the unknown. For texts dealing with the combined use of these techniques in the identification of organic compounds, the reader is referred to a text by Silverstein and Bassler (16).

B. Determination of Nominal Mass

Single-focusing mass spectrometers determine the m/e of an ion to the *nearest* mass unit for ions with masses as high as 1000. This section deals with identification and structure studies performed with mass spectrometers with resolution of 600–2500. Although the advent of high-resolution instruments opened up new avenues in qualitative analysis (Section 10-III), the principles of spectral interpretation are essentially the same in both cases.

The simplest method for mass determination is the *counting* of peaks. This technique can be used only when the spectrum contains a limited number of peaks, when reference masses can readily be found, and when peaks are relatively widely separated. Peak counting normally starts at m/e 12 ($^{12}C^+$) and peaks are successively identified at each mass unit. For magnetic and time-of-flight mass spectrometers the m/e scale is usually proportional to $(m/e)^{1/2}$, while cyclotron resonance and quadrupole instruments normally have linear scales.

At lower masses the m/e 17 and 18 peaks of water and the omnipresent 28 and 44 peaks may be used as "landmarks." When mercury diffusion pumps are employed, isotopes of both singly and doubly charged mercury ions may conveniently be used. For higher masses, calibration with reference compounds is the most straightforward way to obtain a mass scale. The use of a reference mass scale assumes that mass scanning and chart paper speed are highly reproducible. It is customary to construct a paper mass ruler from the known peaks of the reference compound, but day-to-day instrumental and temperature variations may cause errors unless corrections can be made conveniently (e.g., control of scanning speed). For masses up to about 140, n-octane may be used as a reference compound, while for higher masses perfluorokerosene is frequently used. The base peak of this compound is at m/e 69 (CF_3^+) but many peaks of useful abundance appear up to mass 700. Figure 10-5 shows a section of a perfluorokerosene spectrum in the mass 650 area. The $C_{14}F_{25}^+$ peak has an abundance of about 0.05% with respect to the base peak. In spite of its obvious limitations, peak counting with reference spectra is universally employed and is adequate for many purposes. Metastable ions are relatively easy to recognize by their widths, and multiply charged ions of

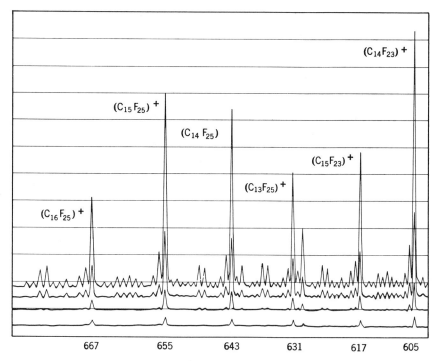

Figure 10-5. A section of the spectrum of perfluorokerosene (courtesy Perkin-Elmer Corp.). Magnetic scan, 1.5 kV constant acceleration voltage, secondary electron multiplier detector.

non-integral masses are normally of such low abundance that they do not impede the use of the mass scale.

According to the basic mass spectrometer equation, m/e is related to the radius, the magnetic field, and the acceleration voltage. Mass can be determined by measuring any of these parameters or a quantity proportional to them, while keeping the others constant. The radius is normally constant because the collector plate is fixed; thus mass may be measured either by determining magnetic field at constant electrostatic acceleration voltage or by measuring precisely the acceleration voltage at constant magnetic field.

Magnetic field can be measured by induction coils, by the field-dependent resistance of bismuth wire (17), or by nuclear magnetic resonance (18). The high precision with which magnetic fields can be measured, particularly with NMR, cannot be fully utilized due to field inhomogeneities and hysteresis effects. The accuracy in mass measurement with magnetic field determination is not better than $1:10^4$. A mass measuring technique widely used in low-resolution spectrometry is based on the precise measurement of the acceleration voltages, at constant magnetic field, at which the unknown and known masses of the reference compound appear. Acceleration voltages are determined with a high quality potentiometer (precision of 1 part in 10^6), and several measurements are made to increase accuracy. Masses can be measured to a precision of $1:10^5$.

Several commercial mass spectrometers feature *mass markers* of various designs. Mass markers are usually based on a Hall generator which uses the Hall effect to produce an output voltage proportional to the magnetic field strength. The Hall effect is a phenomenon involving the development of a voltage between the two edges of a current-carrying metal strip (or semiconductor) whose faces are perpendicular to a magnetic field. Other types of mass markers measure the magnetic field with a rotating coil dynamometer. When the coil is rotated in the magnetic air gap with a constant velocity, an ac voltage proportional to the magnetic flux density is obtained. The sample voltage is compared to a reference voltage and the difference voltage is amplified, properly shaped, and finally used by means of a programmed converter to display the m/e value. Mass scanning may be indicated on a special meter with a linear scale calibrated in terms of mass units, and there are optional markers that indicate every tenth mass number on the chart paper. The accuracy of such markers is usually not better than 0.5 atomic mass unit.

C. Identification of Molecular Ions

The first step in the analysis of an unknown spectrum is the recognition of the molecular peak. Assuming that there are no impurities present, and also that a molecular peak does exist in observable abundance, it will be a member of the group of peaks at the highest m/e in the spectrum. Due to the presence of isotopes (Section 9-VI) there will be peaks present one, two, or more units higher than the molecular mass (the one containing the most abundant isotopes). A general rule regarding the molecular peak states that compounds of even-numbered molecular weight must contain an even number of nitrogen atoms, or none at all, and, conversely, the presence of an odd number of nitrogens results in an odd-numbered molecular weight. This "nitrogen rule," which is valid for the majority of organic compounds, is based upon the observation that the most abundant stable isotope of most elements of even valency has an even mass number, while the corresponding isotope of those of odd valency is of odd mass. Nitrogen is a notable exception with the odd valency of the ^{14}N isotope. A corollary to this rule is that, assuming that all available nitrogen (if any) is part of a fragment, cleavage of a single bond gives an odd-numbered ion fragment from an even-numbered molecular ion, and vice versa.

Besides the nitrogen rule, there are several other tests to prove the identity of the all-important molecular peak. The decrease in intensity of the molecular peak is less than that of other fragment peaks when scanning is repeated at lower and lower ionizing electron energies, and thus molecular peaks may be distinguished from fragment peaks of an impurity of higher molecular weight. Sometimes other peaks in the spectrum may provide a clue; e.g., when two peaks are 3 mass units apart, they are likely to both be fragment peaks, resulting from the loss of water and a methyl group from an alcohol. A quick, superficial analysis of the entire spectrum is very helpful to ascertain that the peak believed to be the molecular ion is compatible with the rest of the spectrum.

Ion–molecule collisions (Section 9-IX) may result in $M + 1$ peaks, which interfere with molecular peak identification. These peaks have intensities proportional to the square of the sample pressure and may also be recognized by changing their residence time in the ion source by varying the repeller voltage. These $M + 1$ peaks of protonated molecules (not to be confused with the $M + 1$ peaks originating from heavy isotopes) may be extremely useful for molecular weight determination when the parent peak is too weak for detection. McLafferty determined molecular weights of nitriles, esters, etc. by this technique (19). It is indeed fortunate that many compounds with a low intensity molecular peak exhibit large $M + 1$ peaks.

When the molecular weight cannot be determined, because neither the parent peak nor the $M + 1$ peaks are visible, other techniques must be resorted to for molecular weight determination. Often, however, careful evaluation of the fragment ions furnish considerable information as to the type of compound involved; e.g., primary alcohols exhibit a weak parent peak but a strong $M - 18$ peak.

Other techniques for molecular weight determination include low-energy ion bombardment (Section 10-I-E), field ionization (10-II-F), micro-effusiometry, and chemical pretreatment. When the sample is introduced from the sample reservoir into the ion source through a molecular leak (Section 6-II), the *rate of effusion* of a compound is inversely proportional to the square root of its molecular weight. On this basis (Graham's law) molecular weight may be determined by comparing the peak height decay per unit time of a peak (any peak) from the unknown to that of a reference compound introduced simultaneously. This method is often useful in cases where no parent peak is visible. It may also be used for mixture analysis: the decay rate of several peaks in the spectrum is measured, and when the logarithms of peak intensities are plotted against time, sets of parallel lines are obtained with each set (peaks from a particular compound) having a different slope. Microeffusiometry has its limitations, but it is very simple and results are often surprisingly accurate (20,21).

Compounds with no molecular ion peak may often be *chemically pre-treated* to obtain a derivative with a more pronounced parent peak. This may be important, for example, when the compound of interest is being used for reaction mechanism studies by stable isotope incorporation (Section 12-IV). For example, a simple oxidation of secondary alcohols to ketones makes it possible to greatly enhance the molecular ion abundance; the molecular weight becomes two mass units lower. Chemical pretreat-ment, incidentally, may also be used to increase sample volatility by re-moving such polar groups as hydroxyl, carboxyl, amido, etc., to degrade large molecules into smaller ones more suitable for mass spectral analysis (pyrolysis, partial hydrolysis), or to convert the molecule into a derivative whose spectra may provide information on fragmentation, etc. Many examples are cited in books on organic mass spectrometry.

D. Spectra and Structure

In the latest available compilation of mass spectral data, published in the fall of 1966 (22), cracking patterns for some 5000 compounds are listed and cross-indexed by reference number, molecular weight, molecular formula, and fragment ion values, including the molecular ion. Such an enormous amount of experimental information on individual compounds provides a natural starting point in the search for common features in the mass spectra of different compound types, and many papers have been published during the last 10 years attempting to empirically relate spectra to structure. On the basis of these studies, several general rules for predicting prominent peaks in a spectrum have been established. These correlations are discussed at great length in many textbooks on organic mass spectrometry and will not be detailed here.

Silverstein and Bassler (16) list nine basic rules for fragmentation. These rules are arrived at by combining experimental evidence with chemical rationalization, using such concepts as resonance, hyperconjugation, polarizability, and inductive and steric effects. The list that follows is reproduced from the reference given:

1. The relative height of the parent peak is greatest for the straight-chain compound and decreases as the degree of branching increases.

2. The relative height of the parent peak decreases with molecular weight in a homologous series.

3. Cleavage is favored at branched carbon atoms; the more branched, the more likely is cleavage. This is a consequence of the increased stability of a tertiary carbonium ion over a secondary, which is in turn more stable than a primary. Generally, the largest substituent at a branch is eliminated most readily as a radical, presumably because a long-chain radical can achieve some stability by delocalization of the lone electron.

4. Double bonds, cyclic structures, and especially aromatic (or heteroaromatic) rings stabilize the parent ion and thus increase the probability of its appearance.

5. Double bonds favor allylic cleavage and give the resonance-stabilized allylic carbonium ion,

$$CH_2 \overset{+}{\cdot} : CH \overset{\frown}{-} CH_2 - R \xrightarrow{-R \cdot} \overset{+}{CH_2} \overset{\frown}{-} CH = CH_2 \rightleftharpoons CH_2 = CH \overset{\frown}{-} \overset{+}{CH_2}$$

6. Saturated rings tend to lose side chains at the α-bond. This is merely a special case of branching (Rule 3). The positive charge tends to stay with the ring fragment,

$$Ⓢ - \overset{\cdot+}{CH_2} - R \rightarrow \overset{+}{Ⓢ} + \cdot H_2C - R \qquad Ⓢ = \text{saturated ring}$$

7. In alkyl-substituted aromatic compounds, cleavage is very probable at the bond beta to the ring, giving the resonance-stabilized benzyl ion or, more likely, the tropylium ion directly,

8. C—C bonds to a heteroatom are frequently cleaved, leaving the charge on the fragment containing the heteroatom whose nonbinding electrons provide resonance stabilization

9. Cleavage is often associated with elimination of small stable neutral molecules such as carbon monoxide, olefins, water, ammonia, hydrogen sulfide, hydrogen cyanide, mercaptans, or alcohols. These cleavages often take place with rearrangement (Section 9-VII).

Regarding the notation employed, it is recalled from Section 9-VII that the regular arrow denotes a two-electron shift, while the "fishhook" refers to a one-electron movement. In homolytic bond fissions, only one fishhook is normally used for brevity. The term "α-cleavage" denotes a fission of a bond originating at an atom which is adjacent to the one assumed to bear the charge. When the site of the positive charge cannot be properly identified, the entire molecular structure is enclosed in brackets with the positive charge outside. This method of notation, advanced by Shannon (23), enables one to make a distinction between odd- and even-electron species. For example, the loss of a methyl radical from a molecular ion is shown as $[RCH_3]^{+\cdot} \longrightarrow [R]^+ + CH_3{}^\cdot$. The *dot* indicates a radical, the symbol $^{+\cdot}$ refers to a radical ion. Ionization normally involves the removal of an electron; thus the ions formed, either molecular or fragment, will possess an unpaired electron. Such ions are called "odd-electron" ions. "Even-electron" ions, in which the outer shell electrons are fully paired, are symbolized by $^+$.

It is emphasized again that the rules of fragmentation should serve only as guidelines, and there are many exceptions. For example, Rule 2 apparently does not apply to fatty esters. Nevertheless, the rules may be applied to the various chemical classes of organic compounds with the objective to arrive at clues which will assist first in the identification of the general class into which the unknown compound belongs, and then in the elucidation of its structure. Several hundred papers have appeared during the last decade on the subject of mass spectral correlations. Only a few examples are mentioned below.

A considerable amount of work has been done on *hydrocarbons* of interest to the petroleum industry. Rules 1 and 3 apply generally and Rule 6 is also evidently observed. In saturated aliphatic hydrocarbons a homologous series of pairs of peaks, corresponding to C_nH_{2n-1} and C_nH_{2n+1}, is observed with maximum abundance around four carbon atoms. When the series of peaks is more abundant than its neighbors at $n + 1$ and $n - 1$, chain branching is indicated. Major peaks are 14 units apart (CH_2), and peaks at m/e 43 ($C_3H_7^+$) and 57 ($C_4H_9^+$) are usually large. Olefinic hydrocarbons show similar patterns to saturated hydrocarbons, but peaks are shifted two mass units lower. Rule 5 is obeyed, especially by polycyclic compounds.

Alcohols are characterized by parent peaks of low intensity. Since Rule 8 applies, primary alcohols usually exhibit a rather strong peak at m/e 31 (CH_2OH^+), while secondary alcohols show strong peaks at m/e 45 (CH_3-$CHOH^+$), 59 ((CH_3)$_2COH^+$), etc. Rule 3 is obeyed. A prominent peak is observed at $P-18$ (loss of water), especially with primary alcohols. *Ethers* are isomeric with alcohols containing the same degree of unsaturation. The presence of oxygen is seen from strong peaks at m/e 31, 45, 49, etc., representing RO^+ and $ROCH_2^+$ fragments. The parent peak of aliphatic ethers is usually weak. The two main fragmentation pathways of ethers are cleavage of the C—C bond next to the oxygen atom, and a C—O bond cleavage with the charge remaining on the alkyl fragment. Aromatic ethers have strong parent peaks (Rule 4). The parent peak of most *ketones* (aliphatic, cyclic, aromatic) is usually quite prominent. In aliphatic ketones, similarly to alcohols and ethers, many fragmentation peaks result from the cleavage at the C—C bonds adjacent to the oxygen atom, the charge remaining with the oxygenated fragment. Peaks at m/e 43 (CH_3CO^+) and 71 ($C_3H_7O^+$) are typical examples. Cyclic ketones have their base peak at m/e 55 (CH_2-$CHCO^+$), while the most intense peak in the spectra of aromatic aralkyl ketones is usually a characteristic $ArC{\equiv}O^+$ fragment.

Other classes of organic compounds, including aldehydes, carboxylic acids, carboxylic esters, amines, nitro compounds, sulfur compounds, and halogen-containing compounds, have been thoroughly investigated and adequately described. *Sulfur*-containing compounds can often be recognized by the contribution of the ^{34}S isotope to the $P+2$ peak. In aliphatic mercaptans the parent peak is normally intense enough to permit the measurement of the $P + 2$ peak. The modes of cleavage resemble those of alcohols. For example, in primary mercaptans pronounced peaks appear at $P-34$ (splitting of H_2S), while a peak at $P-33$ (loss of HS) is usually present in secondary mercaptans. Peaks at mass 33 and 34 ($^{32}SH^+$ and $^{32}SH_2$) are also common. The fragmentation of aliphatic sulfides is similar to that of ethers. It is noted, in passing, that halogen-containing compounds are usually easy to identify on the basis of isotopic peaks (Section 9-VI). In aliphatic chlorides, for example, the P, $P + 2$, and $P + 4$ peaks have characteristic ratios depending on the number of chlorine atoms, and peaks at $P - 35$ and $P - 36$ are also frequently present (loss of Cl and HCl). There is a similar situation in bromine compounds, but, of course, the bromine isotopes of 79 and 81 have a 1:1 ratio. Iodides are often characterized by the m/e 127 peak, and by the large intervals between prominent peaks in the spectrum. A strong peak at m/e 19 indicates fluorine, and prominent peaks are also present at m/e 69 (CF_3^+), 119 ($C_2F_5^+$), etc. in perfluorinated saturated hydrocarbons.

E. Interpretation Procedures

No hard and fast rules can be given for the interpretation of the spectrum of an unknown, and a good knowledge of organic chemistry is just as important as familiarity with the "mass spectrometry" of the analytical procedure, if not more so. Authors of the various books on organic mass spectrometry have their favored step-by-step outlines for the procedure of interpretation, each following a logical argument. These outlines should only serve as guidelines since practically every unknown requires a different approach. Moreover, depending upon the experience of the mass spectroscopist with the particular type of compound the unknown belongs to, certain steps in the interpretation may be omitted as features become obvious in a superficial evaluation.

A rather general procedure for interpretation is suggested by McLafferty in his recent book on the interpretation of mass spectra (24). It is immediately seen from Table 10-6 that certain steps are quite mechanistic and can be performed at a glance by the experienced spectroscopist, while

Table 10-6   Standard Interpretation Procedure (From McLafferty (24))

---

1. Study all available information (spectroscopic, chemical, sample history). Give explicit direction for obtaining spectrum.
2. Verify masses; determine elemental compositions, rings plus double bonds.
3. Mark abundant odd-electron ions. Test molecular ion identity.
4. Study general appearance of spectrum; molecular stability, labile bonds.
5. Identify all low mass ion series.
6. Identify the neutral fragments accompanying high mass ion formation (including "metastables").
7. Postulate structures for abundant ions.
8. Postulate molecular structures; test against reference spectrum, against spectra of similar compounds, or against spectra predicted from mechanisms of ion decompositions.

---

others require much thought and even the development of new mechanisms and concepts.

The first step requires little explanation. Every little bit of information on the unknown helps, and it is indeed surprising how much can be learned about unknown samples just by simply interrogating the person who submits a "completely" unknown material for analysis. The analytical procedure for obtaining the spectrum is dictated by such parameters as sample volatility, quantity of sample available, expected purity, etc. Sometimes, of course, the result of the analysis provides the information on how the analysis should have been made.

The next two steps consist mainly of mass determinations and verifications. The importance of the positive identification of the molecular peak cannot be overemphasized since it leads to the molecular weight of the compound. Together with the parent peak the neighboring isotope peaks should also be identified and their relative abundances measured. Empirical formulas in low-resolution mass spectrometry are frequently determined by utilizing Beynon's tables (25) which enable one to select a number of likely empirical formulas from the knowledge of the parent peak and the abundances of the $P + 1$ and $P + 2$ peaks. Let us assume, for example, that an unknown sample has a parent peak at m/e 158, and the abundances of the $P + 1$ and $P + 2$ peaks are 7.9% and 1.0%, respectively. From Beynon's tables we may select the likely empirical formulas within an arbitrary abundance range of 7.0–9.0 for the $P + 1$ peak.

Formula	$P + 1$	$P + 2$
$C_5H_{10}N_4O_2$	7.17	0.63
$C_6H_8NO_4$	7.15	1.02
$C_6H_{10}N_2O_3$	7.52	0.85
$C_6H_{12}N_3O_2$	7.90	0.68
$C_6H_{14}N_4O$	8.27	0.50
$C_7H_{10}O_4$	7.88	1.07
$C_7H_{12}NO_3$	8.26	0.90
$C_7H_{14}N_2O_2$	8.63	0.73
$C_7H_{16}N_3O$	9.00	0.56

Several of these formulas can immediately be eliminated on the basis of the nitrogen rule, and the $P + 2$ peak appears to indicate that $C_7H_{10}O_4$ fits best, but the rest of the spectrum must also be studied for additional evidence.

After the molecular formula has been found, the number of rings and/or double bonds may be evaluated (24). These calculations are based upon valence equivalency for nitrogen, halogens, silicon, etc. with respect to a general formula $I_yII_nIII_zIV_x$, where I = H, F, Cl, Br, I; II = O, S, III = N, P; and IV = C, Si, etc. The number of total rings plus double bonds for odd electron ions is obtained from the formula: $x - \frac{1}{2}y + \frac{1}{2}z + 1$.

The general appearance of the spectrum often provides several clues, since classes of compounds have characteristic patterns. Construction of a bar diagram (Section 9-I) is a useful tool to obtain an overall impression of the spectrum. The abundance of the molecular ion with respect to that of fragments indicates relative stability of the molecule; aromatic compounds often have large parent peaks while high branching strongly reduces parent peak abundance. The presence of large doubly charged peaks indicates stable parent molecules, usually due to π-electron systems or organometallic compounds. It has been repeatedly pointed out by experienced mass spectroscopists that the higher the mass of a peak in a spectrum the more significant it is in providing clues to the structure, even if its abundance is relatively low with respect to other peaks at the low mass end. Large peaks at the low end should not, of course, be neglected, as their structural implications are very useful in the evaluation of the higher mass peaks. Metastable ions are usually easy to detect and they provide important clues to possible ionic degradation mechanisms.

An important step in structure evaluation, one that is often neglected in the hurry to establish a probable structure for the unknown, is the postulation of ion structures for the abundant ions. A compilation which is extremely useful for this step has been published by McLafferty (26), it lists possible elemental compositions and structures for many observed peaks, and also relative probabilities for the postulations. These numbers were determined by studying thousands of compounds, in order to establish the number of times that an ion of particular structural significance appears as a large peak. It is expected that this book of mass spectral correlations will be enlarged as more information becomes available on larger molecules, but even in its present form it is an indispensable tool in structure studies.

The final step, the postulation of the molecular structure, requires the synthesis of all the information obtained in the course of the study. All possible structures must be listed, and all must be evaluated critically against experimental evidence. Ultimate proof, as mentioned already, requires comparison to known compounds. When reference spectra are not available, the suspected compound should be synthesized, though this suggestion may, on occasion, involve a difficult task. Sometimes the reference spectrum of a compound with a structure similar to that of the unknown may be available, and in such cases Biemann's "shift technique" (27) may be employed: A functional group is added to the molecule in such a fashion that the molecular weight of a particular fragment is increased without changing the general pattern of the spectrum. The technique has been particularly successful in the study of complex indole alkaloids, such as ibogamine, ibogaine, and ibogaline.

The increased rate of production of mass spectral data, particularly with gas chromatograph–mass spectrometer combination systems (Section V in this chapter) necessitated the development of computer-based techniques for both the acquisition and processing of mass spectral data.

The basic principles of *data acquisition* systems are the same for both low and high resolution instruments with electrical detection. The analog output from the electrometer or electron multiplier must be converted to digital form. In one technique, the analog-to-digital conversion is performed directly at the mass spectrometer, the digital numbers are recorded on magnetic tape, and the tape is transferred to the computer for processing. A somewhat more economical approach is to record the electrical signal on FM analog tape which is then digitized and processed at the computer. The intermediate recording on magnetic tape can be eliminated by directly connecting a process control computer to the mass spectrometer. In low resolution mass spectrometry it is adequate to determine only the nominal mass of the ions, and this can be accomplished by determining the time at which particular peaks appear in the course of scanning of the spectrum.

From the measured times the mass is determined by comparison with a time scale obtained in an earlier run using a known compound. The reproducibility of the recording system is usually adequate for this purpose, but corrections, if necessary, may also be included in the general programming of the computer. In high resolution mass spectrometry, of course, accurate mass measurement is desired, and reference peaks must be provided (Section IV in this chapter).

In low resolution mass spectrometry the identification of an organic compound is generally accomplished by a comparison technique, namely, by comparing the cracking pattern of the unknown with that of a reference compound. The need for computerized *data processing* is obvious, particularly with the recent explosion of available data.   First of all, data compilations such as the one by Cornu and Massot (22) must be stored by the computer. Name, molecular weight, mass value, relative intensity, origin and serial number of the spectrum, type of instrument employed, electron energy used, temperature of the ion source, and temperature of inlet system are important characteristics which must be supplied. The process of identification of an unknown compound is performed by using various "keys" which are preselected on the basis of the information desired and the complexity of the program. Next, the computer searches the entire library and extracts those spectra where the "disagreement index" is within the limits specified.

Computer-based data acquisition and data processing of low resolution mass spectra is still in its infancy, but publications describing new and improved techniques are appearing at an increased rate. Several recent papers are listed in reference 28.

F. Field Ionization Spectra and Structure

There are three kinds of peaks in field ionization spectra (14). Parent peaks, obtained at relatively low field strengths, are sharp and symmetrical (Fig. 10-4). Diffuse peaks at non-integral masses represent metastable decompositions, and are, similarly to electron impact spectra, always of low intensity. At higher field strengths fragment peaks appear which are unsymmetrically broadened in the direction of lower mass numbers. These peaks result, according to Beckey et al. (14) from "field dissociation," and their intensity strongly depends upon the field strength employed.

The possible application of field ionization in structure determinations requires the empirical study of the spectra of the various compound classes, similar to the study of electron impact spectra. Many such studies are currently in progress, and a number of general rules have already been formulated. It has been shown, for example, that for paraffins the dissociation of C—H bonds occurs at much lower intensities at low and medium field strengths than dissociation of C—C bonds; at high field intensities both bonds dissociate at about the same intensity level. Another generalization states that the rings of aromatic and heterocyclic compounds are never field dissociated; however, they exhibit doubly ionized peaks with relative abundances as high as 1%. The number of papers on spectral correlations in field ionization is rapidly increasing with the availability of commercial ion sources. The electron impact–field ionization combination sources certainly provide a powerful means for structure studies, particularly for compounds where the detection of parent ion is difficult by electron impact.

III. HIGH-RESOLUTION MASS SPECTROMETRY

The application of high-resolution mass spectrometry in qualitative organic analysis is based upon the accurate determination of the empirical formulas of molecular and fragment ions. Thus, once high resolution is made possible by the geometry of the instrument, techniques must be developed for accurate mass measurement. It has already been pointed out several times that atomic weights differ slightly from integral numbers. These *mass excesses* should not be confused with the *mass defects*; the latter represent differences between the weights of the atoms and the sum of the weights of the constituent subatomic particles (Section 13-I-A). The mass excess of hydrogen, for example, is $+7.82$ mmu in the ^{12}C system, while the same for ^{79}Br is -81.65 mmu.

In low-resolution mass spectrometry, masses are measured to the nearest $1/2$ mass unit by peak counting. In modern high-resolution instruments mass measurements can be carried out with a precision of at least 10 ppm; 2–3 ppm can readily be obtained. The first part of this section deals with techniques of mass measurement. This is followed by applications with special emphasis on combined gas chromatography–mass spectrometry techniques.

A. Determination of Mass

1. Photographic Plates

The mass of an unknown species is accurately determined by measuring the position of the lines on the photoplate with respect to two other lines of known masses. This "bracketing method" has been known since the early days of mass spectrography. Distances are measured with a horizontally moving vernier attachment on the densitometer; positions can be determined to within $\pm 1\ \mu$ or better. The mass of an unknown line M_x, is calculated in a Mattauch-Herzog geometry instrument from (Fig. 10-6)

$$M_x = \left[\frac{l_x}{l_0} (\sqrt{M_2} - \sqrt{M_1}) + \sqrt{M_1} \right]^2 \qquad (10\text{-}5)$$

where l_x is the distance between a known mass line M_1 and the unknown, and l_0 is the distance between two known mass lines, M_1 and M_2. Accuracy of mass measurement depends on how close the known lines are located to the unknown one. In extremely precise atomic mass determinations the reference lines are selected to form a close doublet with the unknown. In most organic applications reference lines are selected to be within 5% of the unknown mass.

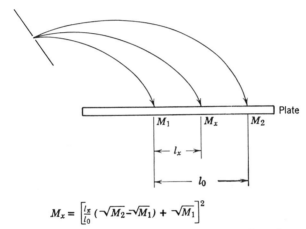

$$M_x = \left[\frac{l_x}{l_0} (\sqrt{M_2} - \sqrt{M_1}) + \sqrt{M_1} \right]^2$$

Figure 10-6.   Mass determination on photoplates.

In spark source spectrography there are usually a number of known lines (matrix, isotopes, etc.) that can serve as references. In organic mass spectrography, perfluorokerosene or some other fluorocarbon is used as the reference compound. Fluorocarbons have adequate vapor pressure and produce a large number of ions (both positive and negative) with which the whole mass scale can be covered. In addition, since fluorine is mass deficient (isotope mass: 18.998402), reference lines can readily be resolved if they occur in doublets or multiplets, e.g.,

CF_3	68.9952
C_3HO_2	68.9976
C_4H_5O	69.0339
C_5H_9	69.0714

Since there are usually several hundred peaks to be measured—perfluoro-kerosene itself gives some 100 reference lines—line position measurement and mass assignment on photoplates is best automated and computerized. Mass identification and line intensity measurement, as discussed shortly, are performed simultaneously. Accuracy of mass measurements on photoplates is of the order of 0.001 mass unit in routine determinations.

2. Acceleration Voltage Measurement

In instruments with electrical detection, mass measurements may be made by accurately determining the acceleration voltage at constant magnetic field. The acceleration voltage is varied, and the deflection voltages, which are proportional to the acceleration voltages, are measured with a precision potentiometer for the unknown peaks and for reference peaks. The unknown mass is determined from the resistance ratio.

3. Peak-Matching Technique

A simple, elegant, and very accurate method of mass measurement has been developed by Nier and his co-workers (29) on the basis of Smith's technique (30) employed with his mass synchrometer. This technique is now more or less standard in high-resolution organic mass spectrometry and all commercial instruments feature the required components. The method is based on a theorem according to which an ion of mass km may be forced to travel the same path in an instrument as mass m (at constant magnetic field) when the electric fields are changed by a factor of $1/k$. Thus, the mass difference in a doublet can be determined by measuring the $\Delta V/V$ necessary to cause the two ions to travel alternately on the same orbit.

Figure 10-7 shows a schematic of Nier's peak-matching circuit (31,32). A small coil is placed in the magnet gap and the fixed magnetic field is modulated at 30 cycles/sec by a sawtooth wave form applied on the auxiliary coil; the sawtooth voltage is in phase with the horizontal sweep of an oscilloscope. The Y-plates of the scope are connected to the amplified output of the ion detector. Multipliers must be used to provide great sensitivity and fast response. By restricting the amplitude of the sawtooth voltage, only a single peak appears on the screen. A pulse synchronized with the flyback of the sawtooth actuates three contact relays. Two relays cause the acceleration voltage and the deflection voltage on the electrostatic analyzer, respectively, to switch between two preselected values. The third relay properly adjusts the gain of the amplifier to assure that both peaks appear with approximately the same intensity. In this way

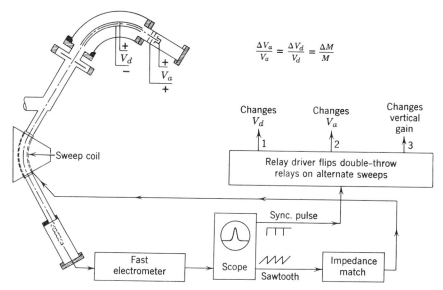

$$\frac{\Delta V_a}{V_a} = \frac{\Delta V_d}{V_d} = \frac{\Delta M}{M}$$

Figure 10-7. Schematic drawing showing the method of determining the width of a mass doublet by peak-matching method (31).

one member of the doublet appears on odd sweeps, while the other appears on even sweeps, and on a long-persistence oscilloscope both members of the doublet can be seen simultaneously. The voltage on the deflection plates (V_d) is adjusted until peak coincidence is obtained. Acceleration voltage (V_a) is changed automatically to provide ions of proper energy. The required deflection voltages are obtained from a circuit shown in Figure 10-8,

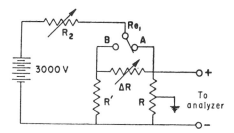

Figure 10-8. Circuit employed for producing precise change in voltage across deflection plates of electrostatic analyzer (31). The source of potential consists of seventy 45-V heavy duty radio "B" batteries connected in series and carefully insulated from one another and ground. ΔR consists of a 100 Ω 10-turn helical potentiometer in series with a decade resistance box having 10, 100, 1,000, and 10,000 ohm steps. R has a value near 3 MΩ. As a result of careful calibration $\Delta R/R$ is known to an accuracy better than 1/50,000 for most measurements made.

while a similar circuit is used to switch the accelerating voltage. The required $\Delta V/V$ is obtained when $R = R'$, and at this point

$$\Delta M/M = \Delta R/R \tag{10-6}$$

where ΔM is the mass difference to be measured and M is the lighter mass in the doublet. In Figure 10-8, Re_1 is the relay that is switched between A and B synchronously with the sweep frequency. ΔR is a calibrated decade resistance with which the peaks are made to coincide. With R_2 fine adjustments can be made when the lighter mass is brought into focus.

Figure 10-9 shows how the two members of a doublet are superimposed. The first and second traces show the doublet when the relay is in positions A and B, respectively. The third trace shows the superimposed peaks when the relay is switching back and forth.

The accuracy of voltage ratio measurement in the peak-matching technique is a few parts per million. The error in mass measurement may be expressed by

$$\text{ppm} = \frac{\text{True value} - \text{Measured value}}{\text{True value}} \times 10^6 \tag{10-7}$$

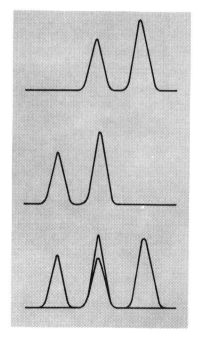

Figure 10-9. Composite photographs of doublets showing how superposition of top and middle figures result in lower figure from which accurate mass measurements may be made (31).

Millimass units (mmu) = (True value − Measured value) $\times 10^3$ (10-8)

Conversion from ppm to mmu units:

$$ppm = mmu \times 10^3/\text{True value} \tag{10-9}$$

The presence of systematic errors can be checked by measuring known doublets and plotting the observed ratio versus the error in ppm. When correction is needed, the multiplying factor is obtained from a graph that must be determined for every instrument. It is stressed again that the accuracy of mass measurement greatly depends upon the selection of the reference mass. Commercial peak ratio attachments usually permit a ratio as high as 2 between the masses of the unknown and the reference, but for accurate results the reference should be within 5% (better still, 1%) of the unknown.

As in photoplate detection, perfluorokerosene is one of the most frequently used reference compounds. For very high molecular weight studies, a compound of formula $C_{72}H_{24}O_8F_{128}N_4P_4$ (molecular weight 3628) has recently been suggested by Fales (33) as a reference compound. This is, incidentally, the heaviest molecule ever studied by mass spectrometry. A section of the spectrum is shown in Figure 10-10. A technique for the controlled introduction of the reference compound is referred to in Section 7-VI-A.

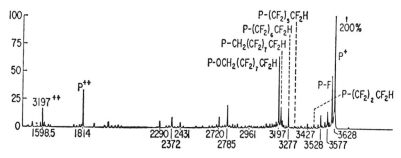

Figure 10-10. High mass end of the spectrum of formula $C_{72}H_{24}O_8F_{128}N_4P_4$ with molecular weight of 3628 (33).

B. Use of Variable Resolving Power

It is usually not necessary to determine the precise mass of every peak in the spectra. An important feature of commercial high-resolution mass spectrometers is the provision for fast and simple changing from high to low resolving power by adjusting the source and collector slits by means of calibrated micrometer controls located outside the vacuum system.

Since the time required to scan through a spectrum is proportional to the resolving power used, it is a waste of time to use resolution greater than necessary. It is best to first scan the whole spectrum rapidly at a resolving power of, say, 1000, and then rescan important portions of the spectrum at high resolution. A disadvantage of this technique is that important doublets may be overlooked. Sensitivity also decreases at high resolution (narrow slits), and in the case of exceptionally small samples resolution may have to be sacrificed in order to detect the sample at all.

Low- and high-resolution scans of a cholesterol sample are shown in Figure 10-11 (34). The low-resolution scan shows that there is an impurity at $P + 14$, but it took a high-resolution scan to reveal that the peak was actually a triplet. The identification of the impurities in this sample is discussed later in the section on applications.

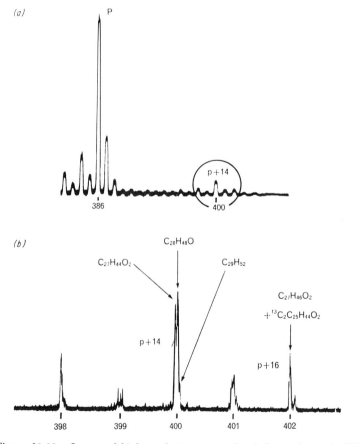

Figure 10-11.   Low- and high-resolution scans of a cholesterol sample (34).

C. Data Handling

1. Element Maps

In the high-resolution mass spectra of complex organic compounds there are hundreds of lines (photographic detection) or peaks (electrical detection). To fully utilize all the information available it would be desirable to measure the exact mass of every fragment, including those with low abundances. This is a formidable task, and until recently analysis usually consisted of low-resolution analysis combined with precise mass measurements on selected portions of the spectra. In addition to solving the problems of *data reduction*, there has been a need to develop new techniques for *data presentation*. A large number of data must be presented in a concise and clear form, preferably in a way that is directly usable by the organic chemist. Biemann and his co-workers developed a new, and indeed revolutionary, technique for the presentation of high-resolution organic spectra (35,36). In the so-called *element mapping* technique, the exact mass of most ions in the spectra is determined and all *possible* elemental compositions are calculated, and the spectrum is interpreted on the basis of the *elemental formulas* of the molecular fragments. The technique has been spectacularly successful during the last few years.

The first step is to determine the mass and approximate intensity of all ions in the spectrum. Biemann's method was originally developed for photographic detection using a Mattauch-Herzog type double-focusing mass spectrograph, but the element mapping technique is equally valuable with electrical detection. Semiautomatic and automatic techniques are now being developed for initial data handling and these will be briefly discussed in the following section. Presently it is adequate to say that the masses and approximate intensities of all fragments are first determined and fed in some way into a computer. The elemental formulas are next sorted by the computer into columns according to heteroatom content; this is, indeed, the heart of the technique.

An element map of deoxydihydro-N_b-methylajmaline (36) is shown in Figure 10-12. The first column shows nominal m/e masses. The second column represents all ions containing only carbon and hydrogen; the first number 7/11 means the formula $C_7H_{11}^+$. The next number indicates the difference of found and calculated mass, which in this case is zero mmu. The relative abundance of this ion is indicated by the number of asterisks that follow. Abundance is measured on a logarithmic scale—asterisks number from 1 to 10—so that even small peaks are characterized. All the ions in the first column represent fragments consisting of only carbon and

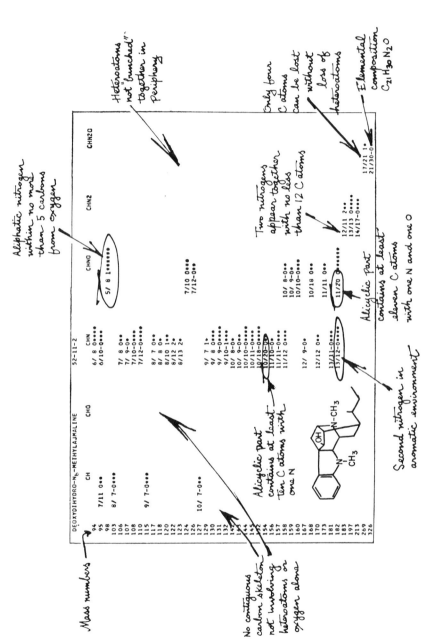

Figure 10-12.   Element map of deoxydihydro-N_b-methylajmaline (36).

hydrogen, in this case $C_8H_7^+$, $C_9H_7^+$, and $C_{10}H_7^+$. The heteroatom content increases from left to right: the third column shows that fragments containing only one oxygen do not occur in measurable quantity. The fourth column indicates many peaks of the general formula C_xH_yN (only one nitrogen). It follows that the molecular ion, if present, should appear as the entry furthest down and to the extreme right. In the present case the formula is $C_{21}H_{30}N_2O$.

The fact that there is a fragment of $C_{17}H_{21}N_2O$ present suggests that a side chain of a C_4H_9 group must be present. The peaks in the CHN2 column show the loss of oxygen and seven, eight, or nine carbon atoms. The base peak is $C_{11}H_{20}NO$. Biemann continues to explain the various fragments and employs both empirical relations and rationalizations based on common organic chemical concepts to justify the proposed final structure. At this point mass spectrometry is left and the experience and ingenuity of the organic chemist is needed to arrive at a final structural formula.

Many other features of the spectra are revealed by the element mapping technique. For example, multiplets appear as two entries at the same nominal mass. Since all peaks are measured, fragmentation patterns not indicated in low resolution readily appear. Hundreds of compounds have already been studied by Biemann's technique and improvements are now reported almost daily on both the experimental part and on shortcuts in interpretation. The method may be considered as a systems approach and utilizes both direct conclusions based on precise mass measurements and mechanistic considerations based on accumulated information and the application of theoretical concepts. In summarizing, the main features of the element mapping technique are the utilization of all information available in the spectra by accurately measuring the masses of all ions, by approximately measuring their relative intensities, and the ordered presentation of the data according to elemental formulas of the fragments. A topographic element map as a display for high-resolution mass spectra has been proposed by Venkataraghavan and McLafferty (37). Data are automatically plotted out in three dimensions with the number of carbon atoms or the number of hydrogens as subdivisions on the x axis, heteroatom content on the y axis, and relative abundance on the z axis.

2. Automatic Data Acquisition Systems

The relative merits of photographic and electric detection in high-resolution mass spectrometry are discussed in connection with gas chromatographic eluent analysis in the next section of this chapter. Automatic data acquisition systems are now being developed for both types of detection. Several systems have been reported during the last few years; they differ mainly in the sophistication of automation. Due to the extremely rapid current progress in both instrumentation and programming techniques, only some basic principles are reviewed here. The reader is, however, urged to follow developments in this field.

On photographic plates the mass of ions is obtainable from the positions of the various lines, while ion abundance is represented by blackness. An automatic plate reader must, therefore, first determine the peak shape, then locate its center and measure its exact location. At the center of the line blackness must also be measured. Plates are usually moved by means of a precision drive screw and a synchronous motor. Blackness is determined by a photomultiplier, the signal of which may be transferred through a high-speed digital voltmeter to a storage register. The information from the automatic plate reader flows into a digital magnetic tape, and from there into a digital computer, according to Scheme 1. The computer receives data for both the unknown and reference lines, computes exact masses and empirical formulas, and prints out partial or complete element maps. Details of a rather elaborate version of such a system have recently been published by McLafferty and his co-workers (38). This paper also discusses computer techniques for improving mass-measuring accuracy and resolution.

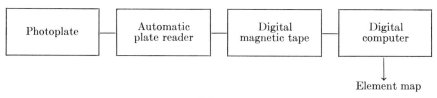

Scheme 1

When electrical scanning is employed, a wide band amplification system permits the use of fast scanning even at high resolving power. The analog signal is recorded on an analog tape recorder from which it can be transferred via an analog-to-digital converter into a digital computer which calculates masses and prints out element maps (Scheme 2). An alternative technique might digitize the analog signal of the detector and record it on digital magnetic tape, from which it may be transferred automatically or manually to a digital computer. Automatic handling of data from electrical detection appears at the present time somewhat more involved than that based on photoplates. There is, however, considerable development work in progress to improve both techniques.

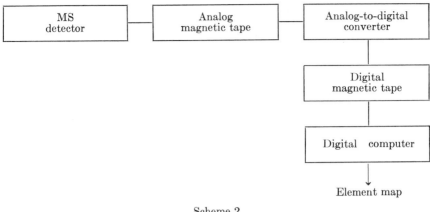

Scheme 2

The next logical step, after obtaining an element map, is to program the computer to carry out as much of the interpretation as possible. The objective, according to Biemann and his co-workers, is to print out "ion types" in addition to the exact mass and elemental composition of the ions detected. Results of recent work in the application of computer interpretation of high-resolution mass spectra of peptides are briefly described in Section 13-III.

D. Applications

The potentials of high-resolution mass spectrometry in composition determination via precision mass measurement have been demonstrated by Craig et al. (39), who analyzed an artificial mixture of tridecylbenzene ($C_{19}H_{32}$), phenyl undecyl ketone ($C_{18}H_{28}O$), 1,2-dimethyl-4-benzoyl naphthalene ($C_{19}H_{16}O$), and 2,2'-naphthyl benzo-β-thiophene ($C_{18}H_{12}S$). The parent peak spectrum is shown in Figure 10-13. The mass measurements, which were made with respect to the m/e 238 peak of dibromobenzene, the reference compound, were all within less than 0.5 mmu of the calculated masses. These compounds could, therefore, have been analyzed had they been present as unknowns.

In Figure 10-11 it was shown that an impurity peak detected in the low-resolution scan of a cholesterol sample was revealed by high resolution as a triplet. When mass measurements were made with respect to the parent cholesterol peak ($C_{27}H_{46}O$), all three peaks were identified, and it was shown that $C_{27}H_{44}O_2$ and $C_{28}H_{48}O$ were impurities, while $C_{29}H_{52}$ was a hydrocarbon introduced in sample handling. A third impurity was also determined at $P + 16$. In this example the impurities were in the low per

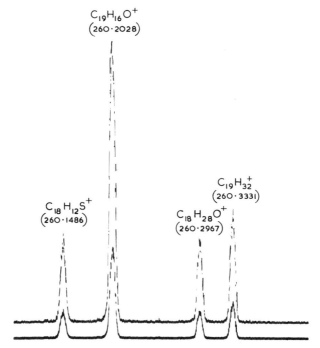

Figure 10-13. Parent peak spectrum of tridecylbenzene, phenyl undecyl ketone, 1,2-di-methyl-4-benzoyl naphthalene, 2,2′-naphthyl benzothiophene (39). (Upper tracing is recorded at $3 \times$ the sensitivity of the lower record).

cent concentration range, but it is clear that even trace level impurities may be detected by the technique.

A second area of application is the study of structures. The principles of structure determination are basically the same as discussed in low-resolution mass spectrometry, but here there is an opportunity to make identifications much more accurately. Indeed, several cases have been reported where high resolution revealed that former interpretations, based on low-resolution measurement, were misleading or false.

The possibility of hydrocarbon *type* analysis by high-resolution mass spectrometry has been reported by Lumpkin (40). A concentrate of trinuclear aromatic fraction (347–360°C boiling range) of a complex petroleum mixture was studied both at high (10,000) and low resolving power, and also at both high (70 eV) and low (8 eV) ionizing voltages. Fourteen different classes of compounds could be identified by combining precise mass measurements of molecular ions with mass measurements on fragment ions to prove or disprove possible assumed structures. Quantitative analysis was made on the basis of low-voltage sensitivities of available pure compounds. For example, it was established that dibenzothiophenes

(C_nH_{2n-12}) constituted 39.7% of the total, with C_nH_{2n-16} type compounds (mainly fluorenes and phenanthrenes) providing an additional 31.3% contribution. The power of combination of low-voltage technique with high resolution in identifying and characterizing complex mixtures is shown in Figure 10-14. At 80 eV only the molecular ion of the substituted carbazole appears at m/e 195. As the electron energy is increased fragment ions of the other compound types appear in order of increasing appearance potential. At the normal operating level of 70 eV, three peaks due to sulfur-, oxygen-, and nitrogen-containing molecules are seen in addition to the hydrocarbon peak.

Quantitative group-type analysis of high boiling petroleum fractions without a preliminary silica gel separation has been accomplished by Gallegos et al. (41), who employed a group-type calibration technique with high-resolution mass spectrometry. The results confirmed, among other findings, the high paraffin, low cycloparaffin content of the Arabian crude versus the low paraffin, high cycloparaffin content of the California and Louisiana crudes.

Direct sample insertion probes (Section 7-VI) make possible analysis of microsamples of involatile and thermally unstable materials; monoenergetic electron sources and photoionization attachments permit the study of the mechanism of ionization; measurement of metastable peaks reveals details of fragmentation mechanisms; in short, high-resolution mass spectrometry is applicable to all problems that could be studied by low resolution. The all-important technique of combination gas chromatography–mass spectrometry is discussed in the next section.

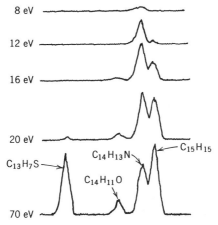

Figure 10-14. High resolution mass spectra at m/e 195 at various electron energies (40).

IV. GAS CHROMATOGRAPHY–MASS SPECTROMETRY COMBINATION

A. Principles

Coupling gas chromatography with mass spectrometry resulted in one of the most powerful tools in instrumental analysis. Gas chromatography provides a highly effective means to separate the constituents of a mixture while mass spectrometry is unique in its specificity, sensitivity, and speed in providing both compositional and structural information on the components separated by the gas chromatograph. The word "spectacular" must be used for some of the recent results obtained by GC-MS. The components of the "vapors" of coffee, strawberry, deteriorating fish, human sweat, etc. have been separated and identified, body fluids are analyzed routinely, the structure of lipids and related compounds is being elucidated, etc.

There are several common problems in the application of GC-MS to such analyses:

1. The components of interest may be present in various proportions, ranging from major constituents to traces, requiring great dynamic range for the analytical technique.

2. The relative volatility of the compounds in question may vary widely. Both gases and liquids with boiling points above 200°C must be handled. This requires temperature programming of the gas chromatograph and poses problems for the mass spectrometer inlet system.

3. Many different types of compounds, possibly with complex (and unknown) structure may occur. High-resolution gas chromatography is thus needed for separation, and high-resolution mass spectrometry must often be employed to facilitate structure evaluation.

4. Available sample size is normally small (0.1 μg or less), making high sensitivity mandatory.

In addition to these problems which are related to the samples, one also has to solve the requirements of the instrument coupling. Most of the difficulties in the GC-MS combination lie in two areas: (*a*) Sample transfer from GC to MS: gas chromatographs operate at atmospheric pressure while mass spectrometers are vacuum instruments. (*b*) Scanning and recording the mass spectra: fast scanning is needed to utilize the separating power of modern chromatographs, but this must be achieved without sacrificing the resolving power of the mass spectrometer. Before discussing these instrumental problems, it is desirable to review the relevant principles of gas chromatography.

B. Separation by Gas Chromatography

Separation in a gas chromatograph is achieved by distributing the components between two phases, one of which is a stationary bed of large surface area, while the other is a fluid percolating through (or along) the bed. In gas–liquid chromatography (GLC) the sample is partitioned between a moving vapor phase (carrier gas) and a stationary liquid phase held on a solid support. In gas–solid chromatography (GSC) the stationary phase is a solid adsorbant. The sample is injected into the moving carrier gas as a "plug," and the various components in the sample are selectively retarded by the stationary phase as the sample moves along. While the process of retention may be based upon adsorption, chemical bonding, solubility, polarity, or molecular filtration, depending upon the kind of column material employed, the end result is always the same: some components are retarded longer than others, and the separated compounds emerge from the column in the inverse order of their retention. The emerging components enter a detector (thermal conductivity, cross section, flame ionization, etc.), the selection of which depends upon the type of column employed and the sensitivity required. In GC-MS combinations the mass spectrometer serves as the detector, although a fraction of the eluent is often diverted into an auxiliary detector.

There are two basic types of columns: *packed* columns normally consist of coiled tubings of copper or stainless steel, of about 5 mm inside diameter and 1.5 m length. The solid support is a form of diatomaceous earth in GLC, and silica gel, molecular sieve, or activated charcoal in GSC. Liquid phases in GLC may be silicone oils or greases, polyglycols, phthalate derivatives, etc. The solvents are used in concentrations of 2–30% by weight, and are normally conditioned by heating in vacuum. The term *column bleeding* is used to describe contributions to the spectra from column contaminants or components. *Capillary* (Golay-type) columns are metal or glass columns of 0.25 mm inside diameter and 50–300 m long. An inside coating of 1 μ thickness with the liquid phase is provided by passing a dilute solution of the liquid in a high vapor pressure solvent (e.g., chloroform) through the capillary. Helium is used almost exclusively as carrier gas in GS-MS applications, partly because its interference with MS analysis is minimal, and partly because of the possibility of sample enrichment by techniques to be described later. In Europe, hydrogen is preferred as carrier gas because of the high cost of helium.

The theory of GC retention and resolution is complex and not well understood. It is sufficient to say that the separating power of a column may be evaluated similarly to that of distillation columns, in terms of the required column length which is equivalent to one theoretical plate (HETP = height equivalent to a theoretical plate). The main parameters determining resolution are the length and diameter of the column, the thickness of the liquid phase, and the speed of the carrier gas. The thickness of the liquid phase is limited by efficiency requirements (small sample size). The number of theoretical plates may be estimated by dividing the length of the column by its radius. With capillary columns HETP values up to 5×10^5 may be achieved, while the limit for packed columns is at the 10^3 level.

The maximum amount of material that can be used without overloading is a few micrograms in capillary columns and a few milligrams in packed columns. The minimum sample quantity needed is determined by the sensitivity of the mass spectrometer under the required resolution and scanning conditions.

A compromise between the two basic types of columns is the support-coated open-tubular column, in which the stationary phase is a thin film on a micron-sized inert solid support deposited on the wall of the tube. These columns accept, due to increased surface area, relatively large sample charges, but still retain the high resolution of the open-tubular columns.

C. Instrumental

1. Sample Transfer

There are two basic techniques for sample transfer from GC to MS: peak trapping for batch-type analysis and direct introduction into the ion source. A detailed review has recently been given by McFadden (42).

In the simplest collection technique both the component effluent and carrier gas are collected in an evacuated glass or metal bulb (volume ~ 250 cc) connected to the exit port of the chromatograph. Such a collection system may consist of a stainless steel manifold needle valve with a by-pass manifold for the connection, and a soap-film flowmeter for flow regulation. When the GC detector indicates the elution of the peak of interest, the flow is directed into the collection flask. The collected component may be concentrated by freezing into the small removable collection tip of the flask. In another technique, eluted components are traped in a melting point capillary connected to the heated exit port of the chromatograph. When a nondestructive GC detector is employed the whole fraction can be condensed and collected on the walls. For high boiling fractions, room temperature is adequate, for components that boil below 150°C (at 1 atm)

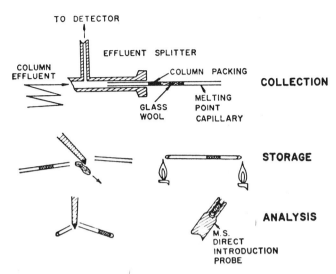

Figure 10-15. Steps in collecting a gas chromatographic fraction and introducing it into the mass spectrometer (43).

Dry Ice or liquid nitrogen cooling is necessary. Many variants of this technique have been reported, including commercial fraction collectors.

Individual submicrogram size samples emerging from a gas chromatograph can be nearly quantitatively collected by the technique of Amy et al. (43). A capillary melting point tube (Fig. 10-15), filled with the GC packing material including liquid substrate, acts as an extension of the GC column. The capillaries are removed (and subsequently flame-sealed for storage) at the proper time, as indicated by a detector through which a small fraction of the effluent is diverted. The capillary is next inserted into a solid introduction probe (Section 7-VI-B) and volatilized inside the mass spectrometer source. The use of activated charcoal to trap GC fractions is suggested by Damico et al. (44).

Individual fraction collection has several advantages: (*1*) there is no decrease in sensitivity due to dilution by the carrier gas or due to sample loss in the enrichment process (see below); (*2*) high MS resolution can be utilized fully since there is no need for fast scanning; (*3*) no quick decisions have to be made during analysis; and (*4*) samples can be taken at any place, at any time. Disadvantages are that microgram size samples must be transferred without loss, requiring much skill, high-resolution capillary columns cannot be utilized conveniently, and the technique becomes awkward when many GC peaks are to be studied.

A technique intermediate between separate trapping and direct effluent monitoring involves the piping of the effluent into a manifold of appropriate valves and traps which serves as a batch inlet system for the mass spectrometer. A paper by Miller (45) discusses the problems involved in using peak trapping in GC-TOF combinations. A manifold inlet system, based on a design by Ebert (46), is available from the Bendix Corporation (Model 1076 GC Manifold).

Direct, high-speed monitoring of GC effluents offer many obvious advantages: improved sample utilization (high sensitivity), full usage of the resolving power of GC, peak profile identification, possible usage in process monitoring, and convenience in sample handling when many peaks are to be analyzed. The main disadvantage, apart from complex instrumentation, is the possibility for "memory." It takes a certain amount of time to remove a sample from the mass spectrometer. The memory is probably solely a function of the polarity of the molecules and is relatively independent of vapor pressure.

A straightforward connection from GC to MS has been used by Gohlke (47), who employed an oscilloscope screen (oscilloscope photography) for recording, thus fully utilizing the fast scanning opportunities of TOF mass spectrometry.

The objective of any *coupling system* between GC and MS is to reduce the atmospheric pressure of the GC effluent to the operating pressure of $\approx 10^{-5}$ torr of the MS with only minimal loss in the partial pressure of the component of interest. In other words, it is desired to get rid of as large a portion as possible of the carrier gas during the short time the effluent spends between GC and MS. At present, there are five techniques available for sample enrichment (48).

In Ryhage's method (49) the effluent from the GC column is fed through two "molecule separators" connected in tandem. These molecule separators (50) operate on the jet principle. The GC effluent is compressed to a narrow beam and is accelerated to about the velocity of sound at which the lighter weight helium atoms (carrier gas)scatter, while the heavy organic molecules tend to remain on the beam. Figure 10-16a shows the operation of the jet molecular diffusion separator (42), while Figure 10-16b is a schematic drawing of a complete GC-MS system (49). A commercial version of the molecule separator is available from LKB Instruments, Inc.

The sample-enrichment system of Watson and Biemann (51) is shown in Figure 10-17. This device is based upon differences in diffusion rates of helium and the organic molecules across the pores (1 μ size) of a fritted glass tube. The tube is pumped to 0.1 torr on the "other "side. The sample thus enriched passes through a second capillary constriction (0.1 mm) into

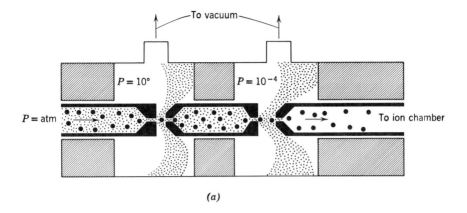

(a)

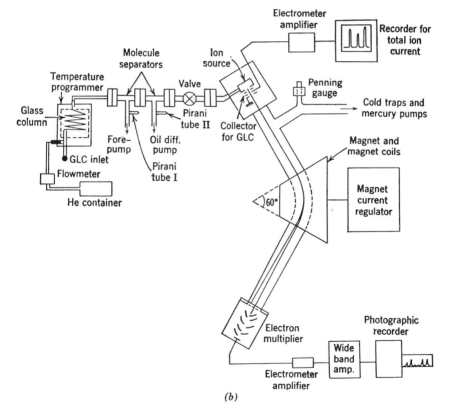

(b)

Figure 10-16. (a) Ryhage-type molecular separator (42); (b) Ryhage's technique for GC-MS combination (49).

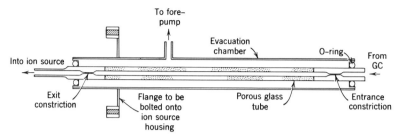

Figure 10-17. Watson-Biemann pressure reduction system linking GC and MS (51).

the ion source. Experimental conditions must be carefully regulated depending upon the molecular weight of the eluent. This technique has been found particularly useful in double-focusing mass spectrometers using photographic detection. Several commercial versions are available.

The use of Teflon as an interface between GC and MS has recently been reported by Lipsky et al. (52). Teflon membranes exhibit such high selectivity for the permeation of helium relative to that of organic vapors that a single stage system appears adequate for most purposes. Once the membrane thickness and surface area are chosen, temperature becomes the major parameter that controls efficiency. No "memory" effect has been noticed over prolonged use.

In the "Llewellyn-type" separator (marketed by Varian Associates), the eluted materials pass through a silicone rubber diaphragm. The separation is based upon the low permeability of helium through the silicone rubber due to its very slight solubility in the rubber.

Finally, a carrier gas separator based on the preferential diffusion of helium through a porous metal membrane has been developed by General Electric Company.

The ability of a separator to allow organic material to pass into the ion source of the mass spectrometer is characterized by the separator *yield,* which is defined as the ratio of the amount of sample entering the spectrometer to that entering the separator. The yield is usually expressed as a per cent value, and yields may be as high as 90%. The *enrichment* factor is obtained by dividing the sample-to-carrier gas ratio on the mass spectrometer side of the separator by the sample-to-carrier gas ratio on the gas chromatograph side. Numerical values are usually in the 1–20 range. It is important to realize that the enrichment factor is of no importance if the absolute amount of sample reaching the ion source is too small to produce a spectrum which is adequate for identification. Additional performance characteristics include lag time, i.e., time for the sample to pass through the separator possible peak distortion, required operating temperature, and usable flow, rate.

The only comparative study published to date is that by Grayson and Wolf (48), who evaluated the performance of the fritted glass tube and the heated porous Teflon tube separators using temperature, carrier gas flow rate, and molecular weight of the sample as variables. These authors appear to favor the fritted glass separator.

2. Scanning Speed Considerations

Scanning speed assumes great importance in GC-MS combinations, particularly in connection with the problem of selecting the "proper" detection system for high-resolution mass spectrometers. It is recalled that photographic plates are composite detector-recorders. Here there is no need for scanning as the entire preselected mass range is recorded simultaneously while the acceleration voltage and the electrostatic and magnetic fields are kept constant. In electrical detection, on the other hand, scanning is accomplished by varying either the acceleration field voltage or the magnetic field strength, and individual ion currents are measured by an amplifier system, connected to a suitable recorder. In the more sophisticated versions of such systems, the output of the electron multiplier is fed into a scope or magnetic recorder. The scope display is simple and straightforward, but such obvious disadvantages as limited dynamic range, inaccurate mass measurement, and problems in obtaining permanent records prevent its general use. Tape recording is desirable from many points of view, but the equipment is complex and expensive.

When comparing the usefulness of the detector systems in GC-MS combinations, the best starting point is to define required performance goals. Undoubtedly, the most important requirement is to maintain the ability to measure masses to at least 5 ppm, so that precise molecular masses can be determined. When this ability cannot be retained, the utility of the high-resolution analyzer system is lost. The second piece of information desired is ion abundance. The main goal here is sensitivity, an ability to ascertain the presence of a very small number of ions. While mass measurements must be precise, some concession can be made as to the precision of abundance measurements in GC-MS work. As far as dynamic range in a single measurement is concerned, electrical recording is obviously advantageous over photoplates.

The importance of fast scanning is appreciated when it is considered that a hundred peaks may emerge from a high-resolution gas chromatograph within 30–60 min. In addition, concentration changes during the course of the elution. Required scanning time is thus determined by the duration of the GC peaks: one would like to scan the whole mass range in a time interval less than the half-width of the GC peak. In practice, a scanning speed of 1–3 sec over the mass range 30–300 is normally required. Occasionally a speed as high as 0.2 sec per decade (factor of 10 in mass) is needed. Variable scanning speed is a standard feature in most commercial mass spectrometers.

The influence of fast scanning on resolution and sensitivity has been instructively illustrated by McFadden and Day (53), who studied the spectra of mercury isotopes at different scanning speeds on a low-resolution mass spectrometer. Figure 10-18 shows several spectra obtained at increasingly faster scans. The natural frequencies of the galvanometers used were 1600 cps (top trace) and 1000 cps (bottom trace), respectively; the different attenuation is of no significance. The instrument had been equipped with an electron multiplier so that the time constants of the recording galvanometers presented the limiting factors. The first scan (0.5 sec/octave) covered the mass range of m/e 24–200 in 1.5 sec, while it took only 0.6 sec to scan the same range at 0.2 sec/octave speed. The adjusted resolution of 400 (10% valley) is retained reasonably well by both galvanometers in the 0.2 sec/octave runs (two are shown to illustrate reproducibility), although the slow galvanometer is seen to indicate a slight lagging. As the scan speed is further increased in the next three scans, not only is resolution reduced, but also "peak clipping" (reduction in peak height) is observed. While these spectra are still usable, peak distortion becomes severe at 0.04 sec/octave scanning speed. Extrapolating these results to a resolution of 10,000, McFadden and Day conclude that a mass peak would be scanned in 10^{-4} to 10^{-5} sec (assuming a tape recorder with 5×10^4 cps response), resulting in a current sensitivity limit of only 10^{-12} to 10^{-13} A (for 100 ions). For better sensitivity they recommend photoplate recording, which presumably has a limit close to 10^{-14} to 10^{-15} coulombs. Thus, although the photoplate has a much lower intrinsic sensitivity, continuous recording of all ions for 10–100 sec more than makes up for this deficiency.

Merritt et al. (54) employed a high-resolution mass spectrometer with electrical tape recording to experimentally determine the effects of fast scanning. They demonstrated that when a uniform sample consumption rate of 10^{-7} g/sec is provided, the number of ions of the base peak arriving at the collector, at a scan rate of 10 sec/decade and at a resolution of 10,000 is of the order of 10^3. Only 10–15 ions are needed to ascertain the presence of a peak and it was also shown that, theoretically at least, mass measure-

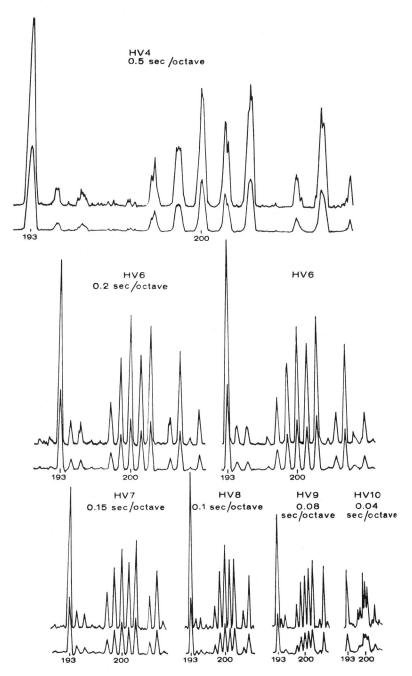

Figure 10-18.   Fast-scan mass spectra of mercury isotopes (53).

ment with a precision greater than 10 ppm can be performed if that number of ions is present during the time in which the peak is scanned. It is concluded that as long as 10^{-7} g/sec of GC eluent is available, high resolution can be used and masses can be measured with adequate precision to facilitate identification.

It is clear from the above discussion that the detector selection for GC-MS combination is presently subject to considerable controversy. Brunnèe et al. (55) discuss the measuring limits due to statistical fluctuations; Banner (56) investigated the distortion of peak shapes in fast-scanning mass spectra; Green et al. (57) describe results with fast-scanning electrical detection; Campbell and Halliday (58) consider the accuracy of mass measurement from low-intensity ion beams; McMurray et al. (59) report some operating parameters in fast-scanning GC-MS combinations.

It is concluded that both photographic and electrical detections can be used in GC-MS combination assuming that proper sophistication in instrumentation is available. For equal performance, electrical detection appears to require more complex and expensive recording equipment—at least at the present time. Photographic detection certainly has the required sensitivity, presents relatively few problems in mass measurement, provides a permanent record, and allows the data to be processed independently. Problems include variabilities in emulsion performance and time requirements in processing. When electrical recording is employed, very fast scanning prevents the use of the peak-matching techniques (unless peak trapping is used) and tape recording becomes absolutely necessary. Required equipment is expensive but real time data processing of the recorded scan may closely be approached. As mentioned already, this is one of the fastest growing areas of instrumentation in mass spectrometry, and the reader is repeatedly urged to follow future developments.

D. Applications

Since its inception in 1957, GC-MS has traveled a long way, and now it is considered one of the most important tools in organic analysis. Today instrument manufacturers feature the required hardware for GC applications, and indeed a considerable portion of the mass spectrometers now being designed are specifically intended for GC-MS use.

Both low- and high-resolution instruments have been combined with gas chromatographs, and each has its own respective merits. For the analysis of mixtures consisting of relatively simple compounds for which cracking patterns are readily available, the direct sampling method and low-resolution mass spectrometry are recommended. High resolution is needed where precise mass measurements must be made and where structure studies are performed. If possible, the GC peak should be trapped out.

Important parameters to be considered in the selection of a proper GC column for a particular application include required GC resolution, maximum operating temperature, desired carrier gas flow rate (separator used), and column bleeding. Column bleeding must be kept at a minimum to avoid interfering contributions to the cracking pattern of the unknown. The extent of column bleed for various stationary liquids has been described by Teranishi et al. (60) in a revealing article in which many other important aspects of GC-MS combination are also discussed.

Increased GC resolution is obtained with capillary or support-coated capillary columns. The coupling of 0.01-in. capillary columns to the MS may be accomplished either by stream splitting or by operating the column exit directly in the vacuum system (60). The coupling of 0.02–0.03-in. capillary columns is also quite straightforward, and as much as 90% of the total effluent can be utilized with the aid of a separator. The dynamic range for such systems may be extended to 10^{-4} to 10^{-10} g (42).

Figure 10-19 illustrates the use of low-resolution GC-MS combination in the analysis of a hydrocarbon mixture. All 53 peaks were identified in the spectrum (61). Analyses of this kind are now being routinely made in many laboratories using standard patterns for identification.

GC-MS is an extremely powerful tool in flavor research. Both low- and high-resolution mass spectrometers have been used for identification and structure work. Figure 10-20 shows the gas chromatogram of an imitation peach flavor (62) and the mass spectrum of peak *4*, identified as cinnamaldehyde. This peak, hardly visible on the GC trace, was known to be present to less than 0.1% concentration, and since the sample size was 0.1 μl, the total amount of the component separated by the GC and hence available for analysis was less than 0.1 μg. A 0.02-in. i.d. support-coated open tubular capillary column (Carbowax 1540) was employed in this study. The temperature of the column was kept at 100°C for 5 min, followed by an 8°/min increase to 190°C. The effluent from the GC was introduced into the MS via a Biemann-Watson separator (0.1 cc/min delivered to MS). Nominal instrument resolution was 2500, scanning speed employed was such that the 12–300 mass range was covered in 3 sec.

Other interesting applications in flavor research include the analysis of strawberry oils (63), volatile hydrocarbons from oranges (64), the aroma fractions of cheese (65), and the volatile components of Jamaica rum (66).

High-resolution gas chromatography, employing a 900-ft. long, 0.03-in. i.d. stainless steel column coated with Polysev (*m*-bis-*m*-(phenoxy-phenoxy)

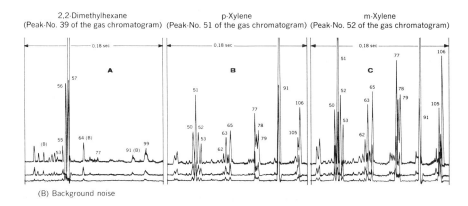

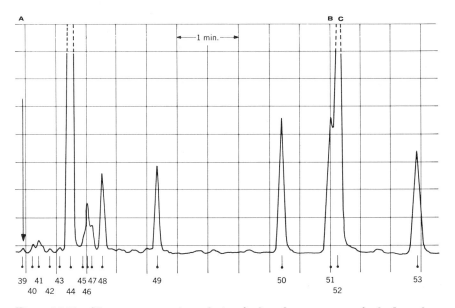

Figure 10-19. **Mass spectrometric analysis of selected components of a hydrocarbon mixture separated by gas chromatography (61).**

phenoxybenzene), was used by Oro' et al. (67) to analyze the pyrolysis products of isoprene and methane. A large number of both aliphatic and aromatic reaction products were separated by GC and many were identified by subsequent low-resolution MS analysis. At low pyrolysis temperatures (300–400°C) aliphatics predominate, while at higher temperatures (600–1000°C) mainly aromatics are formed. This study illustrates the usefulness of the GC-MS technique for the analysis of complex hydrocarbon

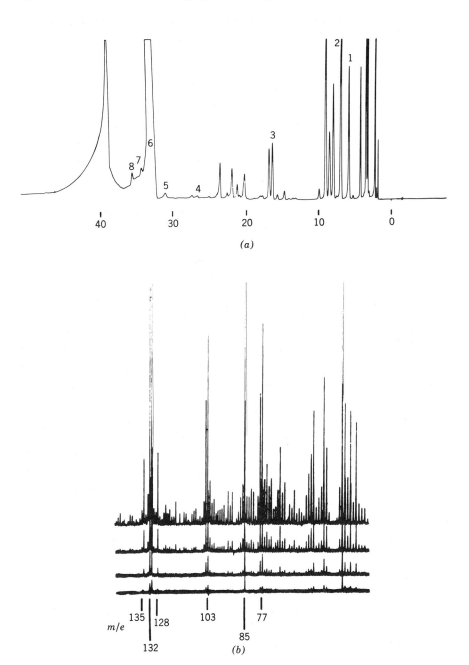

Figure 10-20.   Use of GC–MS combination in flavor research (62). The lower spectrum is an analysis of peak *4* of the upper spectrum. Peak is identified as cinnamaldehyde (mol. weight = 132).

mixtures. Practical applications include the compositional analysis of pyrolysis products involved in air pollution, fuel combustion, and meteorite research.

The composition of unresolved or partially resolved mixtures in GC eluents can be determined with the aid of the "accelerating voltage alternator" (AVA) accessory developed by Sweeley et al. (68). A continuous oscillographic recording of the changes in the intensities of two m/e values with a single collector is achieved by rapid switching of the acceleration voltage (at constant magnetic field) using a time-actuated relay and a voltage-dividing circuit. Quantitative analyses are made from the plot of peak intensities versus time by measuring the areas under the curves with an integrator. An important application of this technique is the analysis of submicrogram quantities of isotopically labeled (e.g., deuterated) organic compounds of biochemical interest.

The application of open-tubular columns with high-resolution mass spectrometers presents a number of problems involving the fast scanning of narrow eluted GC peaks. A compromise is the use of support-coated capillary columns; the peaks are somewhat wider, contain more sample, but the high resolution obtained with regular capillary columns is still retained.

McMurray et al. (59) have reported using an 8 sec/mass decade magnetic scan with a 10,000–20,000 resolution range to monitor (electric detection) the fatty acid effluents emerging from a 50-ft, 0.02-in. i.d. support-coated (Apiezon L) capillary column. Rapid scanning in high resolution instruments with photoplate detection is discussed by Henneberg (69). A plate scan is accomplished by continuously moving the plate during the emergence of a peak. This causes a variation of intensity for the whole spectrum which is proportional to the partial pressure of the substance in the ion source. This technique is not applicable to the analysis of mixtures where peaks are eluted in very short intervals.

V. COMBINATION OF MS WITH THERMAL METHODS

In *mass spectrometric thermal analysis* (MTA) the sample is heated at a constant rate in a miniature furnace placed as close to the ionizing electron beam inside the source as possible, and the gaseous products resulting from phase changes, etc. are observed with the mass spectrometer. In most studies reported thus far, TOF mass spectrometers had been employed but, in principle, the technique is adaptable to any instrument; e.g., an RF-type mass spectrometer is available commercially from Nuclide Corp. for thermal studies. Thermal behavior may be investigated either iso-

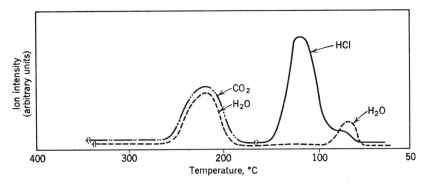

Figure 10-21.   Mass spectrometric thermal analysis trace of $H_4Y \cdot 2H_2O \cdot 2HCl$ (70).

thermally or by programming the temperature change. The main improvement of the method over conventional thermogravimetric analysis is that all volatile reaction products can be identified (with high sensitivity). Figure 10-21 (70) shows the MTA spectrum of $H_4Y \cdot 2H_2O \cdot 2HCl$ (Y = ethylenediaminetetraacetic acid). It is seen, among other things, that H_4Y does not form a stable anhydride since H_2O and CO_2 are released simultaneously.

The limitations of MTA, posed by the difficult-to-control pressure change in the ion source, are overcome in *mass spectrometric differential thermal analysis* (MDTA), where an external commercial differential thermal analyzer (DTA) is combined with effluent gas analysis (71). Figure 10-22 shows simultaneous mass spectrometric and differential thermal analysis of magnesium chloride hexahydrate ($MgCl_2 \cdot 6H_2O$). The sharp peak at 320°C in the MTA scan is MgOHCl, a hydrolysis product. This technique is still in its infancy, but its potential is clearly shown, particularly in studies of pressure-dependent thermal reactions.

Pyrolysis followed by mass spectrometric examination of the volatile products has been increasingly employed in the study of polymers and other large molecules. Thermal decomposition can be carried out in separate pyrolysis outside the ion source (or even away from the mass spectrometer) or in a furnace incorporated into the source; there are many novel designs. An example of the technique has been presented by Brandt and Mohler (72), who showed that polyvinyl chloride decomposes in two stages: between 127 and 220°C the evolving gas is HCl while at higher temperatures (max. 389°C) a variety of aromatic hydrocarbons appear up to the highest mass of 596. Copolymers of 1-pentene and 4-methyl-1-pentene have been studied by Wanless (73) using pyrolysis–mass spectrometry combined with NMR assay. With the advent of direct sample

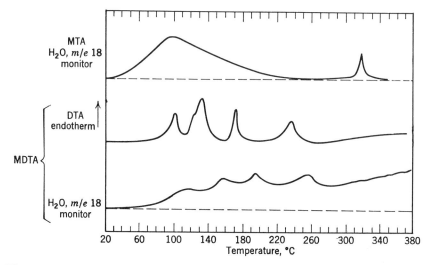

Figure 10-22. Simultaneous mass spectrometric and differential thermal analysis of $MgCl_2 \cdot 6H_2O$ (71).

introduction systems the application of mass spectrometry to polymer studies is expected to become an increasingly active field. A flash evaporation technique, described by Lincoln (74), makes it possible to also study short-lived species, by photographing the spectra that momentarily appear on the scope of a TOF mass spectrometer.

References

1. Washburn, H. W., H. F. Wiley, and S. M. Rock, *Ind. Eng. Chem., Anal. Ed.,* **15**, 541 (1943); **17**, 74 (1945).

*2. Melpolder, F. W., and R. A. Brown, in *Treatise on Analytical Chemistry*, Part I, Vol. 4, I. M. Kolthoff and P. J. Elving, Eds., Interscience, New York, 1963, Chapter 40.

3. Crout P. D., *Am. Inst. Elec. Engrs. Trans.,* **60**, 1235 (1941).

4. Daigle, E. C., and H. A. Young, *Anal. Chem.,* **24**, 1190 (1952).

*5. Dudenbostel, B. F., and P. J. Klaas, in *Advances in Mass Spectrometry*, J. D. Waldron, Ed., Pergamon, New York, 1959, p. 232.

6. Johnsen, S. E. J., *Anal. Chem.,* **19**, 305 (1947); Ruth, J. M., *ibid.* **40**, 747 (1968).

7. Washburn, H. W., in *Physical Methods in Chemical Analysis*, Vol. 1, W. G. Berl, Ed., Academic Press, New York, 1950, p. 587.

8. Brown, R. A., *Anal. Chem.,* **23**, 430 (1951).

*9. *Book of ASTM Standards*, Part 18, 1966, p. 555.

10. Crable, G. F., G. L. Kearns, and M. S. Norris, *Anal. Chem.*, **32**, 13 (1960).

11. Field, F. H., and H. S. Hastings, *Anal. Chem.*, **28**, 1248 (1956).

12. Biemann, K., *Mass Spectrometry, Organic Chemical Applications*, McGraw-Hill, New York, 1962, p. 164.

13. Lumpkin, H. E., and T. Aczel, *Anal. Chem.*, **36**, 181 (1964).

*14. Beckey, H. D., H. Knöppel, G. Metzinger, and P. Schulze, in *Advances in Mass Spectrometry*, Vol. 3, W. L. Mead, Ed., Institute of Petroleum, London, 1966, p. 35.

15. Mead, W. L., *Anal. Chem.*, **40**, 743 (1968).

*16. Silverstein, R. M., and G. C. Bassler, *Spectrometric Identification of Organic Compounds*, 2nd ed., Wiley, New York, 1967.

17. Conn, G. K. T., and B. Donovan, *Rev. Sci. Instr.*, **28**, 7 (1951).

18. Packard, M. E., *Rev. Sci. Instr.*, **19**, 435 (1948).

19. McLafferty, F. W., *Anal. Chem.*, **29**, 1782 (1957).

20. Eden, M., B. E. Burr, and A. W. Pratt, *Anal. Chem.*, **23**, 1735 (1951).

21. Zemany, P. D., *J. Appl. Phys.*, **23**, 924 (1952).

22. Cornu, A., and R. Massot, *Compilation of Mass Spectral Data*, Heyden, London, 1966.

*23. Shannon, J. S., *Proc. Roy. Australian Chem. Inst.*, **1964**, 328.

*24. McLafferty, F. W., *Interpretation of Mass Spectra*, Benjamin, New York, 1966.

*25. Beynon, J. H., *Mass Spectrometry and Its Application to Organic Chemistry*, Elsevier, New York, 1960, p. 294.

*26. McLafferty, F. W., *Mass Spectral Correlations*, American Chemical Society, Washington, D.C., 1963.

27. Biemann, K., *Mass Spectrometry, Organic Chemical Applications*, McGraw-Hill, New York, 1962, p. 305.

28. Hites, R. A., and K. Biemann, *Anal. Chem.*, **39**, 965 (1967); Petterson, B., and R. Ryhage, *ibid.*, **39**, 790 (1967); Barber, M., and W. A. Wolstenholme, *Sci. J.*, Sept. 1967, p. 76; Abrahamsson, S., *Sci. Tools* (LKB instrument journal), **14**, ?9 (1967); Crawford, L. R., and J. D. Morrison, *Anal. Chem.*, **40**, 1464, 1469 (1968).

29. Quisenberry, K. S., T. T. Scholman, and A. O. Nier, *Phys. Rev.*, **102**, 1071 (1956).

30. Smith, L. G., and C. C. Damm, *Rev. Sci. Instr.*, **27**, 638 (1956).

*31. Nier, A. O., in *Nuclear Masses and Their Determination*, H. Hintenberger, Ed., Pergamon, New York, 1957, p. 185.

32. Johnson, W. H., and J. L. Benson, *13th Ann. Conf. Mass Spectrometry*, ASTM Committee E-14, St. Louis, Missouri, 1965, *Proc.*, p. 289.

33. Fales H. M., *Anal. Chem.*, **38**, 1058 (1966).

34. Barber, M., R. D. Craig, R. M. Elliot, and B. N. Green, Publication 2032-78, Associated Electrical Industries, Ltd., Manchester, England, 1964.

35. Biemann, K., *Pure Appl. Chem.*, **9**, 95 (1964).

*36. Biemann, K., P. Bommer, and D. M. Disederio, *Tetrahedron Letters*, **26**, 1725 (1964); also Interview with K. Biemann in *Chem. Eng. News*, **42**, 42 (1964); also K. Biemann, P. Bommer, D. M. Desiderio, and W. J. McMurray in *Advances in Mass Spectrometry*, Vol. 3, W. L. Mead, Ed., Institute of Petroleum, London, 1966, p. 639.

37. Venkataraghavan, R., and F. W., McLafferty, *Anal. Chem.*, **39**, 278 (1967).

*38. Venkataraghavan, R., F. W. McLafferty, and J. W. Amy, *Anal. Chem.*, **39**, 178 (1967).

39. Craig, R. D., B. N. Green, and J. D. Waldron, *Chimia*, **17**, 33 (1963).

40. Lumpkin, H. E., *Anal. Chem.*, **36**, 2399 (1964).

41. Gallegos, E. J., J. W. Green, L. P. Lindemann, R. L. LeTourneau, and R. M. Teeter, *Anal. Chem.*, **39**, 1833 (1967).

*42. McFadden, W. H., *Separation Science*, **1**, 723 (1966).

43. Amy, J. W., E. M. Chait, W. E. Baitinger, and F. W. McLafferty, *Anal. Chem.* **37**, 1265 (1965).

44. Damico, I. N., N. P. Wong, and J. A. Sphon, *Anal. Chem.*, **39**, 1045, (1967).

45. Miller, D. O., *Anal. Chem.*, **35**, 2033 (1963).

46. Ebert, A. A., *Anal. Chem.*, **33**, 1865 (1961).

47. Gohlke, R. S., *Anal. Chem.*, **31**, 535 (1959); **34**, 1332 (1962).

48. Grayson, M. A., and C. J. Wolf, *Anal. Chem.*, **39**, 1438 (1967).

*49. Ryhage, R., *Anal. Chem.*, **36**, 759 (1964).

50. Becker, E. W., in *Separation of Isotopes*, H. London, Ed., Newnes, London, 1961, Chapter 9.

*51. Watson, J. T., and K. Biemann, *Anal. Chem.*, **36**, 1135 (1964); **37**, 844 (1965)

52. Lipsky, S. R., C. G. Horvath, and W. J. McMurray, *Anal. Chem.*, **38**, 1586 (1966).

*53. McFadden, W. H., and E. A. Day, *Anal. Chem.*, **36**, 2362 (1964).

54. Merritt, C., P. Issenberg, M. L. Bazinet, B. N. Green, T. O. Merron, and J. G. Murray, *Anal. Chem.*, **37**, 1039 (1965).

55. Brunnèe, C., L. Jenckel, and K. Kronenberger, *Z. Anal. Chem.*, **189**, 50 (1962); **197**, 42 (1963).

*56. Banner, A. E., *J. Sci. Instr.*, **43**, 138 (1966).

57. Greene, B. N., T. O. Merron, and J. G. Murray, *13th Ann. Conf. Mass Spectrometry*, ASTM Committee E-14, St. Louis, Missouri, 1965, *Proc.*, p. 204.

58. Campbell, A. J., and J. S. Halliday, *13th Ann. Conf. Mass Spectrometry*, ASTM Committee E-14, St. Louis, Missouri, 1965, *Proc.*, p. 200.

*59. McMurray, W. J., B. N. Greene, and S. R. Lipsky, *Anal. Chem.*, **38**, 1194 (1966).

*60. Teranishi, R., R. G. Buttery, W. H. McFadden, T. R. Mon, and J. Wasserman, *Anal. Chem.*, **36**, 1509 (1964).

61. Bulletin M-1113b, Atlas Mess und Analysen Technik, GmbH, Bremen, Germany, 1964; see also Ref. 55.

62. Struck, A. H., and V. J. Coates, *14th Ann. Conf. Mass Spectrometry*, ASTM Committee E-14, Dallas, Texas, 1966, *Proc.*, p. 721.

63. Teranishi, R., J. W. Corse,W. H. McFadden, D. R. Black, and A. I. Morgan, *J. Food Sci.*, **28**, 478 (1963).

64. Teranishi, R., R. E. Lundin, W. H. McFadden, T. R. Mon, T. H. Schultz, K. L. Stevens, and J. Wasserman, *J. Agr. Food. Chem.*, **14**, 447 (1966).

65. Day, E. A., and L. M. Libbey, *J. Food Sci.*, **29**, 583 (1964); Day, E. A., and D. F. Anderson, *J. Agr. Food Chem.*, **13**, 2 (1965).

66. Maarse, H., and M. C. ten Noever de Brauw, *J. Food Sci.*, **31**, 951 (1966).

67. Oro', J., and J. Han, *J. Gas Chromatog.*, Sept. 1967, p. 480; Oro', J., J. Han, and A. Zlatkis, *Anal. Chem.*, **39**, 27 (1967).

*68. Sweeley, C. C., W. H. Elliott, I. Fries, and R. Ryhage, *Anal. Chem.*, **38**, 1549 (1966).

69. Henneberg, D., *Anal. Chem.*, **38**, 495 (1966).

70. Gohlke, R. S., and H. G. Langer, *Anal. Chem.*, **37**, 25A (1965).

71. Langer, H. G., R. S. Gohlke, and D. H. Smith, *Anal. Chem.*, **37**, 433 (1965).

72. Brandt, P., and F. L. Mohler, *Res. Natl. Bur. Std.*, **55**, 323 (1955).

73. Wanless, G. G., *J. Polymer Sci.*, **62**, 263 (1962).

74. Lincoln, K. A., *Anal. Chem.*, **37**, 541 (1965).

Chapter Eleven

Analytical Techniques: Inorganic

I. TRACE ANALYSIS: GENERAL

Almost all applications of mass spectrometry in inorganic analysis deal with the identification and quantitative determination of trace constituents. The term "trace" normally refers to impurities present in a matrix in an amount smaller than 100 parts per million (ppm); levels below 1 ppm are sometimes called ultratrace. It is the objective of this chapter to survey the various analytical techniques available, discuss the problems and limitations involved, and review representative applications. First, the analysis of gases is considered briefly. Demand for ultrapure gases continuously increases in industrial, space, and research applications, and great efforts are being made to improve purification techniques in manufacturing. Requirements for gases with known amounts of impurities (doping gases, calibration gases) are also becoming increasingly stringent. For inorganic gas analysis, conventional mass spectrometers with electron bombardment sources are normally employed and efforts are concentrated mainly on improving sensitivity (preconcentration). Mass spectrometers are also employed in the study of gases in metals, where experimental problems revolve around the extraction and transfer of the gases rather than their detection.

The spectacular advances in solid state physics, chemistry, and technology during the past 15 years created a great demand for new analytical methods for the detection and measurement of trace impurities in solids. The result is what is loosely called "solids mass spectrometry." Certain electric, magnetic, and mechanical properties can be drastically altered by changing the concentration of selected impurities. The role of trace impurities in many branches of the physical sciences has been reviewed by Hannay (1).

There are three major types of problems in general trace analysis:

1. Direct determination of a trace impurity in the presence of a large matrix. The main requirement is high *concentrational* sensitivity, and preconcentration techniques are often needed to improve detection limits and/or eliminate matrix interference.

2. Identification and qualitative determination of impurities and/or the matrix itself in very small samples. Here high *absolute* sensitivity is needed, and sample handling becomes a major problem.

3. Qualitative and quantitative analysis of localized impurities in a large matrix. Examples are inclusions in semiconductors, meteorites, etc. Every problem normally requires the development of special analytical techniques; several are discussed in connection with the spark method.

II. TRACE ANALYSIS OF INORGANIC GASES

A. Preconcentration Techniques

1. Need for Preconcentration

Gas chromatography and mass spectrometry are used in a complementary fashion in inorganic gas analysis, with a mass spectrometric scan up to m/e 200 providing assurance that no important impurity is overlooked. Gas chromatographs designed for inorganic gas analysis have thresholds in the low parts per million region for most permanent gases and in the parts per billion region for many important organic impurities such as hydrocarbons.

Sensitivity limits of the order of ppm can be achieved routinely in commercial mass spectrometers with electrometer detection. Increasing ionization current (up to 150 μA) and sample pressure (up to 750 μ or more behind the molecular leak) are the paths frequently followed to increase sensitivity. Reproducibility in such measurements is not better than $\pm25\%$ of the value reported, and results around the threshold are often reliable only within a factor of 2. Sensitivity may be increased by a factor of 100 with electron multipliers; however, accuracy suffers. Parkinson and Toft (2) determine inorganic trace impurities in various permanent gases using high analyzer pressure and liquid nitrogen freezeout. High sample pressure (mm range) and increased ionization current (100 μA) are employed by Suttle et al. (3) to analyze inorganic impurities in helium. Low concentrations of carbon dioxide in nitrogen may be determined by a high pressure (1 mm in inlet system) technique developed by Hughes and Dorko (4). Concentrations of 180–380 ppm were determined with an accuracy better than 1% by comparing the 44/28 mass ratio of the sample to that in a series of carefully analyzed standards.

Impurities in gases must normally be determined in the presence of a large matrix, i.e., sensitivities refer to concentrational sensitivities. Pre-concentration is employed to improve sensitivity and/or accuracy by making a large absolute quantity of the impurity available for analysis, by eliminating (or reducing) certain adverse effects caused by the presence of the matrix, or by eliminating (or reducing) the need for instrument background (blank) correction.

The concentration factor or enrichment factor, $S_{I/M}$, of impurity I in matrix M is given by

$$S_{I/M} = (Q_M^0/Q_I^0)/(Q_M/Q_I) \tag{11-1}$$

where the Q values refer to quantities after separation, and Q^0 values to quantities in the original sample. The reciprocal of the enrichment factor is the separation factor ($S_{M/I}$). Recovery or yield, R_I, of the desired trace element is given by

$$R_I = (Q_I/Q_I^0) \times 100 \quad (\%) \tag{11-2}$$

The higher the enrichment factor and the closer the yield to 100%, the better the preconcentration. The enrichment factor in the preconcentration techniques to be described can be increased almost indefinitely by using large sample quantities. Available sample quantity is usually not a limiting factor in industrial gas analysis; methods can often be modified for ultra-small samples. Enrichment factors of the order of 10^6–10^7 enable one to extend sensitivities into the fractional ppm region, a level which appears adequate for current requirements.

There are four adverse effects of the matrix gas that can be eliminated by preconcentration: wall effect, linearity effect, matrix interference, and matrix incompatibility.

To obtain instrument response of at least 1 chart div/ppm (electrometer detection 1 mV potentiometric recorder), sample pressure must be of the order of 500 μ. When the sample is introduced into the analyzer, the pre-vailing pressure of 10^{-7} torr increases into the 10^{-5} torr range and various gases adsorbed on the walls may become displaced and appear as virtual impurities; nitrogen is the worst offender. The magnitude of this "wall effect" can be evaluated and results corrected by employing a "zero gas," i.e., a sample believed to be free of impurities. The limitations imposed on accuracy by such techniques are obvious. At high analyzer pressure, linearity between measured ion current and partial pressures of trace components deteriorates and becomes dependent on the nature of the gases. This "linearity effect," known for a long time, is not well under-stood; it is probably connected to changes in the work function of the filament. The maximum pressure that can be tolerated without distortion is about 600 μ, but for greatest accuracy inlet pressure should be kept below 150 μ.

"Matrix interference" is a problem that appears only for certain impurities in specific gases. The interfering peak may be a fragment peak from the matrix (e.g., N_2 in CO_2), or the molecular ion peak of the matrix itself (e.g., N_2 in CO or vice versa). Available resolution is usually not adequate for separating such interfering peaks, and even with adequate resolution it is impossible to detect small impurity peaks due to the peak broadening caused by the matrix. "Matrix incompatibility" refers to the case where the sample gas has certain properties which make it undesirable to introduce large quantities into the mass spectrometer. The most important example is the analysis of ultrapure oxygen; in the same class are corrosive and radioactive matrices.

Background correction can usually be eliminated, or at least reduced, with preconcentration. This is important, since nitrogen, oxygen, hydrogen, and carbon dioxide are almost always present as residuals, and the detection of a few chart divisions of impurity in the presence of a much larger background is not reliable. Thus, high sensitivity is useful and desirable not only in the analysis of ultrapure gases but also in the case of "ordinary" purity. The wall and linearity effects can largely be eliminated by the use of secondary electron multipliers, but matrix interference and incompatibility still often call for preconcentration.

Three techniques are briefly described in the following to illustrate matrix removal by diffusion, freezeout, and chemical reaction.

2. Impurities in Hydrogen

Here there is no peak interference, and the objectives are to increase sensitivity and accuracy. The preconcentration technique, developed by Roboz (5), is based on the removal of the matrix hydrogen by diffusion through a palladium membrane. The selective diffusion of hydrogen (and also deuterium) through hot palladium, or Pd-Ag alloys, is frequently employed for both the large-scale commercial production of pure hydrogen and the introduction of pure hydrogen into ultrahigh vaccum systems (6). For impurity analysis Pd membranes are employed in a reverse manner: hydrogen is pumped away from the sample gas and the remaining impurities are analyzed by the mass spectrometer. As long as there is no chemical reaction between the hot Pd and the impurity, enrichment factors of any desired magnitude can be obtained and at the same time adverse effects of the matrix gas are eliminated.

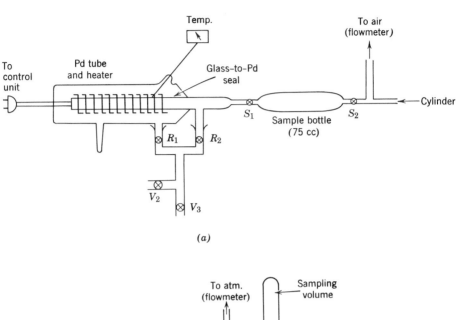

(a)

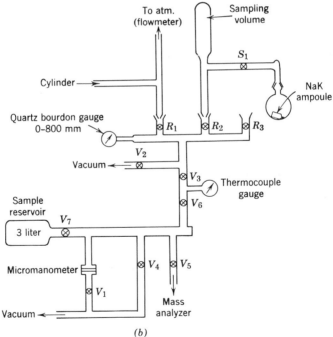

(b)

Figure 11-1. (*a*) Palladium leak assembly for preconcentration of impurities in hydrogen; (*b*) NaK assembly for oxygen removal (5,10). Details of inlet system shown in (*b*) also apply to (*a*).

Figure 11-1a shows a Pd leak assembly mounted on the inlet system of a CEC Model 21-130 mass spectrometer. The Pd tube is heated to 500°C by heater coils embedded in a ceramic tube; the temperature is measured by a chromel–alumel thermocouple. Pumping speed is variable from 0 to 10^4 μl/sec. Hydrogen is pumped away from a known sample quantity (e.g., 75 cc at 1 atm pressure) through the heated leak via valves R_1 and V_2 with all other valves closed. Pumping time is 20 min. The concentrated contaminants are introduced into the sample reservoir by opening R_2, V_3, V_6, and V_7, with all other valves closed (Fig. 11-1b). From here analysis proceeds in a routine manner.

Table 11-1   Determination of Trace Impurities in Hydrogen (ppm/div)[a]

Input resistor	Regular determination		Pd leak method	
	Normal	High	Normal	High
Nitrogen	0.6	0.07	0.02	0.003
Argon	0.5	0.06	0.02	0.002
Helium	1.2	0.15	0.05	0.006
Neon	1.9	0.23	0.08	0.009
Krypton	0.9	0.11	0.04	0.005
Xenon	2.5	0.32	0.10	0.013

[a] Ionization current: 80 μA.

Impurity recovery was found to be complete in linearity measurements with respect to both sample size and impurity concentration. Table 11-1 shows sensitivities for both the regular and the Pd leak techniques. The instrument amplifier system has been modified to include "normal" (10^{10} Ω) and "high" (10^{11} Ω) input resistors. Sensitivity can, if necessary, be further improved by increasing sample size or pressure. Reproducibility is better than $\pm10\%$ of the value reported. Cryogenically purified hydrogen normally contains <4 ppm He and N_2, <2 ppm Ne, while Ar may be as low as 0.01 ppm.

3. Impurity Freezeout

A preconcentration technique for the analysis of high purity helium has been reported by Hickam (7). Concentration of impurities is accomplished by pumping the sample through a coil cooled to 4.2°K in liquid helium. Permanent gas impurities can be determined in the 0.03–500 ppm range.

Determination of nitrogen in carbon dioxide is hampered by a strong matrix interference. It is seen from Table 9-3 that the m/e 28 contribution from CO_2 would be much too strong at high sample pressure to permit analysis of nitrogen. Threshold limit is as high as 30–50 ppm when the m/e 14 peak is used for nitrogen evaluation. Matrix carbon dioxide can be removed by chemical reaction, physical adsorption, or freezeout. The latter appears best (5) because no foreign material is involved.

A freezeout tube (copper or stainless steel) is connected through suitable valving to R_2 (Fig. 11-1). The tube is filled with carbon dioxide at measured pressure, the matrix is frozen out with liquid nitrogen, and the unfrozen portion of the sample is expanded into the sample reservoir for analysis. Experimental conditions must be established with care to prevent impurity loss by occlusion or adsorption. The efficiency of freezeout is such that an original 1 atm pressure of the matrix is reduced to the 5 μ level. Sensitivity depends on the initial sample quantity. Values similar to those presented in Table 11-1 for traces in H_2 can be routinely obtained with reproducibility better than $\pm 10\%$. The freezeout technique offers no advantage for organic impurity analysis in CO_2 by mass spectrometry. A typical ultrapure carbon dioxide contains about 1 ppm nitrogen, while all other impurities are below 0.5 ppm.

The same freezeout technique may be easily adopted for the analysis of oxygen, nitrogen, etc. in chlorine. The matrix gas is completely removed, and corrosive effects are eliminated.

4. Impurities in Oxygen

Gas chromatographic thresholds for nitrogen and argon in oxygen are poor (4 and 10 ppm, respectively) on account of inadequate resolution. Mass spectrometry cannot be routinely applied for trace analysis of oxygen for two reasons. First, oxygen markedly shortens the life of the filament in the ion source. The effect is less severe when rhenium is used instead of tungsten (8). A second adverse effect, a lethal one in trace analysis, is the formation of carbon monoxide in the ion source, originating from a reaction between oxygen and the carbon content of filaments and heaters (9).

Oxygen diffusion through silver tubes, similarly to the Pd technique, could be used for matrix removal, but the technique is inefficient due to the low pumping speed of the silver leak. A method developed by Roboz (10) is based on the removal of the matrix oxygen by chemical reaction with sodium-potassium alloy (NaK). Potassium and sodium are miscible in all portions and the alloy, in concentrates of 40–90 wt % K, is liquid at room temperature.

Figure 11-1 shows the mass spectrometer inlet system and the NaK assembly. A small ampoule of NaK, containing about 1 g of alloy under protective argon filling (commercially available), is placed into a glass bulb with an extra thick bottom. The bulb is filled with argon, and the ampoule is opened by vigorous shaking, with S_1 closed. The alloy is evenly distributed around the walls, the system is reconnected to the inlet line at R_2, and both the sample bulb and the bulb containing the NaK are evacuated. The oxygen sample to be analyzed is introduced at a known pressure into the volume enclosed by R_2 and S_1. After opening S_1, the sample gas is expanded into the previously evacuated bulb which contains the NaK. Reaction starts immediately as evidenced by a darkening of the metal film. Reaction time depends on sample quantity, but it usually takes only a few minutes to reduce the original pressure of oxygen to the level where there is a contribution of only a few divisions to the basic mass spectrometer background. Matrix removal is thus quantitative. Next, the remaining gases are expanded into the instrument reservoir and are analyzed routinely after closing valve V_6. After about 15 analyses oxygen capacity ceases; the NaK bulb is removed and is simply washed out with water after exposure to moist air.

It is important to prove that impurity recovery is complete. No chemical reaction is expected between NaK and nitrogen and the rare gases, but it has also been shown (10) that no impurity is lost due to adsorption or occlusion.

A procedure for calibration is suggested by the constant argon content of air. Another method, one that eliminates the presence of a large amount of nitrogen, is based on the fact that commercial oxygen contains about 0.3% argon. This much argon can be analyzed directly at low sample pressure and without interference, and concentration factors can be obtained for the NaK method. Table 11-2 summarizes the various methods

Table 11-2   Impurity Thresholds in O_2 (ppm)

	N_2	Ar	Kr	Xe	CO_2	N_2O	Total hydroc.
GC/TC[a]	4	10	0.05	0.02	0.1	0.05	—
MS[b]	0.03	0.02	0.05	0.15	—	—	—
IR	—	—	—	—	0.2	—	0.2
GC/Flame	—	—	—	—	—	—	0.3

Individual hydrocarbons: 0.001 by GC/TC and GC/Flame
Chemical detector tubes: 1 ppm CO and 0.1 ppm oxides of N_2

[a] TC = Thermal conductivity detector.
[b] MS = Mass spectrometry–NaK technique.

presently available for oxygen analysis. It is seen that the NaK technique fills the gaps for argon and nitrogen so that complete coverage for all important impurities can be provided. In a typical ultrapure oxygen sample the largest impurity is krypton, at the 8–10 ppm level; nitrogen is below 5 ppm, while argon and xenon are present in quantities less than 1 ppm.

A method for the determination of 10–20 ppm of carbon dioxide in oxygen has been reported by Ross (11). Pure oxygen is admitted alternately with the sample gas, the instrument is focused on the m/e 44 peak, and the CO_2 content is determined by making suitable corrections for the pure oxygen contribution.

B. Vacuum Fusion Mass Spectrometry

The most frequently employed method for the determination of gases in metals is vacuum fusion and/or extraction. In vacuum fusion the metal is heated in a graphite crucible, often in the presence of a metal bath, to 1500–2300°C, and the gases released are quickly removed and transferred to a detector. At the high temperature hydrogen is released, and certain nitrides are decomposed, while reaction of impurity oxides with the graphite results in carbon monoxide formation (some carbon dioxide is also formed). Hot extraction refers to heating the samples in vacuum below their melting points without the presence of any additional material. Gas evolution at various temperatures can be studied by this method.

The evolved and collected gases are analyzed routinely. Most commercial vacuum fusion analyzers employ gasometric methods or gas chromatography. Mass spectrometric detection has definite advantages: (*a*) positive identification of all evolving gases, (*b*) no need for chemical separation, (*c*) high sensitivity, (*d*) possibility to study the mechanism of gas evolution in dynamic measurements.

A high vacuum, low-blank gas transfer system, combined with a cycloidal mass spectrometer, has been utilized in quantitative hot extraction and vacuum fusion studies by Roboz and Wallace (12). Detection limits were extended into the fractional ppm region, and reproducibility was shown to be limited only by sample inhomogeneity. An example of a dynamic type of measurement is shown in Figure 11-2. It is seen that all hydrogen and carbon monoxide evolved in a matter of minutes (chart speed 1 in./min) after sample introduction. Such measurements are very useful in mechanism studies, and in the investigation of the "residual" hydrogen in metals.

A conventional analytical train of a vacuum fusion apparatus is replaced by a mass spectrometer in Hickam's method (13) for the analysis of carbon, oxygen, and sulfur in copper. Peterson and Bernstein (14) determined oxygen and nitrogen in lanthanum; Martin et al. (15) analyzed nitrogen,

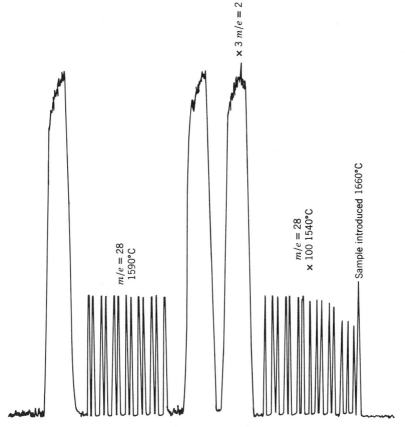

Figure 11-2. Dynamic study of gas evolution from nickel during vacuum fusion (12).

oxygen, and hydrogen in steel, employing mass spectrometric detection. Aspinal (16) describes an instrument and technique for low-level impurity analysis by vacuum fusion–mass spectrometry. Comparative data on various detection methods are presented by Martin et al. (17). The application of isotope dilution technique and spark source mass spectrography to the analysis of gases in metals are briefly discussed later in this chapter.

III. SOLIDS TECHNIQUES

A. Stable Isotope Dilution

1. Principles

In isotope dilution analysis, as the name implies, the concentration of an element in a matrix is determined from the change produced in its natural isotopic composition by the addition of a known quantity of the same element, the isotopic composition of which has been *artificially* altered. Theory has been treated extensively (18), and practical calculations were considered by Hintenberger (19). Several reviews of the technique are available (20,21).

A known amount of the sample mixture is equilibrated with a measured quantity of the isotopically enriched component (called the spike), an extraction procedure is carried out, and the new isotope ratio of the extract is measured. Since the method involves only measurements of isotopic *ratios*, the extraction does not have to be quantitative and sometimes may be omitted altogether. Three isotopic ratios must be measured (or be known): R_n, the ratio in the natural element; R_e, the ratio in the spike; and R_m, the ratio in the extracted mixture. In most cases there are only two isotopes involved, and frequently the available enriched tracer is almost monoisotopic. The concentration of impurity is calculated, in ppm by weight, from

$$E = \frac{R_e - R_m}{R_m - R_n} \times \frac{R_n + 1}{R_e + 1} \times \frac{A}{A_e} \times \frac{W}{M} \tag{11-3}$$

where A and A_e are the atomic weights of the natural and enriched element, respectively (the ratio is 1 for most purposes), M is the weight of the sample in g, and W is the amount of tracer added in μg. It is seen that the more enriched the tracer the larger the ratio difference and the greater the accuracy.

The technique is thus applicable to elements for which at least two stable (or long-lived radioactive) isotopes exist and which are available with an artificially altered isotopic ratio. Seventeen monoisotopic elements, including F, Na, Al, Mn, As, and Bi, cannot be analyzed by stable isotope dilution.

2. Experimental

Both electron impact and thermal ionization are employed in isotope dilution analysis. The former is used for gases and for elements that can be converted into volatile compounds such as CO_2, and tetramethyllead. In most cases, however, thermal ionization is applied (Section 4-III), triple and V-shaped filaments being most widely used. In a novel arrangement (22) the center filament is vertical and parallel to the side filament. Two samples (one may be a standard) can be analyzed without taking the source apart. In addition, improved accuracy is achieved (by a factor of 5) since both samples experience the same conditions and discriminations.

Commercial instruments are readily available for routine use, but for high precision, home-build machines are still common. Most instruments, including the commercial ones are patterned on a design by Nier (23).

Sample and tracer are normally brought to equilibrium in a suitable solution, preceded, if necessary, by chemical conversion into the same form. Next, the element to be analyzed is separated from the sample by extraction. Solvent extraction and ion exchange chromatography are used most frequently (20). As mentioned, the extraction need not be quantitative, and it is omitted altogether when there is no interference from other substances present. Frequently the element is only a few percent (or a few ppm) of the sample which is actually introduced into the mass spectrometer. The solvent is removed from the extract, ordinarily by evaporation on the filament itself, and the isotopic abundance ratio of the mixture is determined. Abundance ratios for the normal material and the tracer are often known but are still determined to cancel instrumental errors. A major experimental problem is how to handle the small quantities of materials without contamination. Much manual skill is needed in this technique.

3. Applications

Isotope dilution is an internal standard technique and, as such, is particularly advantageous where qualitative separation of the components of interest is either impossible or prohibitively laborious. Although the technique is some 30 years old (von Hevesy, 1934), much of the development took place during the last decade due to the ever-increasing availability of enriched stable isotopes (24).

The range, sensitivity, and precision of the method in the field of geochemistry is illustrated in Table 11-3. The first example, the determination of uranium in a stone meteorite, is a classical one. Figure 11-3 illustrates the change in the $^{235}U/^{238}U$ ratio between the $^{235}UO_2$ tracer and the UO_2 sample extracted from the meteorite to which some tracer had been added. Practical sensitivity limits are in the range 10^{-6} to 10^{-12} g and are limited by contamination problems rather than by instrumental factors. In general, an accuracy of 1–5%, and a reproducibility of ±0.1% can be expected in routine isotope dilution determinations in geochemistry. Stable isotope dilution has been used by Schnetzler et al. (25) to determine rare earth elements in rocks and minerals.

The determination of ^{235}U, the plutonium isotopes, and the general problem of "accounting analysis" in chemical processing plants in the nuclear industry have been described in three revealing papers by Webster and his co-workers (26).

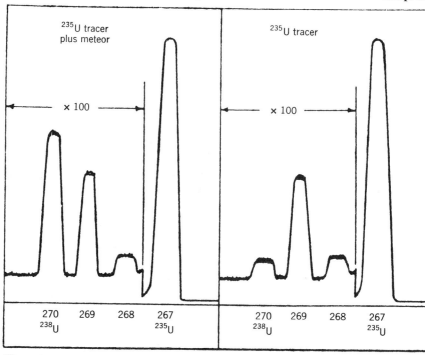

Figure 11-3. Trace uranium analysis by isotope dilution. (After M. G. Inghram, *J. Phys. Chem.*, **57**, 809 (1953).) The apparent mass of uranium is 267 because the element is analyzed as UO_2. Peaks at m/e 268 and 269 are attributable to ^{17}O and ^{18}O contributions, respectively. The change of 500% in the abundance of ^{238}U (left) is caused by the 5 ppb uranium content of the meteorite sample.

Table 11-3 Isotope Dilution in Geochemistry

Element	Sample	Result
Uranium[a]	Stone meteor	5.4 ± 0.2 ppb
	Perthite	0.22 ± 0.03 ppm
	Zircon	2650 ± 40 ppm
	Ammonium nitrate reagent	0.075 ± 0.0004 ppb
Krypton and xenon	Euxenite[b]	10^{-7} ml, STP
Rubidium	Stone meteorite[c]	0.105 ± 0.003 ppm
	Seawater[d]	121.4 ± 1.4 µg/liter

[a] G. R. Tilton, AEC Report, AE CD-3182 (1951); M. G. Inghram, *J. Phys. Chem.*, **57**, 809 (1953).

[b] G. W. Wetherhill, *Phys. Rev.*, **92**, 907 (1953).

[c] R. K. Webster, J. W. Morgan, and A. A. Smales, *Trans. Am. Geophys.* **38**, 543 (1957).

[d] A. A. Smales and R. K. Webster, *Geochim. Cosmochim. Acta*, **11**, 139 (1957).

Methods for the analysis of gases in various metals by isotope dilution were reported by Kirschenbaum and Grosse (27). To accomplish complete exchange between ^{16}O and ^{18}O, a so-called master alloy, containing known weights of both natural oxygen and spike, is prepared and mixed with the sample. The mixture is heated to 1000°C and the evolved gases are circulated into and out of the hot zone by a Toepler pump. Finally isotope ratios in the gas mixture (CO and CO_2) are determined. A wet chemical procedure combined with isotope dilution was developed by Staley and Svec (28) for the determination of nitrogen in metals and alloys. Ammonium sulfate enriched in ^{15}N is added during the process of dissolving the metal in acid and the total amount of combined nitrogen is oxidized. Isotope ratios are then determined in the resulting nitrogen gas. The range of method is from 2 ppm to 6% with an average precision greater than ±5%. Determination of hydrogen in alkali metals, using deuterium as tracer, was reported by Holt (29). Spikes equivalent to 5–1000 ppm are recovered in a 2 g sample with an average deviation of ±2 ppm. A comprehensive review of the use of stable isotope dilution technique in the determination of gases in metals has been published by Masson (30).

The determination of carbon in sodium was reported by Eng et al. (31). Carbon present in the 100 ppm range is capable of carburizing and thus weakening stainless steel; the problem may be serious in thin fuel element cladding. The accuracy of the method is 50 ± 10 μg at the lower end and 150 ± 25 μg at the upper limit.

An interesting application of the method was described by Newton et al. (32) for the determination of traces of boron in silicon. One gram of silicon is dissolved in NaOH and 1 μg of ^{10}B tracer is added in the form of boric acid. Boron is separated, after equilibrium is reached, by a modification of Morrison and Rupp's method (33) involving electrolysis through a cation exchange membrane. Boron is finally determined by isotopic dilution in a thermal source instrument using the peaks at m/e 88 and 89 ($Na_2{}^{10}BO_2{}^+$ and $Na_2{}^{11}BO_2{}^+$). The method is applicable in the 0.1–0.001 μg boron range. Accuracy decreases from ±1% to ±30% as sensitivity is increased.

An application for gas analysis has been reported by Crocker and Hart (34), who determined the fission product xenon distribution in uranium ceramics.

4. Errors

Systematic instrumental errors can largely be eliminated by proper calibration techniques, i.e., by calibrating the tracer itself using isotope dilution. Errors by interference (elements with isobars) are rare and can be avoided (20). If two analyses are made using different quantities of samples but the same amount of reagents, the contamination (from reagents but not from air) can sometimes be estimated from the deviations from proportionality. The elimination of errors due to mass discrimination by the internal standard method is discussed by Dietz et al. (35). A detailed mathematical treatment has been given by Crouch and Webster (36) on the dependence of the errors on the constitution and quantity of tracer. Relations are derived for the proper selection of the most suitable tracer and its optimum required quantity.

5. Discussion

Isotope dilution is the most sensitive and most accurate of all mass spectrometric techniques. The method is specific in that only one (occasionally two) element can be determined at a time. It is particularly attractive in cases where quantitative separation of constituents is problematic. With such high absolute sensitivities as 10^{-6} to 10^{-15} g, concentration procedures are worthwhile even with very small samples, since the method does not depend on chemical recovery. The major limitation is contamination.

B. Spark Source Mass Spectrography

1. Principles

There are four basic assumptions in the use of spark source mass spectrography for trace analysis: (1) solid samples break up completely and evenly in the spark; (2) ionization efficiency is essentially the same for all elements and there is no matrix effect; (3) definite identification of the lines can be accomplished from their position on the photoplates; and (4) the blackening of the photographic emulsion is proportional to ion exposure. None of these assumptions is completely true, and indeed much research work is being currently done to determine the magnitude of the deviations and develop methods for corrections. These efforts are justified by the spectacular success the technique has achieved during the past 5 years in solving analytical problems which could not be attacked by other methods.

The use of the spark source for trace analysis was first suggested by Dempster in 1935. Hannay and Ahearn (37) were the first to apply the method to trace analysis of both bulk and surface impurities in semiconductors, and most current work stems from their pioneering research.

2. Experimental

Commercial instruments are used exclusively, and it was shown in Section 8-V that the available instruments are basically similar, though each one has its respective merits and faults. Modifications by individual researchers most often are aimed at improvement of vacuum conditions in the source to reduce residual gases. Condensible residuals (water, carbon dioxide, hydrocarbons, etc.) can be partially removed by cryopumping utilizing a cold finger attached to the source (38). Harrington et al. employ cryosorption pumping (39), and Roboz suggested (40) sodium-potassium alloy for residual gas removal. A residual gas analyzer, connected through a suitable port, can be of great assistance in evaluating source conditions.

Analytical procedure, apart from special techniques to be discussed later, is simple and straightforward. Two rods of the sample approximately 1 cm long and 0.03 cm^2 in cross section are mounted in the ion source about 0.05 cm apart. Sample handling and cleaning varies widely depending on the objective of the investigation, size, crystalline state, etc. Sometimes presparking is used to remove surface impurities. Another possibility is cathodic etching (glow discharge) instead of chemical cleaning.

Optimum conditions for sparking are established by adjusting spark voltage (up to 100 kV), pulse length (25–200 μsec), and pulse repetition rate (1–10^4 pulses/sec). Exposure can be measured by the duration of sparking, the number of sparks, and by the deposited charge. The first is not reliable, and the second is employed only for short exposures (single sparks), used in the determination of the minimum exposure required for the most abundant isotope of the matrix to become detectable. Exposure is normally measured by a beam monitor which intercepts a certain portion of the total ion beam before it enters the magnetic analyzer. Exposure is measured in coulombs and is increased by steps of $10^{1/2}$. A linear range of 10^7:1 can be covered in 15 exposures on a plate.

The mass range is determined by the accelerating potential and magnetic field intensity employed. There is no mass scanning with photographic detection, and an adequate waiting period is required after a new magnetic field setting to prevent line broadening due to hysteresis. Plate developing techniques have already been discussed (Section 5-II).

3. Qualitative Spectrography

Mass fragment identification is made from the position of the lines on the photoplate. Distances are measured using a vernier attachment on the densitometer. Since distances on the photoplate are porportional to

$(m/e)^{1/2}$ the mass scale can be calculated once two lines are identified. No calibration compounds are necessary (Section 10-III-A) because lines of the major element and its multiply charged fragments are always available for this purpose. Line identification is usually programmed into a computer (see later) so that mass assignment is a routine process. When working with fixed electric and magnetic fields, plates can often be directly compared by experienced operators. Tables are available (41) to aid in rapid mass identification.

4. Sensitivity

Ultimate detection limits depend largely on instrument design, involving such factors as the intensity of the ion beam reaching the photoplate, resolution setting, vacuum conditions in the analyzer, and, of course, the type of photoplate employed. The smallest amount of impurity that can be detected under given experimental conditions depends on how weak a mass line can be detected (visually or instrumentally) on the photoplate. Exposure cannot be increased indefinitely, partially because of the time involved, and, more importantly, because the diffuse background also increases with exposure. The maximum exposure that is still practical is of the order of 10^{-6} coulomb and corresponds to a detection limit of 0.001 ppm (atomic). This value refers to the most favorable cases, i.e., metals, mono-isotopic elements, no interference from matrix lines, good sparking, etc. The limit of detection is poorer by one or even two orders of magnitude in cases where interferences occur.

Many tables have been published and are available from mass spectrometer manufacturers on detection limits of elements in various matrices. Figure 11-4 is a graphic summary (42) of detection limits for some 70 impurity elements in graphite, indium arsenide, copper, and aluminum. Detection for individual elements in aluminum, for example, are as follows:

0.001 ppm:	15 elements, including U, Bi, Au, Ho, Ce, La, Ba, Cs, I
0.002–0.01 ppm:	38 elements, including Pb, Pt, Ag, As, Co, Fe, K, F, B, Li
0.01–0.1 ppm:	19 elements, including Mo, Se, Ge, Ga, Zn, Ni, Cr, Va, Cl, P, S, Si, Mg

These limits are estimated on a 1 μcoul exposure basis, assuming equal relative sensitivities (see later).

An important factor limiting sensitivity is the fogging of plates ("halo") by the heavy ion beams of the matrix element. This is, of course, particularly disadvantageous when the trace sought is near the major component. Various methods of plate splitting and masking (43,44) have been developed to pass the major ion beam through without touching the photo-

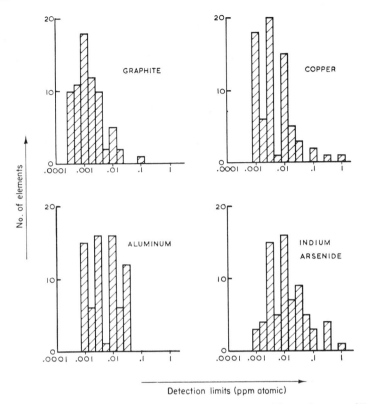

Figure 11-4. Graphical summary of detection limits with spark source (42).

plate. Kennicott (45) reduced fogging by a modified development procedure while Honig et al. (46) reported on a low-fogging gelatin-free photoplate, consisting of a thin film of silver bromide crystals on a glass plate.

Sensitivity is closely related to resolution. Spark source instruments of present-day design operate at a resolution of about 2500 and, as was discussed in Section 8-II, about 35–40% of the elements are resolved from their nearest neighbor at this level.

5. Relative Sensitivities

One of the basic assumptions in spark source mass spectrography is that the method is nonselective, i.e., all elements have approximately the same sensitivity. Due to selective volatilization, non-uniform ionization efficiency, selective transmission of the analyzer, and possibly other processes, the ion beam hitting the photoplates is not truly representative of the sample. The ratio of the apparent concentration found and the true concentration is called the relative sensitivity coefficient (RSC).

Ahearn (47) has normalized and tabulated relative sensitivity coefficients by a number of workers. About 70% of the values are in the 0.5–2.0 range, indicating that the basic assumption of nearly equal sensitivities cannot be in error by more than this factor. Within a factor of 3, the reported coefficients do not reveal any trend suggesting that the coefficients might be functions of the matrix. In a more recent paper Ahearn (48) summarized current work concerning RSC and pointed out the importance of sample inhomogeneity. This leads directly to the problem of calibration standards which is discussed in connection with quantitative spectrography.

6. Quantitative Spectrography

a. *Visual Method.* Semiquantitative estimation of the concentration of an impurity in ppm atomic can be made from the exposures necessary to produce *just detectable* lines of an isotope of a matrix element (or an impurity of known concentration) and an isotope (normally the most abundant) of the impurity element. An internal standard is thus used and the formula, according to Craig et al. (49), is

$$\text{Conc} = \frac{E_s}{E_i} \times \frac{X}{100} \times \frac{I_s}{I_i} \times \frac{S_s}{S_i} \times \frac{A_i}{A_s} \times \frac{M_i}{M_s} \times 10^6 \text{ ppm atomic} \quad (11\text{-}4)$$

where the subscripts s and i refer to standard and impurity, respectively; E denotes the exposure (in coulombs) at which the particular line becomes observable to the naked eye; X is the concentration of the internal standard (%); I is the natural abundance (%) of the isotope on which the estimate is based. Correction factors are A for the area of the spectral line (mm²), S for relative sensitivity, and M for the intensities of singly and multiply charged ions if used.

b. *Microphotometric Method.* Blackness of the spectral lines can be measured by microphotometers (microdensitometers) which actually measure transmissions. Among the many commercial models the comparator types are popular since they permit simultaneous projection of the unknown and a reference spectrum, making line identification convenient. Transmission (transparency) is defined as the ratio of the intensities of the transmitted and the incident light. Other frequently used functions for light attenuation are opacity, the reciprocal of transmission, and density, the $\log_{10}$ of opacity.

To determine impurity concentration, a characteristic curve of the emulsion is obtained by plotting exposure (or log E) versus density (or some function of it) of the major element or a standard. Calibration thus involves the deposition of a known number of ions (beam monitor) and the measurement of the resulting density. The density of the impurity is next measured, and the values for E_i and E_s in equation 11-4 are obtained from the calibration curve. Only the linear portion of the S-shaped calibration curve can be used with good accuracy.

Calibration methods which do not depend on either the validity of the reciprocity law (see below) or the constancy of the ion yield, were developed by Owens and Giardino (50). They make use of the constant abundance ratios of the isotopes of the matrix in plotting the calibration curves. This method is an adaptation of the two-line technique of Churchill (51) used in emission spectrography. The current status of quantitative aspects of spark source spectrography, including discussions of photographic emulsions as ion detectors is given by Owens (52). There are a number of corrections one can make to compensate for errors inherent in photographic ion recording. This is presently an active field and is also covered by Owens' review. Corrections for line width, plate background, ion energy, ion mass, charge distribution, etc. have been made by various investigators. RSC's are, of course, also correction factors. A critical assessment of all available quantitative evaluation methods has been published by Schuy and Franzen (53).

c. Image Effects. Certain image effects that characterize emulsions are, qualitatively at least, similar, regardless of whether the exposure results from electromagnetic radiation or from particle bombardment. The *reciprocity law* states that the photographic response is determined only by the total amount of exposure and the rate of exposure is irrelevant. The validity of the law can be studied by measuring the exposure time required to produce a certain density at different levels of intensity, i.e., at different ion currents. No reciprocity failure has been found with the Q-type and Shumann-type emulsions (37,54).

There are a number of other image effects, the importance of which has not yet been thoroughly investigated. According to the intermittency effect the density produced by a given total energy (exposure) is smaller when the exposure in continuous than when it comes in small installments. The effect of solarization concerns strong exposures. On the horizontal portion of the characteristic curve further exposure produces a partial reversal of the image (43). As a result of the developing process localized changes in density are evidenced (adjacency effect). A weak line between two strong ones may not appear at all as a result of depletion of the developer due to the strong lines (Eberhard effect). A warped shape results when developer inhibition occurs in the region between two lines close together.

d. Computer Data Processing. A very large amount of information is available on the photographic plates and in manual evaluation only portions of it can be utilized. Simple computer programs for line identification were developed almost immediately after the appearance of commercial instruments. Possibilities of data processing for both qualitative and quantitative analysis, however, have been investigated only during the last 2–3 years, following exploratory work on the parameters influencing quantitative spectrography.

Many automatic techniques start with an empirical function, given by Hull (55), which expresses the entire transmission range of the Ilford Q2 emulsions as a function of exposure,

$$E(T) = \left(\frac{1 - T_x}{T_x - T_\infty}\right)^{1/R_x} \tag{11-5}$$

where E_x (ncoul) is "relative exposure" normalized to unity at $E(1) = (1 + T_\infty)/2$, T_x and T_∞ are transmissions of spectral line i as measured and for saturation transmission, respectively, and R_x is proportional to the maximum slope of the photographic response curve (value 1.0–1.5). Woolston's technique (56), which starts with the above equation, also includes corrections for background fog, variable spectral line width and shape, and relative ionization sensitivity. Somewhat different systems for the quantitative evaluation of mass spectrographic plates have been described by Kennicott (57) and Franzen and Schuy (58). As the relative merits of the various parameters influencing quantitative measurements are still strongly debated, significant activity is expected in this field.

e. Precision, Accuracy, and Standards. The average deviation of close isotope ratios was found to be 3–4% (50,53). The deviation is attributed to variations in the ion response of the Ilford Q2 emulsion. This, therefore, represents the best reproducibility that one can expect at present. Relative intensities of lines can be estimated visually to within a factor of 2. The usual reproducibility that can be achieved in microphotometric comparisions is about $\pm 20\%$. In general, an overall factor of 2–3 should be applied in most applications. In the comparative analysis of similar samples, or where standards are available and microdensitometric measurements are employed, a standard deviation, expressed as per cent concentration of impurity, of about 30% can be expected.

In principle, mass spectrometric measurements are absolute and no standards are needed. Ion yields are measured in the form of blackening on photographic plates, and these can be converted to relative ion yields using the basic emulsion calibration curve. Data may then be corrected as described above. In practice, however, particularly at the methods development stage, it is certainly desirable to have standards available for calibration and comparison. Almost everyone working with the spark source has tried, at one time or another, to use materials with known composition. Results of such measurements depend, naturally, on the accuracy with which the sample composition is known.

The main difficulty with standards is connected to sample homogeneity, since only very little material is consumed in an analysis. The volume v (cm³) of a specimen consumed in making an exposure to deposit n ions ($n \simeq 3000$ for a just detectable line, $n \simeq 10^5$ for a measurable line) in order to determine an impurity of concentration p (ppm, weight) in a matrix of density ρ (g/cm³) is given by (59)

$$v = \frac{A_i}{\rho P I_i} \Delta n \times 10^8 N_A \qquad (11\text{-}6)$$

where A_i is the impurity atomic weight and I_i the percentage abundance of the isotope line measured, while N is Avogadro's number. The number of atoms consumed in order to deliver one ion in the photoplate is Δ, which depends upon the specimen material, and upon such experimental parameters as slit dimensions, acceleration voltage, etc. Values for Δ normally lie between 10^{-7} and 10^{-8}.

The disappointing results of cooperative studies using spectroscopically analyzed standard materials called attention to the problem of standards with suitable homogeneity. Considerable work is now being done (48,59) to develop standards with dopant concentrations in the ppm to ppb (atomic) range. This work is indeed needed since the accuracy of present techniques is inadequate in many cases in solid-state physics and chemistry, where a factor of 5 in concentration of an impurity may result in manyfold change in certain properties (1). Recent advances in the precision of the analysis of conducting materials are summarized by Franzen and Schuy (60).

7. Applications

a. Metals, Semiconductors, and Insulators. Analysis of metals and semiconductors in straightforward, and many applications have been reported. In metallurgy, the comparison of alloys and the checking of refining processes (electrolytic, zone melting) are areas in which the technique is ideally suited because of its high sensitivity and its ability to provide wide coverage in a single analysis (49). A typical mass spectrum of a steel sample is shown in Figure 11-5.

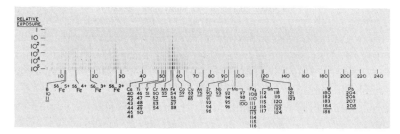

Figure 11-5. Mass spectrum of a steel sample. (Courtesy of AEI/Picker X-ray Co.) Ions covering a mass range of 35:1 are focused simultaneously on the photoplate. All elements from lithium to uranium are analyzed.

Low melting point metals, such as gallium, may be analyzed by Wolstenholme's technique (61), which uses liquid nitrogen to cool the samples during sparking. The main problem is selective volatility of impurities.

The use of the spark source technique in solid-state physics and chemistry has been extensive, and impressive agreements between the impurity determined and the electrical or magnetic properties, etc. have been reported. A detailed review of the mass spectrometry of III–V compounds is given by Brown et al. (62).

Methods for the analysis of insulators were developed by Hannay and Ahearn (37) and Ahearn (63). Ahearn's technique is shown in Figure 11-6; the sketches are self-explanatory.

b. Analysis of Powders. Metal powders have been analyzed since the early days of spark spectrography by applying them as pastes and letting

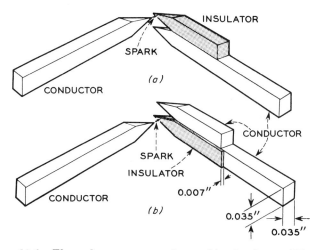

Figure 11-6. Electrode arrangement for sparking insulators (63).

them dry on graphite electrodes, or by using electrodes to which a sufficiently rough powder could adhere. It is obvious that the ratio of sample to supporting material being ionized is erratic in such techniques. The method of "pressed electrodes" is now becoming increasingly popular and the many variations developed are summarized by Owens (64).

In the technique of Brown and Wolstenholme (65), the powder to be analyzed is mixed with an equal amount of pure graphite powder, and the mixture is pressed in a die to form a conducting rod about 2 cm long and 2 mm in diameter. The surfaces are scraped, and the rod is broken in the middle to give two electrodes with freshly broken ends. Rather long presparking is suggested to remove surface impurities. Rods thus produced are electrically conducting and have the required physical strength. Graphite powder of very high purity is now available from several suppliers and most impurities are below 1 ppm. It is claimed that there are no severe problems with interference from carbon lines, and the contribution of the support graphite to the spark has been found reasonably constant and consistent. On long exposures, however, carbon clusters (e.g., C_{16}) might present problems. An interesting application has been reported by Brown and Wolstenholme (65). Two magnesium oxide samples were compared, one that failed as an insulator after 2000 hr, the other an unused material. Portions of the spectra are shown in Figure 11-7. The results indicate that the magnesium oxide had become contaminated with chromium and manganese, the materials in contact with it. The chromium concentration increased from 15 to 1800 ppm, and that of manganese from 4 to 50 ppm. The nickel, cobalt, and iron content remained unchanged.

The choice of conducting powder is limited by the availability of sufficiently pure materials which are also mechanically adequate for pressing. Graphite has several disadvantages (e.g., mechanical strength, carbide formation, carbon clusters) and there is a current search for a substitute; silver might be a possibility (64).

c. Microsamples. Problems with microsamples may be divided into two groups: isolated particles and localized inhomogeneities. Small isolated

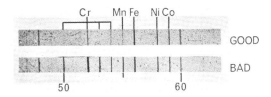

Figure 11-7. Part of mass spectra of magnesium oxide sample (65). (Courtesy of AEI/ Picker X-ray Co.)

particles such as tiny crystals, whiskers, and filings present difficulties in handling and mounting. Every case should be treated individually depending upon the type of the sample and the nature of the information sought. The sample holder must be of a material of high purity, and mounting must be such that the contribution from the support material is minimal. Ahearn (47) employed high purity silicon crystal holders to mount small samples. Holder and sample were held together by a platinum wire, and crystals of any shape could be mounted. Almost every investigator has a favored technique to mount small samples and they differ mainly in the selection of the supporting material.

A technique for exceptionally small materials was developed by Wallace and Roboz (66). The sample is folded into a pure platinum foil strip which serves, after cutting, as both electrodes (Fig. 11-8). A portion of the spectrum using a blank strip of platinum foil without sample is shown in Figure 11-9a; a small gold impurity is also seen in addition to the doubly charged platinum lines. The unknown material, actually scrapings from the anode of a special thermionic material, weighing less than 20 μg, turned out to be zirconium (Fig. 11-9b). Two milligrams of platinum foil was con-

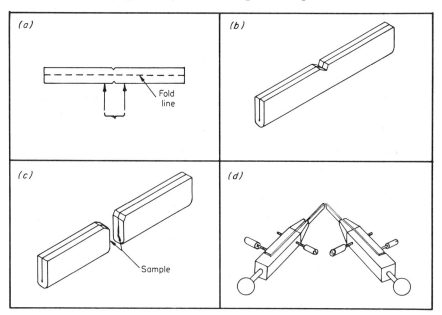

Figure 11-8. Technique for analysis of microsamples in spark source (66). Prepare pure matrix ribbon 0.001 $\times$ 0.050 $\times$ 0.750 in. (a) Notch ribbon at midpoint and fold lengthwise. Open fold and place microsample as close to center notches as practicable. (b) Fold and compress matrix ribbon over microsample. (c) Cut matrix ribbon at index notches to form electrode pair. (d) Finished microelectrodes clamped and ready for use.

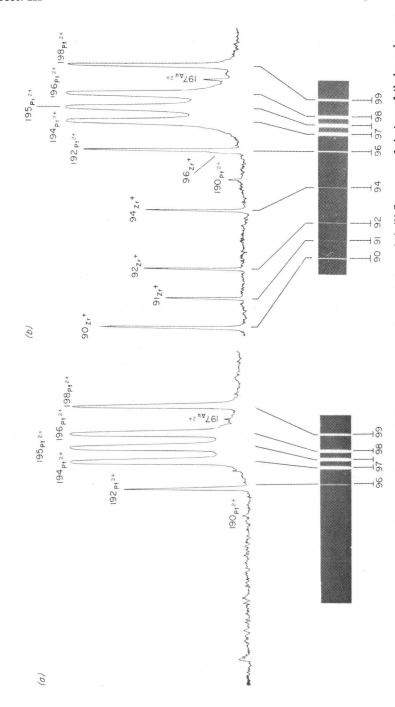

Figure 11-9. Analysis of microsamples in spark source (66). (*a*) Spectrum of pure platinum foil. (*b*) Spectrum of platinum foil plus unknown sample.

sumed. A second sparkling of the electrode confirmed that essentially all the unknown sample deposit had been used up in the initial exposure. In addition to zirconium, which was actually the matrix of the unknown sample, other impurities were also detected. The $^{11}B^+$ isotope, for example, appeared at a density equal to that of the $^{92}Zr^+$ isotope (19% abundance), representing a 0.2% impurity in the microsample.

Inclusions, segregations at grain boundaries, and layer structures constitute the second category of microsamples. Here the main problem is to confine the spark to the localized area. A detailed review of mass spectrographic microprobe analysis is given by Hickam and Sweeney (67), who did much of the development work in this field. These authors discuss problems in probing volumes, surfaces, and depth. In all cases the objective is to obtain under controlled sampling a small volume of material from the specific area of interest.

The isometric view of an electrode structure for "milliprobe" analysis using a point to stationary plane configuration is shown in Figure 11-10. In this technique, developed by Roboz and Wallace (66), an unconfined spark is used to remove material of the order of 1–100 μg, forming hemispherical craters of 100–500 μ diameter. The technique is useful to study impurities in the 1–100 ppm range in striated metallic samples, where the zones of interest frequently have dimensions of a few tenths of a millimeter. The pointed counterelectrode is a high purity metal (gold, silicon, etc.) and the gap between electrode and sample is about 25 μ. Figure 11-11 shows how a crater is formed in the middle of a striated structure. To minimize the size of the craters, Hickam and Sweeney (67) employed "single" sparks,

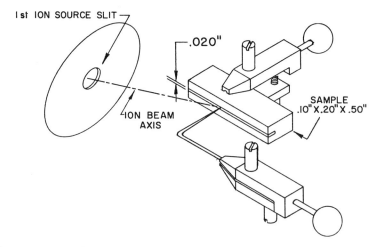

Figure 11-10.   Isometric view of electrode structure for milliprobe analysis (66).

Figure 11-11.   Analysis of layer structures using spark as a probe (69).

consisting of a few 20 μsec pulses. Crater diameters of 25 μ with depths of 3 μ were obtained in iron samples, and the quantity of material removed (10^{14} atoms) was adequate to determine impurities to about 0.1%.

The area of sparkling can be increased, and the depth simultaneously decreased, by slowly translating the sample with respect to the point. Sparking an area of 10^{-4} to 10^{-3} cm^2 to a depth of 3–5 μ enables one to detect impurities of the order of 1 in 10^4 and an equivalent average thickness of 3–5 Å (67). A point-to-rotating plane configuration is shown in Figure 11-12. The sample, in a highly polished disk form of 2 cm diameter, is rotated at 1750 rpm. Near the edge of the disk and 25 μ above it is a stationary point of the probe material. The surface moves with a linear velocity of 2 μ/μsec with respect to the point. Both inorganic and organic materials have been studied by this technique (67). Microprobe methods are still in an early development stage, but the potential has been amply demonstrated and widespread use is expected as soon as commercial instruments feature the required accessories.

d. Surface Contamination.   Bulk impurities and surface contamination can be distinguished by the spark source method. Ahearn has shown (68) aluminum as a bulk impurity in a nickel sample when the density of the aluminum line did not change as the spark eroded into the sample. In a sample of silicon, however, magnesium was found to be a surface impurity, since the density of the magnesium line decreased significantly as the surface layer was removed. Microphotometer tracings are shown in Figure 11-13. The aluminum line shown is that of Al^{2+} at mass 13.5, the magnesium line is Mg^{2+} at mass 12.5. All three exposures were the same

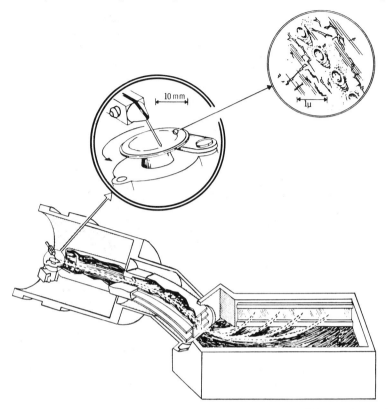

Figure 11-12. Mass spectrographic rotating electrode microprobe. (After W. M. Hickam and Y. L. Sandler in *Surface Effects in Detection*, J. I. Bregman and A. Dravniel, Eds., Spartan Books, Washington, 1965, p. 193.)

in each case. Surface contaminants equivalent to 0.01 monolayer may be detected by this method.

e. Gases in Metals.   Carbon, hydrogen, nitrogen, and oxygen are components in the background gases in every mass spectrometer. They are also adsorbed both as gases and as oxides on freshly etched surfaces. The possibility of determining nitrogen in metals has been explored by Roboz (69), who reported experiments with standard steel and titanium samples.

A more recent study by Harrington et al. (70) proved that carbon, oxygen, and nitrogen can indeed be analyzed in trace quantities in various matrices, including iron, tungsten, copper, and silver. Results compared favorably with combustion–chromatographic, inert gas fusion, and micro-Kjeldahl techniques, but problems involving sample inhomogeneity were clearly evident. The same authors also demonstrated the use of spark source technique for the analyses of these "difficult" elements in various sections of zone-refined metals.

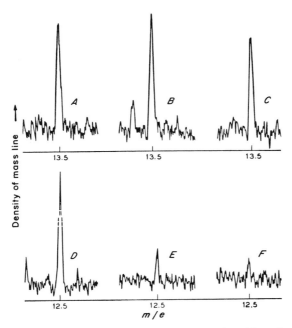

Figure 11-13. Bulk impurity versus surface contaminants (68). Top, aluminum as bulk impurity in nickel. Bottom, magnesium as surface contaminant on silicon.

f. Analysis of Liquids. A sensitive but rather simple technique, based on the ability to determine surface contaminants, has been developed by Ahearn (68) for the analysis of trace impurities in liquids. The impurities are transferred to a high purity electrode by dipping it into a solution, by electroplating, or by placing the liquid on an electrode and evaporating the solvent. This latter technique is greatly improved by the use of specially shaped electrodes and an electric field with which proper localization of the water droplets can be achieved. The practical limit of detection of impurities is 10^{-10} g. The method is used by Chastagner (71) to analyze impurities at the ppb level in heavy water moderator from water-moderated nuclear reactors.

g. Other Applications. Spark source mass spectrography has been applied in many areas of physics, chemistry, metallurgy, and geology, and the reader is directed to reference journals for lists of applications. More unique recent applications include the analysis of blood plasma (72), investigation of organic materials (73), techniques for spark isotope dilution (74), and applications for geochemical analysis (75). Direct current sources have been discussed briefly in Section 4-IV. Although there is considerable current interest in the low-voltage dc sources, relatively few papers have appeared on analytical applications (76).

8. Discussion

Spark source mass spectrography provides a very sensitive means for impurity analysis. The limit of sensitivity is 0.001 ppm (atomic) in a number of cases, and is in the 0.1–0.001 ppm range for most elements in any matrix. The method is applicable to almost all elements in the periodic table with equal sensitivity within a factor of 3–10. An overall coverage of the elements is possible in a single analysis, and there is linearity with concentration. Both bulk and surface impurities can be studied, and samples as small as a few micrograms can be analyzed. The major disadvantage is that quantitative determination with good accuracy is not possible, at least at this time. Sample inhomogeneity appears to be the major problem in developing suitable standards. The undesirable effects of sample inhomogeneity may be reduced by employing an ion *beam chopper* so that the ion beam is allowed to enter the analyzer region for only a fraction of the total sparking time (77). This is a controversial area at the present time. Another area of current interest is the development of electrical detection systems for improved accuracy and convenience in the analysis of individual elements.

C. Ion Bombardment

1. Principles

Sputtering is the removal of surface particles by positive ion (primary ion) bombardment. Those particles which leave the surface as ions (secondary ions) can be directly introduced into a mass analyzer, but neutral particles must first be ionized in an auxiliary electron bombardment source prior to mass analysis. The ejected ions and neutral can thus, in principle, be studied separately by turning on or off the electron beam in the auxiliary source and by applying suitable bias potentials. The behavior of the positive and negative ions can be predicted, qualitatively at least, by the Langmuir-Saha equation (eq. 4-2). Elements with a low ionization potential produce positive ions; elements with a high electron negativity produce negative ions.

The bombarding primary ions, even those with relatively high energy, do not penetrate deeply into the surface. It is a reasonable assumption that the secondary particle emission takes place only from the top few layers, so the method permits identification of species emanating from surfaces. By successive "peeling" of the surface, thin surface films may be analyzed.

2. Experimental

Earlier investigations were confined to secondary ions rather than to sputtered neutrals due to difficulties in obtaining sufficiently strong ion currents from ionized neutrals. This problem is now being overcome with high efficiency sources and sensitive detectors. There are severe vacuum problems in sputtering sources. It is important that the target region be completely separated from the source of primary ions so that a high intensity ion beam can be focused on the target without having an excessive gas pressure caused by an overflow of un-ionized gases. Residual gases will, naturally, be ionized together with the neutral sputtered particles by ionizing electrons. Figure 4-9 shows an ion bombardment source. Targets are of the order of 0.3×0.8 mm in size. Since the objective is to study the surface, proper sample preparation is vital; the method of cleaning must be determined by the aim of the investigation. Analytical procedure, after initial pumpdown, starts with an extended baking to achieve the best possible vacuum (10^{-8} torr level in the target region). Background gases consist of the omnipresent CO and CO_2. Operating pressure increases to 10^{-6} torr level after the primary ion beam is turned on. The purest available gases (argon) are used to form the primary ion beam.

Background is determined by a separate measurement in the Nier source with the accelerating voltage of the primary beam off. A more sophisticated method with which background peaks can be suppressed is the so-called synchronous source-detector (78). Two synchronized electrical choppers are employed. The first one pulses the bombarding ion beam so that the analyzed beam has an ac component in addition to a dc current from the background; the latter originates from the continuously operated Nier source. The dc component is recorded from the modulated current in an ac amplifier, the output of which is rectified by the second chopper. This technique is particularly useful when relatively high (10 mA) electron emission current has to be used in the Nier source to increase the sensitivity for neutrals. Mass analysis is performed in a conventional manner with magnetic scanning.

3. Applications

Until recently most studies were aiming to explore the possibilities and limitations of the method for investigating surfaces, rather than trying to solve specific problems. Data have been reported on the mass, charge, and energy distribution of the sputtered particles as a function of such parameters as the mass, energy, and density of the primary ionization beam for almost every instrument designed for sputtering. High purity, single-crystal copper is a favored target material (79). Studies specifically directed to measure sputtering yield as a function of experimental variables were made for a number of elements, such as copper, gold, graphite, aluminum (80).

Peaks caused by sputtered copper ions are shown in Figure 11-14 (77); the spectra were obtained in a Nuclide-type source (Fig. 4-9). The long tails on the upper mass side (Fig. 11-14a) indicate large initial energy spread (>350 eV), while the sharp low mass side shows that the ions had a definite minimum initial energy. In Figure 14b, the superimposed background gases and sputtered neutrals are shown (Nier source "on"). The purpose of the synchronous source-detector is to eliminate this kind of interfering background.

An advantage of the ion bombardment technique is that the sample is not heated appreciably. To explore this property Honig (81) deposited a monolayer of ethyl radicals on an etched and chlorinated germanium surface. The neutrals obtained included species from C_2 to C_2H_5 in addition to the background CO and CH_4 peaks. The presence of the ethyl radical monolayer was demonstrated and the H/C ratio was found to be close to

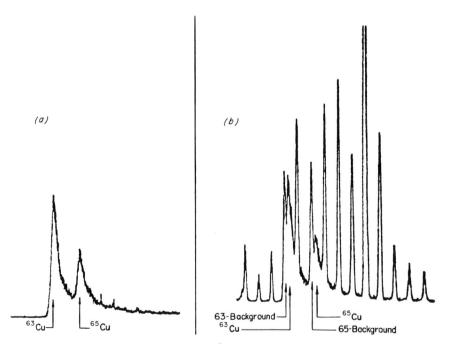

Figure 11-14. Mass spectra of sputtered copper ion peaks (78). (*a*) Spectrum showing the sputtered ion peaks and part of their energy tails (argon ion energy 1000 eV). (*b*) Spectrum showing sputtered copper ion peaks along with background peaks from residual gases in electron impact source.

the expected value. The positive ion spectrum was quite complex including combinations of Ge with C_1, C_2, C_3 and C_4 groups. Peaks observed with "clean" germanium (Ge^+, $GeOH^+$, etc.) were also present. The negative ion spectrum consisted of peaks from elements with high electron affinity (O^-, F^-, C_6^-, etc.). Other targets investigated included coal, diamond, and silicon carbide. Neutral particles detected were mainly adsorbed gases and radicals, and only occasionally were neutrals ejected from the substrate.

A double-focusing mass spectrograph has been used by Beske (82) to utilize a sputtering ion source (Penning-type arc) for trace analytical applications. Photometric comparisons with standards permitted the detection of 10^{-9} g of lithium, 10^{-8} g of aluminum, copper, and arsenic, and 10^{-7} g of cobalt. Concentrational sensitivities as low as a few ppm were achieved for a number of elements. Variation of the concentration of gallium with depth as it diffused into a silicon disk has also been investigated by Beske.

Preliminary results on an ion bombardment microprobe (Fig. 11-15), consisting of a sputtering source mass spectrometer and an ion microscope, were reported by Castaing and Slodzian (83). The secondary ions obtained by bombarding the sample surface with primary rare gas ions are mass analyzed. Next, the focused ion beam is accelerated onto an image converter releasing tertiary electrons which, in turn, are accelerated in the

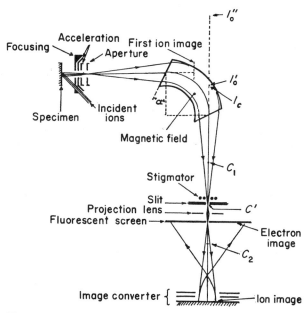

Figure 11-15.   Ion bombardment microprobe analyzer (83).

opposite direction to eventually fall on a fluorescent screen. A mass map of the top few atomic layers of the target is thus obtained. The method, when fully developed (calibration problems), may become a very useful complement to electron microprobe x-ray analysis. An ion microprobe mass spectrometer with improved vacuum system has recently been described by Barrington et al. (84).

4. Discussion

The adjective "preliminary" is frequently used in the previous section. Indeed, the capabilities of the technique as a trace analytical tool are just being explored. The heavy dependence on the work function of the surface and the uncertainty of the apparent temperature make quantitative determinations difficult. Neutral sputtered particles are believed to be better suited for compositional analysis since discrimination is reduced. The spectra consist mainly of gases and radicals adsorbed on the surface. The limiting factor is background; suppression, as described, may be the solution. Honig estimates (81) that surface impurities at a concentration of 1 part in 1000 can be identified by present methods, but there are no standards available. Accuracy and precision cannot even be considered. The main advantage of the method is the possibility of studying surfaces without significantly heating the sample. Disadvantages are poor selectivity and problems in quantitative determinations.

D. Complete Thermal Vaporization

In this method the total amount of the solid sample is evaporated from a crucible which is placed very near to an electron impact source. A source where the crucible is collinear with the ion beam is described by Honig (85), who also evaluated many materials fror crucible use. The maximum evaporation temperature for most materials is 1500°K. The ion current corresponding to the element in question must be integrated over time to eliminate the effects of vapor pressure differences and to obtain total concentrations. Comparison with standards is needed for quantitative measurements. About 1 mg of the properly cleaned sample is placed in a previously outgassed crucible, the temperature is increased in 30°K steps, and the complete spectrum is taken at every level. Attainment of equilibrium is checked by monitoring major peaks, and all ion currents are integrated until the sample is completely consumed.

The method is applicable to about 3/4 of the elements in the periodic table. Elements emitted as positive ions can be detected in the 0.1–10 ppm range, while the limit for those with poor ionization cross section, and for those with background interference, is a few ppm. Elements with low vapor pressure at the crucible temperature (boron, carbon, titanium, etc.) cannot be detected (85).

This technique is rather tedious and is subject to large errors. Fractional distillation may result in a sort of "self-concentration," which may be useful in certain types of problems, such as analysis of arsenic. There is relatively little interest in complete thermal vaporization at the present time. For high-temperature studies, the crucible is replaced by a Knudsen cell, so that equilibrium between solid and vapors is established (Section 12-II).

E. Summary of Solids Techniques

While the spark source technique is most useful for exploring impurities in a sample, isotope dilution is the technique to use when specific impurities are studied in detail. Isotope dilution is the most sensitive and accurate method of the four discussed and rates high in these respects among all trace analytical procedures. Its absolute sensitivity is in the 10^{-6} to 10^{-15} g range, and accuracy as good as 1% can be achieved. The method is almost unique in the ease with which preconcentration may be applied; it does not depend on quantitative chemical recovery. The method is, however, highly selective. Isotope dilution is relatively slow and requires special care.

The vacuum spark method is a survey type technique; it has wide coverage, it is sensitive, it is accurate enough for many problems, and it is fast. Almost all the elements in the periodic table can be analyzed in almost any matrix. Metals, semiconductors, insulators, solids and liquids, powders and microsamples, bulk and surface impurities can be equally well handled. Concentrational sensitivities range from 0.001 to 0.1 ppm for practically all elements; only those appearing in the instrument background are exceptions. Results are reproducible within a factor of 3–10 without sensitivity corrections and standards. In one analysis that may take no more than an hour, a complete, broad, semiquantitative survey can be obtained on the impurities in a sample at the fractional ppm level.

The main features of the ion bombardment source are that it is well suited for surface studies as the temperature of the samples is not appreciably increased. Accuracy is poor. Complete thermal vaporization offers a means of reducing interferences by separating impurities from major constituents in time. It may be used as a tool to study interactions at high temperatures.

The use of focused radiation (laser, arc-image furnace, exploding wire) for ion production has not been employed for analytical purposes. Honig and Woolston (86) and Honig (87) investigated the laser-induced emission of ions, neutral atoms, and electrons from solid surfaces and estimated that impurities to the ppm level can be detected with proper techniques. More recent work on laser-induced ion analysis is reported by Fenner and Daly (88).

References

*1. Hannay, N. B., in *Trace Analysis: Physical Methods*, G. H. Morrison, Ed., Interscience, New York, 1965, Chapter 2.

2. Parkinson, R. T., and L. Toft, *Analyst*, **90**, 220 (1965).

3. Suttle, E. T., D. E. Emerson, and D. W. Burfield, *Anal. Chem.*, **38**, 51 (1966).

*4. Hughes, E. E., and W. D. Dorko, *Anal. Chem.*, **40**, 750 (1968).

5. Roboz, J., *12th Ann. Conf. Mass Spectrometry*, ASTM Committee E-14, Montreal, 1964, *Proc.*, p. 536.

6. Young, J. R., *Rev. Sci. Instr.*, **34**, 374, 891 (1963).

7. Hickam, W. M., *9th Ann. Conf. Mass Spectrometry*, ASTM Committee E-14, Chicago, 1961, *Proc.*, p. 346.

8. Robinson, C. F., and A. G. Sharkey, *Rev. Sci. Instr.*, **29**, 250 (1958).

9. Young, J. R., *Appl. Phys.*, **30**, 1671 (1959).

*10. Roboz, J., *Anal. Chem.*, **39**, 175 (1967).

11. Ross, P. J., *Anal. Chem.*, **38**, 1437 (1966).

12. Roboz, J., and R. A. Wallace, *10th Ann. Conf. Mass Spectrometry*, ASTM Committee E-14, New Orleans, 1962, *Proc.*, p. 199.

13. Hickam, W. M., *Anal. Chem.*, **24**, 362 (1952).

14. Peterson, D. T., and D. J. Bernstein, *Anal. Chem.*, **29**, 254 (1957).

15. Martin, S. F., J. E. Friedline, and G. E. Pellissier, *Trans. AIME*, Aug. 1958, p. 514.

16. Aspinal, M. L., *Analyst*, **91**, 33 (1966).

17. Martin, J. F., R. C. Takacs, R. Rapp, and L. M. Melnick, *Trans. Met. Soc. AIME*, **230**, 107 (1964).

18. Gest, H., M. D. Kamen, and J. M. Reiner, *Arch. Biochem.*, **12**, 273 (1947).

19. Hintenberger, H., in *Electromagnetically Separated Isotopes and Mass Spectrometry*, M. L. Smith, Ed., Butterworths, London, 1956, p. 177.

*20. Webster, R. K., in *Methods in Geochemistry*, A. A. Smales and L. R. Wagner, Eds., Interscience, New York, 1960; and in *Advances in Mass Spectrometry*, J. D. Waldron, Ed., Pergamon, New York, 1959, p. 103.

21. Wilson, H. W., and N. R. Daly, *J. Sci. Instr.*, **40**, 273 (1963); de Bievre, P. J., and G. H. Debus, *Nucl. Instr. Methods*, **32**, 224 (1965).

22. Patterson, H., and H. W. Wilson, *J. Sci. Instr.*, **39**, 84 (1962).

*23. Nier, A. O., *Rev. Sci. Instr.*, **18**, 398 (1947); *Anal Chem.*, **34**, 1358 (1962); also Ridley, R. G., and D. E. Silver, *J. Sci. Instr.*, **38**, 47 (1961).

24. *Catalog and Price List of Stable Isotopes*, Oak Ridge National Laboratory,

Tennessee; *Radioactive Materials and Stable Isotopes*, Isotope Division, U.K. Atomic Energy Commission, Harwell, Berks, England. Both lists are reissued periodically.

25. Schnetzler, C. C., H. H. Thomas, and J. A. Philpotts, *Anal. Chem.*, **39**, 1888 (1967).

*26. Webster, R. K., D. F. Dance, J. W. Morgan, E. R. Preece, L. J. Slee, and A. A. Smales, *Anal. Chim. Acta*, **23**, 101 (1960); Webster, R. K., A. A. Smales, D. F. Dance, and L. J. Slee, *ibid.*, **24**, 371 (1961); Webster, R. K., D. F. Dance, and L. J. Slee, *ibid.*, **24**, 509 (1961).

27. Kirschenbaum, A. D., and A. V. Grosse, *Anal. Chim. Acta*, **16**, 225 (1957); *Anal. Chem.*, **29**, 980 (1957).

28. Staley, H. G., and H. J. Svec, *Anal. Chim. Acta*, **21**, 289 (1959).

29. Holt, B. D., *Anal. Chem.*, **31**, 51 (1959).

*30. Masson, C. R., *Met. Rev.*, **12**, 147 (1967).

31. Eng, K. Y., R. A. Meyer, and C. D. Bingham, *Anal. Chem.*, **36**, 1832 (1964).

32. Newton, D. C., J. Sanders, and A. C. Tyrell, *Analyst*, **85**, 870 (1960).

33. Morrison, G. H., and R. Rupp, *Anal. Chem.*, **27**, 1150 (1955).

34. Crocker, I. H., and R. G. Hart, *Anal. Chem.*, **38**, 781 (1966).

35. Dietz, L. A., C. F. Pachucki, and G. A. Land, *Anal. Chem.*, **34**, 709 (1962).

36. Crouch, E. A., and R. K. Webster, *J. Chem. Soc.*, **1963**, p. 118.

*37. Hannay, N. B., and A. J. Ahearn, *Anal. Chem.*, **26**, 1056 (1954).

38. Socha, A. J., and R. K. Willardson, *14th Ann. Conf. Mass Spectrometry*, ASTM Committee E-14, Dallas, Texas, 1966, *Proc.*, p. 127.

39. Harrington, W. L., R. K. Skogerboe, and G. H. Morrison, *Anal. Chem.*, **37**, 1480 (1965).

40. Roboz, J., *Anal. Chem.*, **38**, 1629 (1966).

41. Heath, R. L., in *Table of Atomic Masses*, J. W. Guthrie, Ed., Sandia Corp., Albuquerque, 1961. E. B. Owens and A. M. Sherman, *Mass Spectrographic Lines of the Elements*, Tech. Rept. 265, Lincoln Laboratory, Mass. Inst. Technol., 1962. A. Cornu, R. Massot, and J. Terrier, *Atlas de Raies*, Centre d'Etudes Nucleaires, Grenoble, France, 1963.

42. Associated Electrical Industries, Ltd., Manchester, England, *Data Sheets* 2030 Series, Nos. A2, A12, A13, and A14.

43. Mai, H., *J. Sci. Instr.*, **42**, 339 (1965); also in *Advances in Mass Spectrometry*, Vol. III, W. L. Mead, Ed., Inst. Petroleum, London, 1966, p. 163.

44. Ahearn, A. J., and D. L. Malm, *Appl. Spectry.*, **20**, 411 (1966).

45. Kennicott, P. R., *Anal. Chem.*, **38**, 633 (1966).

46. Honig, R. E., J. R. Woolston, and D. A. Kramer, *14th Ann. Conf. Mass Spectrometry*, ASTM Committee E-14, Dallas, Texas, 1966, *Proc.*, p. 481.

47. Ahearn, A. J., *11th Ann. Conf. Mass Spectrometry*, ASTM Committee E-14, San Francisco, Calif., 1963, *Proc.*, p. 223.

*48. Ahearn, A. J., in *Symposium on Trace Characterization, Chemical and Physical*, Natl. Bur. Std. Monograph 100, 1967.

*49. Craig, R. D., G. A. Errock, and J. D. Waldron, in *Advances in Mass Spectrometry*, J. D. Waldron, Ed., Pergamon, New York, 1959, p. 146.

50. Owens, E. B., and N. A. Giardino, *Anal. Chem.*, **35**, 1172 (1963).

51. Churchill, J. R., *Ind. Eng. Chem. Anal. Ed.*, **16**, 653 (1944).

*52. Owens, E. B., in *Mass Spectrometric Analysis of Solids*, A. J. Ahearn, Ed., Elsevier, New York, 1966, Chapter III; also *Appl. Spectry.*, **21**, 1 (1967).

*53. Schuy, K. D., and J. Franzen, *Z. Anal. Chem.*, **225**, 260 (1967).

54. Wagner, H., *Ann. Physik*, **7**, 189 (1964); also Kawano, H., *Bull. Chem. Soc. Japan*, **37**, 697 (1964).

55. Hull, C. W., *10th Ann. Conf. Mass Spectrometry*, ASTM Committee E-14, New Orleans, La., 1962, *Proc.*, p. 104; also *Tech. Paper* TP 131, dated 4-65, Consolidated Electrodynamics Corp., Monrovia, Calif.

56. Woolston, J. R., *RCA Rev.*, **26**, 539 (1965).

57. Kennicott, P. R., *14th Ann. Conf. Mass Spectrometry*, ASTM Committee E-14, Dallas, Texas, 1966, *Proc.*, p. 278.

*58. Franzen, J., and K. D. Schuy, *Z. Naturforsch.*, **21a**, 1479 (1966).

59. Halliday, J. S., P. Swift, and W. A. Wolstenholme, in *Advances in Mass Spectrometry*, Vol. 3, W. L. Mead, Ed., Inst. of Petroleum, London, 1966, p. 143.

*60. Franzen, J., and K. D. Schuy, *Z. Anal. Chem.*, **225**, 295 (1967).

61. Wolstenholme, W. A., *Appl. Spectry.*, **17**, 51 (1963).

62. Brown, R., R. D. Craig, and J. D. Waldron, in *Compound Semiconductors*, Vol. 1, R. K. Willardson and H. L. Gering, Eds., Reinhold, New York, 1962, p. 106.

63. Ahearn, A. J., *J. Appl. Phys.*, **32**, 1195 (1961).

64. Owens, E. B., in *Advances in Mass Spectrometry*, Vol. 3, W. L. Mead, Ed., Inst. of Petroleum, London, 1966, p. 197.

65. Brown, R., and W. A. Wolstenholme, *Nature*, **201**, 598 (1964).

66. Wallace, R. A., and J. Roboz, *11th Ann. Conf. Mass Spectrometry*, ASTM Committee E-14, San Francisco, Calif, 1963, *Proc.*, p. 451; Roboz, J., in *Trace Analysis: Physical Methods*, G. H. Morrison, Ed., Interscience, New York, 1965, Ch. 11.

*67. Hickam, W. M., and G. G. Sweeney, in *Mass Spectrometric Analysis of Solids*, A. J. Ahearn, Ed., Elsevier, 1965, Chapter V.

68. Ahearn, A. J., in *6th National Symposium on Vacuum Technology*, Pergamon, New York, 1960, p. 2.

69. Roboz, J., *11th Ann. Conf. Mass Spectrometry*, ASTM Committee E-14, San Francisco, Calif., 1963, *Proc.*, p. 572.

*70. Harrington, W. L., R. K. Skogerboe, and G. H. Morrison, *Anal. Chem.*, **38**, 821 (1966).

71. Chastagner, P., *Chem. Eng. News*, **42**, 79 (July 13), 1964.

72. Wolstenholme, W. A., *Nature*, **203**, 1284 (1964).

73. Chastagner, P., *Appl. Spectry.*, **19**, 33 (1965).

74. Leipziger, F. D., *Anal. Chem.*, **37**, 171 (1965).

*75. Taylor, S. R., *Geochim. Cosmochim. Acta*, **29**, 1243 (1965); also R. Brown and W. A. Wolstenholme, *Nature*, **201**, 598 (1964).

76. Hintenberger, H., and K. D. Schuy, *Z. Anal. Chem.*, **197**, 98 (1962); *Z. Naturforsch.*, **18a**, 926 (1963).

*77. Jackson, P. F., J. Whitehead, and P. G. Vossen, *Anal. Chem.*, **39**, 1737 (1967); Vossen, P. G., *ibid.*, **40**, 632 (1968).

78. Smith, A. J., A. Cambey, and D. I. Marshall, *J. Appl. Phys.*, **34**, 2489 (1963); *Vacuum*, **14**, 263 (1964); Liebl, H. J., and R. F. K. Herzog, *J. Appl. Phys.*, **34**, 2893 (1963).

*79. Woodyard, J. R., and C. B. Cooper, *J. Appl. Phys.*, **35**, 1107 (1964).

80. Walther, V., and H. Hintenberger, *Z. Naturforsch.*, **18a**, 843 (1963).

81. Honig, R. E., in *Advances in Mass Spectrometry*, Vol. 2, R. M. Elliott, Ed., Pergamon, New York, 1963, p. 25.

82. Beske, H. E., *Z. Angew. Phys.*, **24**, 30 (1962).

*83. Castaing, R., and G. Slodzian, *J. Microscopie*, **1**, 395 (1962); also in *Advances in Mass Spectrometry*, Vol. 3, W. L. Mead, Ed., Inst. Petroleum, London, 1966, p. 9.

*84. Barrington, A. E., R. F. K. Herzog, and W. P. Poschenrieder, *J. Vacuum Sci. Technol.*, **3**, 239 (1966); also *Geochim. Cosmochim. Acta*, **29**, 1193 (1965).

85. Honig, R. E., *Anal. Chem.*, **25**, 1530 (1953); *J. Chem. Phys.*, **22**, 1610 (1954); also in *Trace Analysis of Semiconductor Materials*, J. P. Cali, Ed., Macmillan, New York, 1964, pp. 169–206.

86. Honig, R. E., and J. R. Woolston, *Appl. Phys. Letters*, **2**, 138 (1963).

87. Honig, R. E., *Appl. Phys. Letters*, **3**, 8 (1963).

88. Fenner, N. C., and N. R. Daly, *Rev. Sci. Instr.*, **37**, 1068 (1966).

Chapter Twelve

Mass Spectrometry in Physical Chemistry

I. IONIZATION AND APPEARANCE POTENTIALS

A. Definitions

The first ionization potential (I or IP) of an atom or molecule is defined as the energy input required to remove to infinite distance a valence electron from the lowest occupied atomic or molecular orbital of the neutral particle to form the corresponding atomic or molecular ion, also in its ground state. The removal of more strongly bound electrons requires more energy, giving rise to a series of higher ionization potentials. The unqualified term "ionization potential" refers to the first ionization potential. In such 0–0 type transitions, the energy difference between the states is determined by the difference in energy levels of the equilibrium minima of the potential curves of the two states (Fig. 12-1). In contrast to this *adiabatic* ionization potential, the term *vertical* ionization potential refers to a Franck-Condon type transition (Section 9-XI-A), where no change in the inter-nuclear separation occurs during the transition, because the interaction time of the order of 10^{-15} sec is too short compared to the 10^{-12} to 10^{-13} sec vibration intervals. When an electron is removed from a neutral species, the stability of the remaining structure usually decreases, resulting in an increase of equilibrium distances. The process of ionization will thus bring the ion into a vibrationally excited state, as shown in Figure 9-10, curve *2*. When the difference between the internuclear distances of the two states is large, the transition leads to a dissociative state of the ion (curve *3*, Fig. 9-10). Conversely, when equilibrium distances for the neutral and ionized state are nearly equal, the vertical and adiabatic ionization potentials become equal. For polyatomic molecules the potential curves must be replaced by multidimensional potential energy surfaces. Contour diagrams are often used to simplify the presentation of internuclear separations.

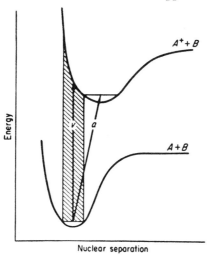

Figure 12-1. Franck-Condon diagram illustrating vertical (*v*) and adiabatic (*a*) transition in diatomic molecule. (After 21, p. 120).

When dissociation accompanies (or rather follows) the process of ionization, fragment ions start to appear, and the amount of energy required for the appearance of a particular fragment ion in its ground state, from the molecule also in its ground state, is designated by the term *appearance potential* (*A* or AP). In the course of fragmentation neutral particles may, of course, also be formed. Ionization potentials can thus be considered as appearance potentials in those cases where ionization occurs without dissociation.

In electron impact experiments both reactants and products may be in excited states, and usually little is known about these excited states. For practical purposes, therefore, the definitions of IP and AP should be modified in such a manner that numerical values obtained are understood to refer to the particular apparatus and technique employed in the study. In fact, the divergence of a newly determined IP or AP from that corresponding to the theoretical definitions given provides a measure of the "quality" of the apparatus and technique used. The true values are often not available, so IP and AP values must always be treated with caution. This warning will become more obvious after the description of the experimental methods available for IP and AP determination.

B. Energetics of Electron Impact Processes

When a neutral diatomic molecule XY (Y being electronegative) is bombarded by electrons of controlled energy, the following dissociative ionization processes may occur:

$$XY + e^- \rightarrow XY^+ + 2e^- \tag{12-1}$$
$$XY + e^- \rightarrow X^+ + Y + 2e^- \tag{12-2}$$
$$XY + e^- \rightarrow (XY^-)^* \rightarrow X + Y^- \tag{12-3}$$
$$XY + e^- \rightarrow X^+ + Y^- + e^- \tag{12-4}$$

It is noted here that ionization may also be caused by photoionization when quanta of sufficiently high energy are available,

$$XY + h\nu \rightarrow XY^+ + e^- \tag{12-5}$$

The energy balances for processes 12-1 and 12-2 are given by

$$A(XY^+) = I(XY) \tag{12-6}$$

and

$$A(X^+) = I(X) + D(X—Y) + E_{kin} + E_{exc} \tag{12-7}$$

where $A(XY^+)$ and $A(X^+)$ denote the appearance potentials of XY^+ and X^+, and $D(X—Y)$ denote the dissociation energy of XY; the I values refer to ionization potentials. In the first process there is no fragment formation, but in the second process the total kinetic energy available is divided between the fragments according to the law of conservation of momentum. The particles may or may not have associated with them kinetic energy of appreciable quantity. In addition to kinetic energy (E_{kin}), fragments may also possess varying amounts of excitation energy (E_{exc}). The excitation energy term includes electronic, vibrational, and rotational components. Although the excitation energy term is usually small, it is not negligible. At any rate, it is immediately clear that the measurement of appearance potentials provide only an *upper* limit for the dissociation energy of the X—Y bond, provided $I(X)$ is known. More will be said about the connection between AP and dissociation energy later.

Processes 12-3 and 12-4 involve negative ion formation. The formation of XY^- is a relatively rare event since it involves resonance electron capture. An example is the formation of SF_6^- from SF_6; this reaction may be used to calibrate the electron-energy voltage scale at low energies. The XY^- ion usually decomposes to form Y^- and a neutral X fragment. The last process is simple ion pair formation. The appearance potentials are given by

$$A(Y^-) = D(X—Y) - EA(Y) + E_{kin} + E_{exc} \tag{12-8}$$

and

$$A(Y^-) = A(X^+) = I(X) + D(X—Y) - EA(Y) + E_{kin} + E_{exc} \tag{12-9}$$

where the EA terms refer to electron affinities.

Numerical values for the various processes can be estimated by considering that the first ionization potential of most atoms lies in the 6–15 eV range, electron affinities are normally between 1.5 and 3.5 eV, while the dissociation energy of most chemical bonds lies in the 3–5 eV range. Available tables are discussed after the description of the experimental determination of IP and AP values.

C. Determination of IP and AP

1. General Remarks

There are three techniques for the determination of IP and AP: ultraviolet spectroscopy, electron impact, and photon impact. In ultraviolet spectroscopy IP's are calculated from the Rydberg series, i.e., a series of adsorption bands resulting when an electron is excited from its ground state into orbitals with increasingly higher energies. The technique yields true adiabatic ionization potentials but, due to the complexity of the spectra, can only be used for simple molecules.

Electron impact methods consist of two basic steps: First, an *ionization efficiency curve* is obtained for the ion under study by measuring the intensity of the ion beam as a function of electron energy. These curves (Fig. 1-8) are *S-shaped* with a nearly asymptotic onset, and have a tendency to level off about 10–20 eV above the onset. Since the neighborhood of the ionization threshold is very sensitive to changes in electron energy, the main instrumental difficulty, as detailed below, is to obtain a monoenergetic electron beam. The second step is the interpretation of the experimental ionization efficiency curve. There are several techniques, each attempting to eliminate, or at least reduce, the uncertainties introduced by the nonhomogeneity of the electron beam.

In photon impact methods the problem of energy homogeneity is replaced by difficulties in obtaining sufficiently intense and continuous photon beams.

2. Electron Energy Spread

Although conventional mass spectrometers normally operate at 70 eV, most instruments feature adjustable electron energy controls. Variations of electron beam energy arise from three sources:

(*a*) Electrons obtained in thermal emission possess considerable kinetic energy. The energy distribution is of Maxwellian nature, and the number of electrons, dN of mass m, emitted from unit area of the filament per second, with kinetic energy between U and dU, can be obtained by considering the electrons in thermodynamic equilibrium with the filament (1)

$$dN(U) = (4\pi mA/h^3) \, U \, \exp\left[-(W + U)/kT\right]dU \qquad (12\text{-}10)$$

where T (°K) and W (erg) are the absolute temperature and work function, respectively, of the filament of surface area A, h is Planck's constant, and k is Boltzmann's constant. The general shape of the curve is shown in Figure 12-2. Work function appears as the determining factor: with tungsten or rhenium filaments (Table 4-2) operating at 2000°K, the energy spread is 0.2–0.5 eV. Thoriated iridium (2) or oxide coated filaments, e.g., rhenium coated with lanthanum hexaboride (3), provide electron beams with somewhat lower energy spread.

(b) The potential difference through which electrons are accelerated before entering the ionization region is measured (with a potentiometer) between the cathode and the final electron collimating slit (Fig. 4-1). The true electron energy, however, is somewhat different due to contact potentials and space charge effects. Normally it is hoped that the use of a calibrating gas (see below) will compensate for these effects. Sometimes the interior of the source is gold-plated to reduce contact potentials and surface effects.

(c) The electron beam traveling through the ionization region toward the electron trap (anode) is subjected to the electric field which is employed to draw out positive ions from the source into the accelerating-collimating region. The drawing-out field is perpendicular to the direction of the motion of the electrons, while the source magnetic field (used for electron collimation) is parallel to it. The result is that this energy superposition is periodic.

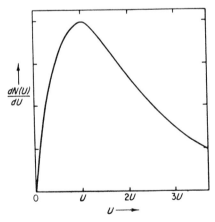

Figure 12-2. The number of electrons in the energy interval between U and $(U \pm dU)$ as a function of U, assuming a Maxwellian distribution.

The magnitude of the excess kinetic energy from this source is about an order of magnitude lower than that originating from thermal emission uncertainties. The effect can be minimized by keeping potentials in the source unchanged during analysis, by employing electric fields as small as possible, and by reducing the thickness of the electron beam (measured in the direction of the repeller field) by using a large magnetic collimating field.

The undesirable effects of energy distribution of the electron beam can be almost entirely eliminated by constructing ion sources which incorporate some kind of electron energy selector. Monoenergetic electron impact techniques are discussed shortly.

3. Conventional Methods

In an indirect approach to the problem of electron energy spread, no significant changes are made on the ion source, but available source potential adjustments are utilized as advantageously as possible, and then appearance potentials are determined by ingenious data handling procedures. All such techniques are based on comparisons with a calibrating gas introduced simultaneously with the unknown. Argon and krypton are frequently employed since their ionization potentials are well established.

The oldest is the *vanishing current* (or *initial onset*) technique: The electron voltage at which the ion just appears is taken as IP or AP. The measured voltage (i.e., the voltage applied to the accelerating electrodes) is, of course, not the true voltage, and correction is made by introducing a calibration gas. An improved version of the method (4) is the *extrapolated difference* technique. After the ionization efficiency curves of both the unknown and known gases (introduced simultaneously) are determined, the ordinate scales are adjusted so as to make the linear portions of the curves parallel (Fig. 12-3). Next, a plot of ΔV versus i is drawn, where the ΔV's are the differences of electron voltages corresponding to various values of the ion current i. The line thus obtained is extrapolated to zero current and the ΔV value corresponding to that point is taken as the difference between the appearance potentials of the unknown and the calibration gas. This technique is simple and often gives remarkably reliable data. It can only be employed with ions that are reasonably intense ($>5\%$ of base peak). The main asset of both the vanishing current and the extrapolated difference techniques is that values obtained are close approximations of the adiabatic ionization (or appearance) potential.

Several logarithmic methods have been proposed, most of them based on the *critical slope method* of Honig (1). This technique starts from equa-

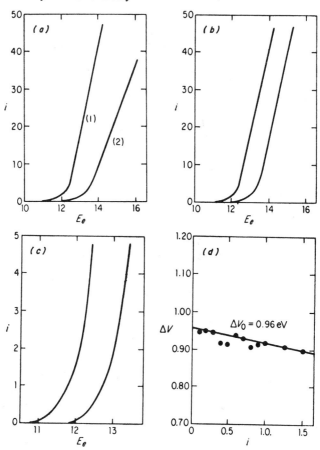

Figure 12-3. Illustration of the Warren method of interpreting ionization efficiency curves. (After 21, p. 170.) First, curve 2 in (a) is multiplied by 1.75 to cause the linear rising portions of the two curves to be parallel as shown in (b). Next, from the expanded portions of the feet of the curves (c) the voltage differences, ΔV, are determined and plotted as a function of the current, i, as shown in (d).

tion 12-10 and assumes that the probability that an electron of energy U produces an ion which is collected is proportional to the square of the electron energy in excess of the appearance potential. The critical voltage is determined eventually by establishing the point on a semilog plot of the ionization efficiency curve which has a slope of $\frac{2}{3} kT$, where T (°K) is the temperature of the filament and k is Boltzmann's constant. Lossing, Tickner, and Bryce (5) modified Honig's method: the pressure of the gases is first adjusted to yield comparable ion currents, and then the logarithms of peak heights as percentages of the abundances at 50 eV are plotted against the uncorrected electron energies. In the region of about 1% of the ion current obtainable at 50 eV, the curves have been found parallel for many substances and values were claimed to be reproducible to as low as 0.01 eV (Fig. 12-4). The difference in ionization potentials between two samples is given by the displacement along the voltage axes of the two curves. Another semilogarithmic method based on Honig's technique has been proposed by Morrison and Nicholson (6). The appearance potential is taken (somewhat arbitrarily) as the point at which the log ion current versus electron energy curve ceases to be a straight line. Still another logarithmic method, termed energy compensation technique, has been suggested by Kiser and Gallegos (7) to quickly determine appearance potentials with TOF instruments. First, ion currents of the unknown and the calibration gas are measured at 50 eV on a dual-channel recorder. Next, the sensitivities of both amplifiers (one for each output) are increased a hundredfold, and the electron energy is reduced until observed ion intensities read the same for each ion as previously at 50 eV. The desired appearance potential is determined from the measured voltage difference and the known IP of the calibrating gas.

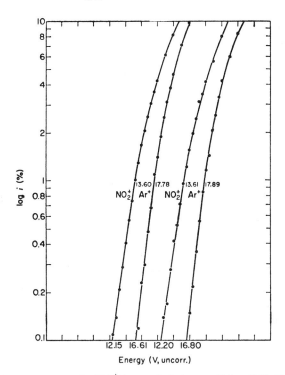

Figure 12-4. Semilog curves for NO_2^+ from ethyl nitrate. (After J. E. Collin, *Bull. Soc. Roy. Sci. Liege*, **32**, 133 (1963).) Argon is used as reference.

In studying mathematical techniques for the interpretation of ionization efficiency curves, Morrison (8) developed the "derivative" method in which differential curves of $\Delta V/\Delta i$ (V = energy of ionizing electrons, i = measured ion current) are evaluated. This technique permits the study of the fine structure of the ionization curves and provides information on the excited states of the ions. For details of this extremely interesting and useful technique the reader is referred to the original papers (8) and several subsequent publications by Dorman (9).

An interesting technique of ionization efficiency curve evaluation is Morrison's "deconvolution" technique (10) in which the use of Fourier transforms and computerized noise removal results in the uncovering of many structural details from the curves. Still another technique, proposed by Winters et al. (11), called the *energy distribution difference* method, reduces the effective electron energy distribution by a relatively simple mathematical transformation to such a degree that the ionization curves obtained in conventional instruments become equivalent to those obtainable with monoenergetic methods. Krypton is employed for calibration. The method has been shown to be applicable to inert gas ions and simple molecular ions such as CO, O_2, and N_2^+ and work is in progress for extension to more complex organic molecular ions as well as fragmentation ions.

4. Monoenergetic Electron Impact Methods

a. RPD Technique. In the *retarding potential difference* (RPD) method developed by Fox et al. in 1951 (12), the energy of the bombarding electrons is confined to a small range and ion formation takes place in a field-free region. In recent years "Fox sources" have become commercially available as accessories, resulting in an increased popularity of the technique.

In Figure 12-5a the normal electron acceleration voltage is V_1; when no retarding potential is employed, electrons enter the ionization region (7 in Fig. 12-5c) with the usual Maxwellian energy spread. When a potential of V_R, which is negative with respect to the cathode, is applied to the retarding electrode (electrode 4), slow electrons cannot pass and the low end of the energy distribution curve is sharply cut. Those electrons with initial energy just equal to eV_R will leave the slit of the retarding electrode with essentially zero velocity, while those with initial energies less than eV_R will not pass at all. The difference in the observed ion current for the given distribution in the ionization chamber and that obtained when V_4 is changed by ΔV_R, keeping V_1 fixed, is the ion current produced by the electrons within the ΔV_R band. In Figure 12-5b, V_{4M} is the largest

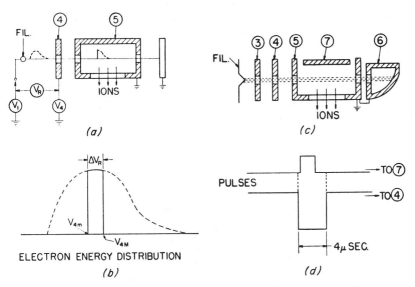

Figure 12-5. (*a*) Simplified electron gun illustrating the effect of the retarding potential in preventing low energy electrons from entering ionization chamber. (*b*) Electron energy band of width ΔV_R utilized in obtaining ionization curves. (*c*) Electrodes in gun assembly. (*d*) Pulses employed to insure that the ionization takes place in a field-free region (12).

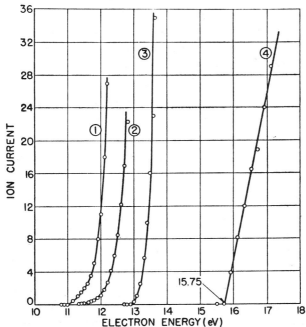

Figure 12-6. Effect of electron energy spread on the observed shape of the ionization curve for argon (12).

and V_{4m} is the smallest in absolute value of the two V_4 settings. The appearance potential is obtained at the point where the current becomes zero in the plot of the difference in ion current versus V_{4M} (keeping V_R constant). The ionization efficiency curve this way corresponds to one obtained with electrons whose energy spread is ΔV_R. In practice ΔV_R is of the order of 0.1–0.2 eV, making it possible to observe the fine structure of the ionization efficiency curves which, in turn, can be used for the study of excited ionic states in correlation with the molecular orbital theory.

Ion formation in a field-free region is achieved by the use of a pulse generator capable of producing two synchronized pulses at a fixed repetition rate (Fig. 12-5c and d): one pulse is used for ion acceleration (electrode 7), the other for electron retardation (electrode 4). When the electrons are passing through 4, no voltage is applied to 7 and thus ions are formed in a field-free region. Next, a negative voltage of sufficient magnitude is applied to 4 to stop the electron beam completely, followed a fraction of a microsecond later by a voltage to electrode 7 to facilitate ion removal. Figure 12-6 shows how the ionization efficiency curve of argon can be improved by the RPD method: Curve 1 is obtained with no retarding potential employed; in curve 2 the amplifier sensitivity was reduced by a factor of 10. In curve 3 retarding potential is employed, and in curve 4 both retarding potential and pulsing fields are used. It is seen that the curvature near the threshold is almost completely eliminated. The appearance potential of argon obtained this way (15.75 V), and also values for several other gases, showed excellent agreement with spectroscopic values.

Cloutier and Schiff (13) modified the RPD method by utilizing the potential minimum that exist in the region between the cathode and the anode of a space charge limited cathode with which they replaced the retarding slit. Fine control in the retarding potential is obtained by changing the anode voltage. The application of the RPD method for TOF instruments is described by Melton and Hamill (14).

b. Electrostatic Electron Selectors. The most straightforward, though certainly not the simplest, method for obtaining monoenergetic electrons is the combination of the mass spectrometer with an "electron spectrometer." Both parallel plate (15,16) and 127° cylindrical condenser energy selectors have been described, the latter type being more popular.

Figure 12-7 shows an energy selector developed by Marmet and Kerwin (17), who considerably improved an earlier basic design by Clarke (18). Electrons originate from filament F and are accelerated by a few volts between the filament and the first slit ($\sim 0.7 \times 4$ mm) of the selector. The

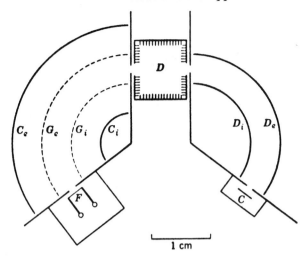

Figure 12-7. Energy selector of Marmet and Kerwin (17). F is the filament, C_e and C_i are collector electrodes, G_i and G_e are the main selector electrodes (made of 90% transparent tungsten mesh), D is the ionization chamber, D_i and D_e are analyzer electrodes, and C is the collector plate.

electrons are directed along the main condensers G_e and G_i and enter into the ionization chamber D after accelerated to the desired final energy by a potential difference between the exit slit of the selector and the entrance slit of D. The inside deflection plate is usually at a set voltage (of the order of +5 V) while the potential on the outer deflection plate is variable to about 4 V. Those electrons which pass through the main selector electrodes, which are made of 90% transparent tungsten mesh, are collected on the collector electrodes C_e and C_i which are maintained at a positive potential. The energy distribution of the electron beam is determined in a second section, where D_e and D_i are the "analyzer" electrodes and C is a collector plate. Problems in electron selectors include space charge effects, secondary electrons, and spurious reflections, in addition to the major problem of obtaining the desired electron current density.

An electron beam of about 10^{-7} to 10^{-8} A intensity with an energy spread of about 0.03 eV can routinely be obtained with electron selectors. This enables one to study the fine structure of the ionization efficiency curves and several such studies have been reported; the references quoted describe several applications. Brion et al. reported investigations on the ionization of oxygen and the rare gases by monoenergetic electrons (19). The photograph of a commercially available electron selector is shown in Figure 12-8.

Figure 12-8. Photograph of electron energy selector (courtesy of Nuclide Corp.).

5. Photoionization Methods

When quanta of sufficiently high energy are available, ionization may proceed according to equation 12-5 and appearance potentials may be determined using a photon beam. In fact, as mentioned in Section 4-VII, where instrumentation is briefly discussed, the main use of photoionization spectroscopy is in molecular energetics studies. The source shown in Figure 4-12 is a typical example of recent designs incorporating a vacuum UV monochromator and a mass spectrometer. The presence of a mass spectrometer (20) enables one to unequivocally identify the ions produced in the photoionization process, and provides considerable assistance to the study of the upper energy states of polyatomic molecules. The limiting factor in instrumentation is the sensitivity of the detection device; energy spread of less than 0.05 eV is now routinely obtained. Other problems include required photon density, upper energy limit, and variations in photon density with λ.

Figure 12-9 shows the photoionization efficiency curve of ethylene obtained with the apparatus shown in Figure 4-12. The curve is a plot of the photoionization yield as a function of photon energy measured in wavelengths and expressed in energy units. The photoionization yield (ions/

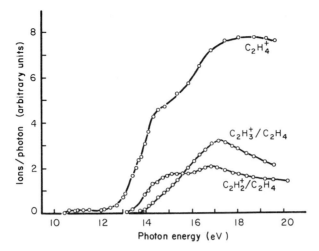

Figure 12-9. Photoionization efficiency curve of ethylene and fragments. (From ref. 25 in Chapter 4; also see Fig. 4-12.)

photons), is obtained as the ratio of the ion current, measured on the collector of the mass spectrometer, and the light intensity, measured on a photomultiplier.

6. Theoretical Calculations

Theoretical calculation of appearance potentials from basic physico-chemical properties is an extremely difficult task which has been attempted only for simple molecules. For diatomic and polyatomic molecules various semiempirical methods have been developed based on equivalent orbitals or group orbitals. For a relatively elementary review and additional references the reader is directed to reference 21. Calculated and experimental ionization potentials have often shown excellent agreement.

D. Significance of IP and AP

1. Numerical Values

Ionization potentials for a number of atoms, molecules and radicals are summarized in Appendix II. Almost all the atomic ionization potentials have been determined by vacuum ultraviolet spectroscopy which is certainly the most reliable method for such measurements. The molecular ionization potentials quoted are determined either by electron impact or by photoionization. There is an immense amount of data available in the literature on appearance potentials. The National Bureau of Standards has recently established a service to provide information on published appearance potentials and related data (see Section 14-IV).

The reliability of IP and AP data is an open question. Some of the published values were obtained in long and involved studies, some in superficial measurements. Indeed, appearance potentials may be obtained on most commercial mass spectrometers in a matter of minutes. It happens much too frequently that the method of obtaining data is only briefly indicated and the "accuracy" quoted is only a measure of reproducibility. In general, ionization potentials are accurate within 0.1–0.3 eV for abundant ions, and within 0.3–0.6 eV for fragment ions assuming the use of a calibration gas and a reasonably "normal" ionization efficiency shape. The accuracy of the monoenergetic techniques is, of course, better, although there are many obviously erroneous values in the literature.

In Table 12-1 are summarized IP values obtained for ethylene by various methods as given by Frost et al. (See Figs. 4-12 and 12-9 and reference 25 of Chapter 4.)

Table 12-1   **Ionization Potential of Ethylene by Various Methods**[a]

Method	IP (eV)	Method	IP (eV)
Spectroscopic	10.51	Photoionization	10.52
Electron impact	10.48	Photoionization	10.50
		Theoretical	10.75

[a] Data from reference 25 of Chapter 4.

2. Calculation of Bond Energies

Most chemical reactions involve the breaking of old bonds and the making of new ones. The knowledge of energy and enthropy changes in these processes is vitally important in the establishment of relations between molecular structure and chemical reactivity. Bond dissociation energy is defined as the energy required to dissociate a molecule into known fragments, each in a known state. Mass spectrometric methods have been popular in determination of bond dissociation energies due to their relative simplicity and wide applicability compared to kinetic studies, although the latter generally provide more reliable values.

As shown in equation 12-7, the appearance potential of an ion X^+, formed from a molecule XY by the process of equation 12-2, is equal to the sum of the ionization potential of X and the dissociation energy of the X—Y bond, assuming the kinetic and excitation energies of the process to be

known or absent. A mass spectrometric method for the determination of
kinetic energies of fragment ions has been developed by Hagstrum (22),
who designed an instrument incorporating electrodes with which various
retarding potentials can be applied to the ion beam. Kinetic energy is
determined from a plot of ion current I_p versus retarding potential V_p, or
from the asymmetry (about $V_p = 0$) of the dI_p/dV_p versus V_p derivative
curves. Another instrument, the Lozier tube (23), is also based upon applying retarding potentials to ion-discriminating plates, but here these is no
provision for mass analysis. Both techniques have been applied extensively
for negative ion studies. Other methods to determine excess kinetic energy
include deflection techniques (24) and the study of peak shapes (25).

According to *Stevenson's rule* (26), in a dissociation process that can
proceed in two directions,

$$X\text{—}Y + e^- \begin{array}{l} \overset{(1)}{\nearrow} X^+ + Y + 2e^- \\[2ex] \underset{(2)}{\searrow} X + Y^+ + 2e^- \end{array} \qquad (12\text{-}11)$$

the products of the first process will be in their lowest states or without
kinetic energy when $I(X) < I(Y)$. In this case

$$A(X^+) = I(X) + D(X\text{—}Y) \qquad (12\text{-}12)$$

and the bond dissociation energy can be determined accurately. The second
process of equation 12-11 would, according to the rule, result in the formation of fragments with excitation energy. The rule is based upon observations of the behavior of paraffin hydrocarbons, but subsequent studies have
proven it to be of wide applicability although there are many apparent
exceptions. The rule has no theoretical explanation as of this time.

A simple example of the direct application of the technique is the determination of the bond energy of the CH_3—H bond (27) in the formation
of the CH_3^+ ion from methane. The appearance potential of the CH_3^+ is
measured as 14.39 eV and the ionization potential of the methyl radical is
9.86 eV, yielding the dissociation energy of the CH_3—H bond in methane
as $14.39 - 9.86 = 4.53$ eV. This value was shown to agree well with that
obtained in the kinetic study of the bromination of methane (28), proving
that no kinetic energy correction was needed.

Dissociation energies may be derived in an indirect way from electron
impact data in conjunction with available thermochemical data. The bond
energy $D(R_3$—$R_4)$ is determined in the following manner: First, determine
the appearance potential of R_1^+ in two different processes from molecules

R_1R_2 and R_1R_3. These can be arbitrarily (and conveniently) selected as long as there is no excess energy involved in the processes,

$$R_1R_2 + e^- \rightarrow R_1^+ + R_2 + 2e^- \quad (+A_1) \qquad (12\text{-}13)$$

$$R_1R_3 + e^- \rightarrow R_1^+ + R_3 + 2e^- \quad (+A_2) \qquad (12\text{-}14)$$

Then,

$$R_1R_2 - R_1R_3 = R_2 - R_3 + A_1 - A_2 \qquad (12\text{-}15)$$

Assuming that the heats of formation of R_1R_2 and R_1R_3 and those of R_2R_4 and R_3R_4 are known, and also that the value of $D(R_2\text{---}R_4)$ is available, then the heat (ΔH) of the following reaction can be calcuated,

$$R_1R_2 + R_3R_4 = R_1R_3 + R_2 + R_4 + \Delta H \qquad (12\text{-}16)$$

Substracting equation 12-16 from equation 12-15 one obtains

$$R_3R_4 = R_3 + R_4 + D(R_3 - R_4) \qquad (12\text{-}17)$$

i.e.,

$$D(R_3\text{---}R_4) = A_1 - A_2 + \Delta H \qquad (12\text{-}18)$$

where by Hess's law,

$$\Delta H = \Delta H^\circ(R_1R_3) + \Delta H^\circ(R_2) + \Delta H^\circ(R_4)$$
$$- \Delta H^\circ(R_1R_2) - \Delta H^\circ(R_3R_4) \qquad (12\text{-}19)$$

the ΔH° values referring to the heats of formation of the appropriate compounds in their standard states. The compound R_2R_4 is selected arbitrarily (e.g., CH_4, H_2) according to convenience. This technique, introduced by Stevenson in 1942 (29), has been used extensively for hydrocarbon studies. For the determination of the $CH_3\text{---}H$ bond strength by five different processes using the above method, the reader is referred to Stevenson's 1951 paper (29). For further thermochemical applications see texts on electron impact phenomena (Section 14-II).

II. HIGH TEMPERATURE MASS SPECTROMETRY

A. Principles

The term *high temperature chemistry* refers to the chemistry of systems at sufficiently high temperature (1000–3500°K) so that the general chemical behavior of compounds differ appreciably from those at room temperature. The potentials of mass spectrometry in the study of high temperature equilibria become obvious after the classical studies of Chupka and Inghram (30) and Honig (31), who studied the graphite–carbon vapor system. These investigations revealed that near 2500°K the vapor phase contains C_2 and C_3 in addition to carbon vapor. Hundreds of solid–vapor (and also liquid–vapor) systems have been investigated during the subsequent fifteen years and it is now a generally accepted fact that the composition of the vapor phase is quite often different from that of the solid due to thermal decomposition or reaction with the materials with which contact is made. Indeed, it has been said that "at high temperatures everything reacts with everything, and the products may be anything." The need to understand chemical reactions and processes at high temperature is associated to a great extent with defense and space efforts, such as studies on solid propellants, selection of materials for missile components, etc. Characterization of materials under extreme conditions is also needed in glass technology, pyrometallurgy, and other industries.

The most important advantage of high temperature mass spectrometry, compared to classical methods, is the possibility to sample and analyze vapors at high temperature *directly*, i.e., without the intermediate step of condensation. Three types of information are simultaneously provided by mass spectrometry (32): (*a*) the composition of the gaseous phase, (*b*) the pressure of every gaseous species, and (*c*) the variation of each pressure with temperature. From these data both thermodynamic and kinetic studies may be made. Thermodynamic properties that can be determined include reaction enthalpy, heat of atomization of polyatomic molecules, and heat of dissociation of dimers or polymers. In kinetic studies mechanisms and activation energies are determined from the data on the flux of the vaporizing species as a function of such parameters as temperature, pressure, and surface conditions.

There are two basic types of molecular beam sources. A molecular beam which is a representative sample of the vapor in thermodynamic equilibrium with the condensed vapor phase is obtained in a *Knudsen cell*. Most Knudsen cell sources are basically of the same design (Section 7-VI-B), employing a small orifice and a relatively large sample area to maintain the ratio between the effusing and evaporating surfaces much smaller than 1. When the Knudsen cell is replaced by a filament from which the sample is evaporated (Langmuir evaporation), equilibrium is not established, and only the heat of activation of the evaporation can be measured. Double filament arrangements may be employed for reflection and adsorption coefficient determinations. A conventional electron impact type source is commonly employed to ionize the molecular beam. In instruments designed for high temperature investigations, the analyzer is normally a 60° or 90° sector magnetic field with resolving power between 1000 and 2500; the detector is usually a secondary electron multiplier. Recent general purpose commercial mass spectrometers, even those with high resolution, can be modified with relative ease for Knudsen cell studies.

The interpretation of the mass spectrum obtained at a particular temperature consists of two steps: determination of the composition of the gaseous phase and determination of the partial pressure of every component *in the cell.* To determine the composition of the gaseous phase, one has first to identify every peak in the spectrum and then establish the neutral molecule from which they were produced. *Identification* of the various species in the gas phase is accomplished by standard mass spectrometric techniques, i.e., by determining the mass of every peak and establishing isotopic distributions (and cracking patterns where applicable). Mass is commonly determined using a calibrated mass scale (background peaks are excellent landmarks). Acceleration potential or magnetic field measurements may be applied in case of doubt. Available resolution is usually adequate to distinguish between common background peaks and unknowns. The use of the shutter (Section 7-VI-B) provides information on those peaks originating from hot surfaces outside the cell. Appearance potential measurements are often extremely helpful in recognizing particular ions. For example, sodium ions at m/e 23 appear at 4.5 eV, while the BC^+ ion (also at mass 23) has an appearance potential of about 10 eV. In this case, the presence of a peak at mass 22 (boron-10 isotope) also assists in the identification. Once all ions are identified, parent and fragment ions must be distinguished and identification of the neutral molecule antecedent to each of the ions must be made. This may or may not be a problem depending on the complexity of the mixture.

The *partial pressures* of the components inside the Knudsen cell must be determined from the measured peak heights. This involves the establishment of the relationships between measured current (peak height) and pressure in the ion source, and between pressure in the ion source and pressure inside the cell. The first one involves such parameters as the efficiency of ionization, the overall transmission of the mass spectrometer, and the efficiency of the secondary electron multiplier. As far as the second relationship is concerned, it can be shown (33) that the partial pressure of a component inside the Knudsen cell is *proportional* to the height of the parent peak times the absolute temperature, T. For surface ionization the pressure is shown to be proportional to the height of the parent peak times $T^{3/2}$. For *relative* measurements, therefore, the relationships are simple and thermodynamic properties (e.g., heat of sublimation) can be calculated directly from the variation of peak height with temperature. For *absolute* pressure determinations all components of the proportionality constant must be known or determined. To circumvent this, a nonreacting calibration substance (often silver) is evaporated simultaneously with the unknown so that pressures can be compared. In an alternative method, an accurately known quantity of the substance is totally evaporated at constant temperature, and the vapor pressure (in absolute units) is calculated from the Knudsen formula (33) and the measured total evaporation time. The silver calibration method is used more frequently. Often, evaporation of silver standard is the best method to assess the overall performance of an instrument. Figure 12-10 (34) shows the determination of the heat of vaporization of silver in the temperature range 900–1300°K, which corresponds to a pressure variation from 1×10^{-9} to 1×10^{-5} atm. The electron energy was set at 5 V above the appearance potential of silver. Results are plotted as log IT versus $1/T$, where I is the intensity of the

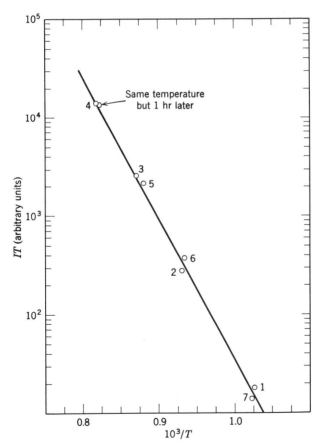

Figure 12-10. Determination of heat of evaporation of silver (34). Numbers beside points indicate sequence of observations. Ionization current: 20 μA, secondary electron multiplier gain: 7×10^4, temperature measurement by Pt/Pt-Rh thermocouple. From the slope, $L_e = 65.3$ kcal/mole.

Ag^+ ion current and T is the absolute temperature of the cell. Ion intensity is normally expressed in arbitrary units. When electron multipliers are employed a silver pressure of 1×10^{-9} atm normally yields a current of about 10^{-13} A, orders of magnitude above multiplier noise (10^{-18} A). The heat of vaporization (L_e) is calculated from the slope of the curve.

B. Thermodynamic Considerations

The relationship between measured ion current, I^+ and partial pressure p for every peak is given by

$$I^+ = (p/kT)\, \sigma\gamma K = pS/T \tag{12-20}$$

where σ is ionization cross section from the process yielding the peak in question, γ is the yield of the secondary electron multiplier, K is an instrumental constant (geometric factor, ionizing electron current, overall transmission), and S is "sensitivity," a grouping of all factors; k and T have their normal meaning of Boltzmann's constant and absolute temperature (of the cell), respectively. From this equation one obtains pressures and the calculation of heats of vaporization may proceed by applying either the second or third laws of thermodynamics.

The second law method is based upon the Clausius-Clapeyron equation,

$$d \ln K_p/d(1/T) = -\Delta H_T^\circ/R \tag{12-21}$$

which describes the variation of vapor pressure with temperature. Here ΔH_T° (at T°K) is reaction enthalpy R is the gas constant, and K_p $(= \pi p n_i)$ is the equilibrium constant at constant pressure; p and n refer to partial pressures and stoichiometric coefficients of products and reactants. Combining equations 12-20 and 12-21 one obtains

$$d \ln I^+T/d(1/T) = -\Delta H/R \tag{12-22}$$

where it is assumed that S is independent of temperature.

The reaction enthalpy can also be determined from the third law,

$$\Delta G_T^\circ = -RT \ln K_p = \Delta H_T^\circ - T\Delta S_T^\circ \tag{12-23}$$

or

$$\Delta G_T^\circ = -RT \ln K_p = \Delta H_0^\circ + T\Delta[(G_T^\circ - H_0^\circ)/T] \tag{12-24}$$

where ΔH_0° is reaction enthalpy at 0°K and S_T° or $(G_T^\circ - H_0^\circ/T)$ are entropy or free energy functions, respectively. These equations combined with equation 12-20 yield

$$K_p = \Pi(I_i^+/\sigma_i\gamma_i)^{n_i} \quad [\Sigma n_i = 0] \tag{12-25}$$

and

$$K_p = \Pi(I_i^+T/G_i)^{n_i} \quad [\Sigma n_i \neq 0] \tag{12-26}$$

The last equation requires the determination of the sensitivity by vaporizing a known amount of the substance and measuring the total evaporation time so that the product $I^+T^{1/2}$ can be obtained. When a silver standard is employed, relative pressures are given by

$$\frac{P_i}{P_j} = \frac{I_i^+ T_i (\sigma\gamma)_j}{I_j^+ T (\sigma\gamma)_i} \tag{12-27}$$

Entropies and free energy functions in the above considerations must be obtained from statistical mechanics or experimental thermodynamics. The relative merits of the second and third law calculations have been argued thoroughly, but this is beyond the scope of this discussion. For further details on thermodynamic considerations the reader should consult reference 33.

C. Applications

Hundreds of elements, compounds, and alloy systems have been investigated during the last decade by the mass spectrometric technique, and applications have been amply reviewed (32,35,36). Many more or less general relationships have been established and an immense amount of thermodynamic data has become available. It has been shown, for example, that the vapor of most metals is essentially monoatomic. Many volatile oxides were found to vaporize without decomposition but exhibiting polymeric molecules in appreciable quantities. Refractory oxides of the M_xO_y type, on the other hand, often vaporize to diatomic MO or triatomic M_2O or MO_2 molecules with frequent decomposition to the elements. Vapor species over borides, carbides, and silicides reveal decomposition to the elements with occasional species of unusual composition.

Determination of thermodynamic properties by the Knudsen cell method using the second law technique is illustrated by a recent study by Margrave and co-workers (37) on the sublimation pressures of the fluorides of scandium, yttrium, and lanthanum. When yttrium trifluoride was vaporized from a Knudsen cell in the temperature range of 1256–1434°K, the main ionic species were YF_2^+, YF^+, and Y^+. Their ionization potentials were found to be 28.0, 21.5, and 13.5 eV, respectively. The heat of sublimation of YF_3 was determined as $\Delta H^\circ_{298} = 115 \pm 5$ kcal/mole from the slope of the plot of log (I^+T) versus $1/T$. To determine absolute pressures, 10^{-3} g of YF_3 was effused at 1352°K through an orifice of 0.14 cm diameter, with continuous monitoring of the YF_2^+; total integration time was 11,200 sec. From these data a least-squares equation could be calculated,

$$\log P_{\text{atm}} = -(2.185 \pm 0.03) \times 10^4/T + 9.77 \pm 0.23 \tag{12-28}$$

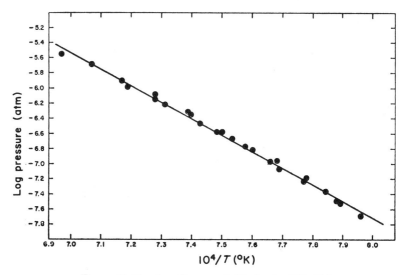

Figure 12-11. Log P versus $1/T$ plot for YF_3 (37).

which is plotted in Figure 12-11. The second-law entropy of sublimation was found to be $\Delta S^\circ_{1332} = 44.7 \pm 1.0$ cal $\deg^{-1}$ mole^{-1}. Heats of atomization and average bond energies, calculated from thermochemical data, are summarized in Table 12-2 (37). Similar studies on other fluorides have also been reported (38). A study of the vaporization of cobalt oxide, based upon silver calibration and third-law calculations, has been reported by Inghram and co-workers (39). Hundreds of other papers are listed in general reviews and in a recent bibliography of the field (40). For a typical example

Table 12-2   **Heats of Atomization and Average Bond Energies of Fluorides at 298°K (37)**

	ΔH_{sub} (kcal mole^{-1})	$^3/_2 D(F_2)$ (kcal mole^{-1})	$-\Delta H^{MF3}$ (kcal mole^{-1})	ΔH^{MF3}_{sub} (kcal mole^{-1})	ΔH_{atm} (kcal mole^{-1})	$^1/_3 \Delta H_{atom}$ (kcal mole^{-1})
ScF_3	82 ± 1	56.5 ± 1.5	380 ± 10	101 ± 5	417 ± 15	139 ± 6 $(6.0 \pm 0.3$ e$V)$
YF_3	100 ± 2	56.5 ± 1.5	410.7 ± 0.8	115 ± 5	467 ± 6	151 ± 2 $(6.6 \pm 0.1$ e$V)$
LaF_3	102 ± 2	56.5 ± 1.5	434 ± 10	108 ± 5	485 ± 15	162 ± 6 (7.0 ± 0.3)
CeF_3	95 ± 3	56.5 ± 1.5	429 ± 10	99 ± 3	482 ± 15	161 ± 5 (7.0 ± 0.3)

of kinetic studies on heterogeneous reactions by high temperature mass spectrometry, reference is made to the extremely interesting work of Mc-Kinley (41) on the surface reaction between nickel and chlorine, bromine, and fluorine.

The application of mass spectrometry to problems in high temperature chemistry has been extremely successful during the last decade and contributed greatly to the present day understanding of the processes of inorganic chemistry in the gas phase. The recent progress in high intensity ion sources and new methods for heating (arc image, laser) will certainly open up new levels of understanding of high temperature chemistry with the likely results of hitherto unknown industrial processes.

III. REACTION MECHANISM STUDIES

The mass spectrometer can be used as a qualitative as well as quantitative analytical tool in kinetic studies. A complete kinetic study of a chemical reaction involves first the determination of the mechanism of the reaction, followed by the measurement of the rate of the reaction. The first step is basically a qualitative one; the objective is to detect and identify all transitory chemical species which occur during the course of the reaction. The sensitivity of mass spectrometers permits the identification of extremely small quantities of gases, while the speed of time-of-flight instruments permits one to follow even shock-wave reactions. In addition, it is often possible to monitor several components simultaneously. The determination of rate constants is a quantitative problem. The role of mass spectrometers is obvious again since the number of molecules (concentration) can be determined in addition to qualitative identification.

The use of mass spectrometry in reaction mechanism studies includes organic reaction path determinations, elucidations of fragmentation processes under electron impact, fast reaction in flames, discharges, shock wave tubes, the study of free radicals, and many other problems involving ion–molecule reactions in the gas phase. Only a few examples will be given in the following to illustrate the principles of the techniques most frequently employed.

A. Isotope Labeling

The principle of isotope labeling is simple and well known: The substitution of a heavier isotope for a normal atom in a particular position in a molecule enables one to follow the compound or its fragments through physical, chemical, or biological processes. For stable isotopes, mass spectrometry is the only method for detection. Here the substitution results in a gain in the mass of every fragment ion containing the isotope atom. The most commonly employed stable isotopes are carbon-13, oxygen-18, nitrogen-15, and deuterium. There are two basic techniques for the analysis of labeled compounds. In the first, the compound is degraded into small gaseous molecules, followed by high precision isotopic analysis. This is often employed in biochemical research and several variations have been reported (42) to increase reliability. In the second basic method the stable isotope in the molecule is left intact and is determined either in the unbroken molecule or in the fragments formed during ionization.

Table 12-3   Cracking Patterns for Deuterated Methanes[a]

$m/9$		CH_4	CH_3D	CH_2D_2	CHD_3	CD_4
1		7.05	5.92	4.40	2.05	3.00
2		0.34	0.68	1.24	1.86	
3			0.11	0.18	0.11	
4				0.04	0.09	0.22
12	C	2.57	2.46	2.39	2.30	2.19
13	CH	8.21	4.90	2.80	1.42	
14	CH_2, CD	16.3	8.80	6.39	6.41	7.23
15	CH_3, CHD	86.1	20.9	9.79	6.75	
16	CH_2D, CD_2, CH_4	<u>100</u>	77.2	30.7	13.2	12.5
17	CHD_2, CH_3D		<u>100</u>	62.4	51.1	
18	CH_2D_2, CD_3			<u>100</u>	27.1	83.0
19	CHD_3				<u>100</u>	<u>100</u>
20	CD_4					

[a] Ref: API spectral data, Serial 455–458.

The cracking patterns of deuterated methanes are shown in Table 12-3. The shifts are due to the incorporation of the heavy isotope. The appearance of the various new peaks and even their relative sizes can be explained on a rational basis by considering the influence of deuterium. The situation becomes more complex with larger molecules and it must be realized that the probability of losing a deuterium in fragmentation may not be the same as that of losing a hydrogen.

A detailed discussion of the principles involved in the experimental techniques of isotope labeling and in the determination of the isotopes in the intact molecule, together with sources of possible errors, is given by Biemann (43). He lists three techniques to incorporate deuterium into a molecule for subsequent reaction mechanism studies: exchange of labile hydrogen atoms, replacement of a functional group with deuterium, and saturation of multiple bonds with deuterium. The mechanism of the reaction must, of course, be known and under complete control. Depending on the degree and specificity of deuteration, the resulting mixture may contain species deuterated to different extents. For methods of quantitative determination of the extent of isotope incorporation reference is again made to Biemann's treatment (43).

The application of isotopically labeled molecules in the study of the mechanisms of organic chemical reactions, electron impact fragmentation processes, biochemical cycles, etc. is a broad field and hundreds of applications have been reported. Only two examples are discussed here. The theory of kinetic isotope effects and uses in physical chemistry are reveiwed by McMullen and Thode (44). Deuterium labeling in the elucidation of the structures of natural products is discussed by Budzikiewicz et al. (45).

As shown in Section 9-VII, a common type of rearrangement involves the intramolecular migration of hydrogen atoms in molecules containing heteroatoms. It was proposed by McLafferty that the rearrangement proceeds through a sterically favorable six-membered cyclic transition state. In the case of ethyl butyrate the γ hydrogen is transferred to the oxygen through a six-membered transition state, and the neutral fragment eliminated (mass 28) contains the β and γ carbon atoms (Scheme I).

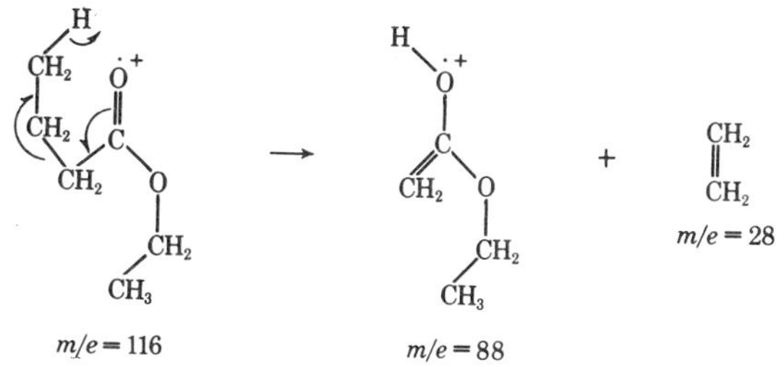

Scheme I

Figure 12-12 shows the mass spectra of ethyl butyrate and various deuterated derivatives (46). The molecular weight is increased due to the incorporation of deuterium, and if the proposed mechanism were correct the structures illustrated in Scheme II should be obtained after deuterium labeling, and peaks should appear at mass 90, 88, and 89, respectively, and also at mass 62, 60, and 61 after subtraction of ethylene. The observed spectra fully confirmed the suggestion that it is exclusively the γ hydrogen that migrates to the oxygen when the β and γ carbons are released as an olefin.

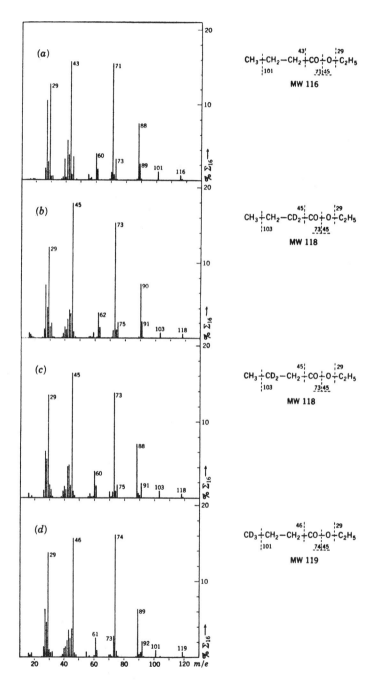

Figure 12-12.   Mass spectra of ethyl butyrate and various deuterated derivatives (46).

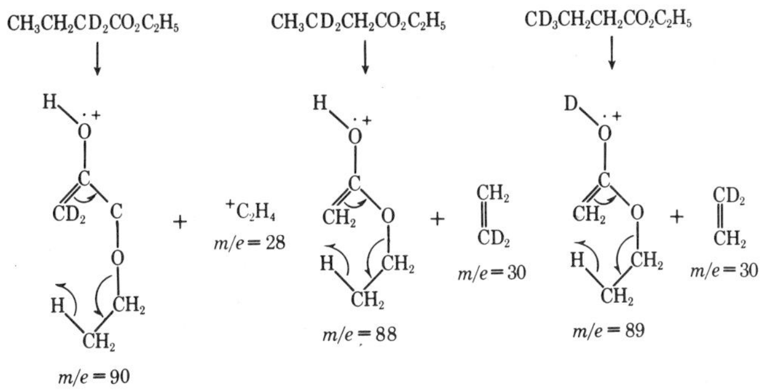

$CH_3CH_2CD_2CO_2C_2H_5$ $CH_3CD_2CH_2CO_2C_2H_5$ $CD_3CH_2CH_2CO_2C_2H_5$

(Scheme II structures with labeled fragments)

$m/e = 90$ $m/e = 88$ $m/e = 89$

$^+C_2H_4$ $m/e = 28$ $m/e = 30$ $m/e = 30$

Scheme II

In the review of the mass spectra of alkylbenzenes, Grubb and Meyerson (47) describe many applications of labeled compounds for the study of ionic reactions. The interpretation of labeling data is based upon calculating label retentions in fragment ions by comparing the spectra of the labeled and unlabeled species of a compound after removing certain irrelevant isotopic contributions. A relative intensity scale is defined by taking the intensity of the molecular ion as 100.

Studies on toluene and xylene indicated that the hydrogen atoms in the $C_7H_7^+$ fragment ion lost their identity. Label retention values with deuterium (and also ^{13}C) have indicated a nearly complete loss of identity of carbon as well as hydrogen atoms implying that the structure of $C_7H_7^+$ is not that of the benzyl ion but tropylium instead, a structure that involves ring expansion. The mechanism and timing of the ring expansion is not known. Assuming the dissociation of alkylbenzenes to proceed through tropylium rather than benzyl ions leads to the explanation of a number of anomalies. The problem of tropylium ion formation is still a subject of much research: a mechanism proposed for several benzyl and methyl-substituted benzyl derivatives (48), for example, cannot explain the formation of $C_7H_7^+$ ions from many other compounds.

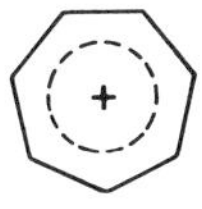

Tropylium ($m/e = 91$)

B. Free Radical Studies

A molecular beam sampling system and the block diagram of the mass spectrometer and ion detector system developed by Foner and Hudson

(49) to study the mechanism of inorganic radical formation are shown in Figures 12-13 and 12-14. The molecular beam traverses the 10 cm distance from the entrance aperture through collimating slits into the center of the source in 230 μsec. Perhaps the most important problem in free radical studies is the transport of the sample from reactor to ion source without change and loss. Discrimination against background is provided by a vibrating reed beam chopper which mechanically interrupts the entering beam at a rate of 170 cps. Mass analysis is performed in a conventional 90° sector analyzer. Every ion arriving at the multiplier causes a pulse of about 10^{-13} coul to be further amplified and counted. The mechanical chopper and the counter are synchronized in such a manner (electronic switch) that one counter registers the beam, the other the background. Best sensitivity is 0.01 ion/sec; average sensitivity is 0.1 ion/sec.

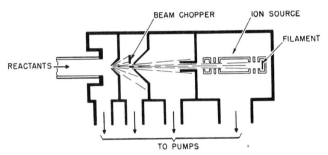

Figure 12-13. Molecular beam sampling system of mass spectrometer for free radical studies (49).

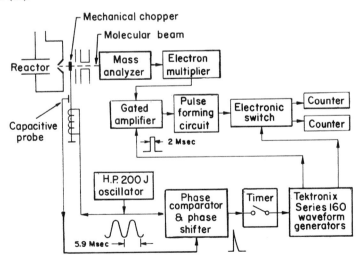

Figure 12-14. Block diagram of mass spectrometer for free radical studies (49).

The yield of HO_2 radicals from an electrical discharge in H_2O_2 as a function of decomposition is shown in Figure 12-15. The species were obtained by using a confined low power electric discharge in a high speed gas stream of hydrogen peroxide; two wire loop electrodes, spaced 1 cm apart, were wrapped around a Pyrex tube, and connected to a radiofrequency generator. By measuring the appearance potential of $HO_2{}^+$, the bond dissociation energy $D(H—O_2)$ could be calculated and further experiments with the HO_2 radicals yielded much additional information concerning the mechanism of the reaction.

The mass spectrometry of organic radicals have been reviewed by Harrison (50) and Lossing (51). Organic free radicals are usually produced in reactors where a neutral molecule RX is heated to temperatures sufficiently high to cause partial thermal dissociation. Both the neutral molecule and the radical R (or X) that form enter the ionization source of the mass spectrometer and undergo ionization,

$$RX + e^- \rightarrow R^+ + X + 2e^- \quad A_1 \geq I(R) + D(R—X) \tag{12-29}$$

and

$$R + e \rightarrow R^+ + 2e \quad A_2 \geq I(R) \tag{12-30}$$

where the A's are appearance potentials. A_1 is greater than A_2 by an amount corresponding to the bond dissociation energy. Consequently, the presence of R^+ ions originating from free radicals can be definitely established when the energy of the applied electrons is between A_2 and A_1.

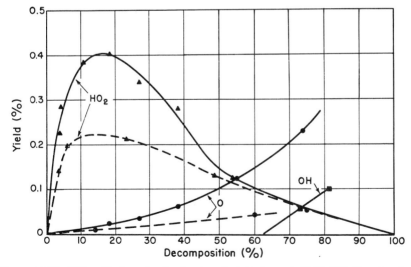

Figure 12-15. HO_2 radical production by electrical discharge in H_2O_2 (49). Reactor: 1 cm i.d. Pyrex tube. Pressure: 120 μ. Sample rate: 122 liter-microns/sec. (———) Concentrations at 0.003 sec. (– – –) Concentrations at 0.006 sec.

Since appearance potential curves increase rapidly with electron energy above the threshold, for maximum detection sensitivity the value of the electron energy used must be carefully selected. The sensitivity of detection can be calculated (52) by considering concentrations and appearance potentials of the species involved, the electron energy, and the temperature. Typically, one methyl radical in 10^2–10^4 molecules of methane can be detected. Sensitivity will increase with monoenergetic electron sources if high electron density is achieved.

A homogeneous thermal reactor designed by Lossing (51) is shown in Figure 12-16. With this instrument the reaction products are bombarded

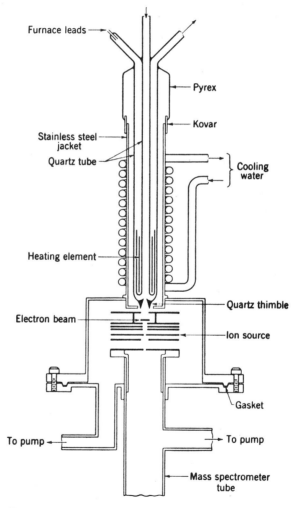

Figure 12-16.   Thermal reactor and ion source of mass spectrometer for study of homogeneous thermal decompositions (51).

with 50–75 eV electrons and the contributions from the stable molecules are subtracted on the basis of separate experiments with the pure compound as in mixture gas analysis. At least 1% concentration of the free radical is necessary. For the determination of the sensitivity for methyl radicals, 6–20 mm pressure of He carried a 8–14 μ pressure of dimethylmercury reactant through the heated quartz tube and into the ionization chamber via a 30 μ diameter quartz leak. Using various assumptions, a carbon balance was established. Results were expressed as methyl/methane ratios in the form of sensitivity coefficients.

A large number of free radicals have been studied. These may have originated in homogeneous or heterogeneous thermal reactions, in flames, or electric discharges. These studies are summarized and reviewed in references 50 and 51.

In addition to the identification of free radicals in various processes, mass spectrometric techniques have been developed for the determination of the rates at which free radicals react in recombination, disproportionation, and decomposition reactions. Considerable work has also been done on appearance potential measurements since AP values of free radicals permit the calculation of radical heats of formation and are helpful in establishing correlations between structure and stability. Experimental methods for AP determination are essentially the same for radicals as for ordinary organic compounds with the added difficulty of providing high enough concentrations on account of low thermal stability. Appearance potentials for a number of radicals are given in Appendix II.

C. Reactions in Flames

The conditions for successful extraction and identification of ions from flames are listed as follows (53): high pumping capacity, rapid quenching of the sample, a minimum of collisions in traversing the system, and a method of focusing the ions while allowing the neutral molecules to be pumped away. An inlet system meeting these requirements is shown in Figure 12-17 (53). The flame burns at 1 atm pressure in front of a 75 μ diameter hole in a 25 μ thick gold sheet. The pressure is progressively reduced to 10^{-6} torr in a series of differentially pumped chambers. The ions formed in the source enter the drift tube of a time-of-flight mass spectrometer and pass through a series of collimating ion lenses followed by a grid system designed to decelerate and bunch the ions. Resolution of about 200 was achieved with a collection efficiency of 1 ion in 10^4. Typical recorder traces of ion spectra in propane–air flame are shown in Figure 12-18.

Both qualitative and quantitative aspects of sampling 1 atm flames have been investigated by Greene et al. (54). They discuss the theory of beam

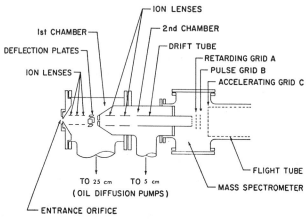

Figure 12-17. Ion inlet system for flame studies (53).

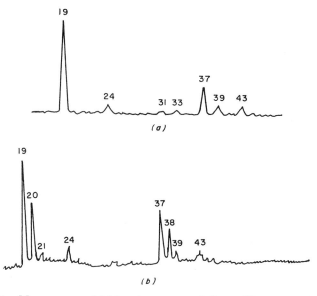

Figure 12-18. Mass spectra of (a) ions in propane–air flame, (b) propane–air flame with deuterium added (53).

formation and describe a relatively simple source which gives molecular beam intensities of 10^{17} molecules cm^{-2} sec^{-1} at 10 cm from the first orifice.

Flame temperatures normally vary between 1000 and 4000°K. Most ions formed in flames result from ion–molecule reactions following the formation of primary ions from neutral species. There is a rather large amount of energy in flames but most of it quickly degrades as thermal energy. The number of ions in a typical laboratory flame is in the range of one part in 10^5 to one part in 10^{13}. The formation of the most prominent positive ions in hydrocarbon flames has been described by two reactions:

$$CH + O \rightarrow CHO^+ + e^- \tag{12-31}$$

and

$$CH^* + C_2H_2 \rightarrow C_3H_3^+ + e^- \tag{12-32}$$

The asterisk denotes possible electronic excitation. The $C_3H_3^+$ peak (at mass 39) is prominent in all hydrocarbon flames and its concentration maximizes in rich flames in contrast with that of other positive ions; therefore it is believed to be produced in a primary ionization step. Other important ions occurring in hydrocarbon flames include CH_3^+, H_3O^+, CH_3O^+, $CH_5O_2^+$, $C_2H_5O_2^+$, etc. The appearance of various ionic, neutral, and excited species and their relative quantities can strongly be influenced by addition of hydrocarbons to a hydrogen/oxygen or hydrogen/air flame (see flame ionization detector in gas chromatography). An extremely interesting field is the study of reaction rates when ion–molecule reactions take place in flames in the presence of metals and halogens. Table 12-4 summarizes rate constants for a number of ion–molecule reactions in flames (55).

Table 12-4   **Rate Constants for Ion–Molecule Reactions in Flames (55)**

Reaction	ΔH (kcal/mole)	k (cm³ molecule⁻¹ sec⁻¹)
$CHO^+ + H_2O \rightarrow H_3O^+ + CO$	$-34,000$	10^{-8}
$OH^- + M \rightarrow OH + e^- + M$	$+41,000$	$10^{-9}e^{-41,000/RT}$
$OH^- + H \rightarrow H_2O + e^-$	$-77,000$	$\sim 10^{-14}(?)$
$Cl^- + H \rightarrow HCl + e^-$	$-18,000$	$\sim 10^{-13}$
$H_3O^+ + Pb \rightarrow Pb^+ + H_2O + H$	$+26,000$	$\sim 10^{-91}$
$H_3O^+ + Li \rightarrow Li^+ + H_2O + H$	$-21,000$	$7.5 \times 10^{-10}(?)$
$LiH_2O^+ + M \rightarrow Li^+ + H_2O + M$	$\sim +50,000$	$10^{-9}e^{-50,000/RT}$
$SrOH^+ + Na \rightarrow Na^+ + SrOH$	~ 0	$\sim 10^{-8}$

D. Shock Wave Studies

The first use of mass spectrometry to study shock waves was reported by Bradley and Kistiakowsky (56), who utilized a time-of-flight mass spectrometer to analyze the effluent from a shock tube. The original equipment has been used to study the thermal decomposition of nitrous oxide, and the polymerization and oxidation of acetylene. Dove and Moulton, from the same laboratory, reported (57) on a completely redesigned version of Bradley's original apparatus. The gas is sampled continuously from the shock tube into the ion source of the TOF mass spectrometer in which samples are taken every 20 μsec. A method was developed to photograph individually a number of consecutive complete mass spectra without using a moving film. This way the change with time in the concentrations of all species present can be followed. In a preliminary reinvestigation of mechanisms proposed earlier for the oxidation of acetylene at high temperature, it was established that the reaction $C_2H + O_2 \rightarrow CO_2 + CH$ is the most important step in the regeneration of hydrogen atoms in the formation of diacetylene.

The schematic diagram of a shock tube–mass spectrometer combination designed by Diesen and Felmlee (58) is shown in Figure 12-19. The basic design is again similar to that pioneered by Bradley and Kistiakowsky. Mass spectra are displayed on an oscilloscope and recorded on a high-speed drum camera. In addition, time intensity profiles for two peaks may be viewed and photographed on a dual-beam oscilloscope. In the first experiments the thermal dissociation of chlorine was studied, primarily to test the apparatus. Subsequent works included studies on the thermal decomposition of hydrazine and thermal dissociation of fluorine (59). It was established that the primary process in the dissociation of hydrazine is the rupture of the N—N bond to yield NH_2 radicals which is followed by radical–radical reactions of various types leading to the final decomposition products of nitrogen and hydrogen gas. Order-of-magnitude values are given for several rate constants. The thermal dissociation of molecular fluorine in inert gases was investigated in the 1650–2700°K temperature

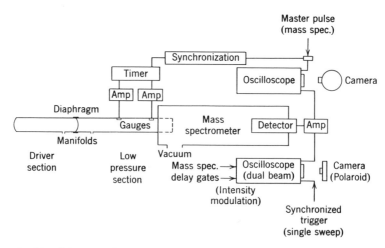

Figure 12-19. Schematic diagram of shock-tube mass spectrometer combination (58).

range. Rate constants were determined for the bimolecular dissociation product, $F_2 + M \rightarrow 2F + M$, where M, the collision partner, was taken to be the diluent gas.

Instrumentation for a shock wave tube–quadrupole mass filter combination has recently been reported by Gutman et al. (60). The advantage of this instrument is the opportunity to obtain detailed continuous ion current records thereby increasing the accuracy of measurements. At the same time, however, ability to observe several species simultaneously is sacrificed. To test the instrumentation as a possible tool for quantitative gas kinetic studies, the thermal decomposition of nitrous oxide diluted in argon has been studied (61). The rate constant of the reaction $Ar + N_2O \rightarrow N_2 + O + Ar$ has been determined, and it was also established that the activation energies of reactions $O + N_2O \rightarrow N_2 + O_2$ and $O + N_2O \rightarrow 2NO$ are the same because the ratio of their rate constants was found to be independent of temperature.

References

*1. Honig, R. E., *J. Chem. Phys.*, **16**, 105 (1948).

2. Morrison, J. D., and A. J. C. Nicholson, *J. Chem. Phys.*, **31**, 1320 (1959).

3. Buckingham, D., *Brit. J. Appl. Phys.*, **16**, 1821 (1965).

*4. Warren, J. W., *Nature*, **165**, 810 (1950); also J. W. Warren and C. A. McDowell, *Discussions Faraday Soc.*, **10**, 53 (1951).

5. Lossing, F. P., A. W. Tickner, and W. A. Bryce, *J. Chem. Phys.*, **19**, 1254 (1951).

6. Morrison, J. D., *J. Chem. Phys.*, **19**, 1305 (1951); also J. D. Morrison and A. J. C. Nicholson, *ibid.*, **20**, 102 (1952).

7. Kiser, R. W., and E. J. Gallegos, *J. Phys. Chem.*, **66**, 947 (1962).

8. Morrison, J. D., *J. Chem. Phys.*, **21**, 1767 (1954); **21**, 2090 (1954); **22**, 1219 (1954); *J. Appl. Phys.*, **28**, 1409 (1957); F. H. Dorman and J. D. Morrison, *J. Chem. Phys.*, **34**, 578 (1961).

9. Dorman, F. H., *J. Chem. Phys.*, **41**, 2857 (1964); **42**, 65 (1965); **43**, 3507 (1965); **44**, 35 (1966).

10. Morrison, J. D., *J. Chem. Phys.*, **39**, 200 (1963); *Bull. Soc. Chim. Belges*, **73**, 399 (1964).

*11. Winters, R. E., J. H. Collins, and W. I. Courchene, *J. Chem. Phys.*, **45**, 1931 (1966).

*12. Fox, R. E., W. M. Hickam, T. Kjeldaas, and D. J. Grove, "Mass Spectroscopy in Physics Research," *NBS Circ.*, **522**, 211 (1953).

13. Cloutier, G. G., and H. I. Schiff, in *Advances in Mass Spectrometry*, J. D. Waldron, Ed., Pergamon, New York, 1959, p. 473.

14. Melton, C. E., and W. H. Hamill, *J. Chem. Phys.*, **41**, 546, 1469, 3464 (1964).

15. Foner, S. N., and B. N. Nall, *Phys. Rev.*, **122**, 512 (1961).

16. Hutchison, D. A., in *Advances in Mass Spectrometry*, Vol. 2, R. M. Elliott, Ed., Macmillan, New York, 1963, p. 527.

*17. Marmet, P., and L. Kerwin, *Can. J. Phys.*, **38**, 787 (1960); also *J. Appl. Phys.*, **31**, 2071 (1960).

18. Clarke, E. M., *Can. J. Phys.*, **32**, 764, (1954).

19. Brion, C. E., *J. Chem. Phys.*, **40**, 2995 (1964); also C. E. Brion, D. C. Frost, and C. A. McDowell, *ibid.*, **44**, 1034 (1966).

20. Hurzeler, H., M. G. Inghram, and J. D. Morrison, *J. Chem. Phys.*, **28**, 76 (1958); W. A. Chupka, *ibid.*, **30**, 191 (1959).

21. Kiser, R. W., *Introduction to Mass Spectrometry and Its Applications*, Prentice-Hall, Englewood Cliffs, N. J., 1965, pp. 176–186.

*22. Hagstrum, H. D., *Rev. Mod. Phys.*, **23**, 185 (1951).

23. Lozier, W. W., *Phys. Rev.*, **35**, 1285 (1930); J. T. Tate and W. W. Lozier, *ibid.*, **39**, 254 (1932).

24. Berry, C. E., *Phys. Rev.*, **78**, 597 (1950); H. I. Stanton, *J. Chem. Phys.*, **30**, 1116 (1959).

25. Lagergren, C. R., *Phys. Rev.*, **96**, 823 (1954); F. L. Mohler, V. H. Dibeler, and R. M. Reese, *J. Chem. Phys.*, **22**, 394 (1954).

*26. Stevenson, D. P., *Discussions Faraday Soc.*, **10**, 35 (1951).

27. Lossing, F. P., K. U. Ingold, and I. H. Henderson, *J. Chem. Phys.*, **22**, 621 (1954); C. A. McDowell and J. W. Warren, *Discussions Faraday Soc.*, **10**, 53 (1951).

28. Kistiakowsky, G. B., and E. R. Van Artsdalen, *J. Chem. Phys.*, **12**, 469 (1944).

*29. Stevenson, D. P., *J. Chem. Phys.*, **10**, 291 (1942); *Discussions Faraday Soc.*, **10**, 35 (1951); *Trans. Faraday Soc.*, **49**, 867 (1953).

*30. Chupka, W. A., and M. G. Inghram; *J. Chem. Phys.*, **21**, 371, 1313 (1953); **22**, 1472 (1954); *J. Phys. Chem.*, **59**, 100 (1955).

31. Honig, R. E., *J. Chem. Phys.*, **22**, 126 (1954).

32. Inghram, M. G., and J. Drowart, in *High Temperature Technology*, McGraw-Hill, New York, 1960.

*33. Boerboom, A. J. H., "High Temperature Mass Spectrometry," and P. Goldfinger, "Mass Spectrometric Investigation of High Temperature Equilibria" in *Mass Spectrometry*, R. I. Reed, Ed., Academic Press, New York, 1965, p. 251 and p. 265.

34. McKinney, C. R., *Tech. Publ.*, **102**, Consolidated Electrodynamics Corp., Monrovia, Pasadena, Calif., 1963.

35. Drowart, J., in *Condensation and Evaporation of Solids*, E. Rutner, P. Goldfinger, and J. P. Hirth, Eds., Gordon and Breach, New York, 1964, p. 255.

*36. Drowart, J., and P. Goldfinger, in *Advances in Mass Spectrometry*, Vol. 3, W. L. Mead, Ed., Inst. Petroleum, London, 1966, p. 923; *Angew. Chem. Intern. Ed.*, **6**, 581 (1967).

*37. Kent, R. A., K. F. Zmbov, A. S. Kana'an, G. Besenbruch, J. D. McDonald, and J. L. Margrave, *J. Inorg. Nucl. Chem.*, **28**, 1419 (1966).

38. Kent, R. A., T. C. Ehlert, and J. L. Margrave, *J. Am. Chem. Soc.*, **86**, 5090 (1964); K. F. Zmbov and J. L. Margrave, *J. Chem. Phys.*, **45**, 3167 (1966).

39. Grimley, R. T., R. P. Burns, and M. G. Inghram, *J. Chem. Phys.*, **45**, 4158 (1966).

40. Redman, J. D., "A Literature Review of Mass Spectrometric–Thermochemical Technique Applicable to the Analysis of Vapor Species Over Solid Inorganic Materials," ORNL, TM 989, 1966.

*41. McKinley, J. D., *J. Chem. Phys.*, **40**, 120 (1964); **40**, 576 (1964); **45**, 1690 (1966).

42. Han, InGun, and G. J. Fritz, *Anal. Chem.*, **37**, 1442 (1965).

*43. Biemann, K., *Mass Spectrometry, Organic Chemical Applications*, McGraw-Hill, New York, 1962, Chapter 5.

44. McMullen, C. C., and H. G. Thode, in *Mass Spectrometry*, C. A. McDowell, Ed., McGraw-Hill, New York, 1963, Chapter 10.

45. Budzikiewicz, H., C. Djerassi, and D. H. Williams, *Structure Elucidation of Natural Products by Mass Spectrometry*, Vol. I, Holden-Day, San Francisco, Calif., 1964, Chapter 2.

46. Biemann, K., *Mass Spectrometry, Organic Chemical Applications*, McGraw-Hill, New York, 1962, p. 121.

*47. Grubb, H. M., and S. Meyerson, in *Mass Spectrometry of Organic Ions*, F. W. McLafferty, Ed., Academic Press, New York, 1963, Chapter 10.

48. Meyer, F., and A. G. Harrison, *J. Am. Chem. Soc.*, **86**, 4757 (1964).

*49. Foner, S. N., and R. L. Hudson, *J. Chem. Phys.*, **36**, 2676, 2681 (1962).

*50. Harrison, A. G., in *Mass Spectrometry of Organic Ions*, F. W. McLafferty, Ed., Academic Press, New York, 1963, Chapter 5.

51. Lossing, F. P., in *Mass Spectrometry*, C. A. McDowell, Ed., McGraw-Hill, New York, 1963, Chapter 11, *Ann. N. Y. Acad. Sci.*, **67**, 499 (1957).

52. Robertson, A. J. B., *Mass Spectrometry*, Wiley, New York, 1954, p. 109.

*53. King, I. R., and J. T. Scheurich, *Rev. Sci. Instr.*, **37**, 1219 (1966).

54. Greene, F. T., J. Brewer, and T. A. Milne, *J. Chem. Phys.*, **40**, 1488 (1964); T. A. Milne and F. T. Greene, *J. Chem. Phys.*, **44**, 2444 (1966).

55. Calcote, H. F., and D. E. Jensen, in *Ion–Molecule Reactions in the Gas Phase* (*Advan. Chem. Ser.*, **58**), R. F. Gould, Ed., American Chemical Society, Washington, D. C., 1966, p. 291.

56. Bradley, J. N., and G. B. Kistiakowsky, *J. Chem. Phys.*, **35**, 256, 264 (1961).

*57. Dove, J. E., and D. McL. Moulton, *Proc. Roy. Soc.* (*London*), *Ser. A*, **283**, 216 (1965).

*58. Diesen, R. W., and W. J. Felmlee, *J. Chem. Phys.*, **39**, 2115 (1963).

59. Diesen, R. W., *J. Chem. Phys.*, **39**, 2121 (1963); **44**, 3662 (1966).

60. Gutman, D., A. J. Hay, and R. L. Belford, *J. Phys. Chem.*, **70**, 1786 (1966).

61. Gutman, D., R. L. Belford, A. J. Hay, and R. Pancirov, *J. Phys. Chem.*, **70**, 1793 (1966).

Chapter Thirteen

Miscellaneous Applications

I. NUCLEAR PHYSICS

Mass spectrometry began with the discovery of the stable isotopes and was mainly concerned with the determination of their masses and abundances for more than two decades. Two of the fundamental properties of atomic species are the mass number, A (the number of protons and neutrons in the nucleus) and the accurate isotopic weight (giving a measure of the total energy released in the formation of the nucleus from Z protons and $N = A - Z$ neutrons). Mass spectrometry yields direct information concerning both quantities. Indirect applications in nuclear physics include the determinations of half-lives of radioactive nuclides, cross sections of neutron-capture processes, and fission yields. Only the basic principles involved and a few representative applications are reviewed here; for details the reader is referred to books and reviews listed in Chapter 14.

A. Isotopes: Identification and Abundance

1. Identification of Natural Isotopes

The identification of the naturally occurring nuclides was practically complete by 1940. White et al. (1) reported in 1955 the discovery of ^{180}Ta with an abundance of 1 part in 10^4 of the abundance of ^{181}Ta. Several subsequent studies with electron multiplier detectors failed to reveal new isotopes. Abundance limits were placed at 1 part in 10^4 in general, with limits as high as 1 part in 10^8 for several elements. In 1963 Leipziger (2) employed a spark source mass spectrograph with photographic detection in a search for hitherto undetected isotopes. Upper limits of 1–5 parts in 10^8 were established for many elements which were not susceptible to mass analysis by thermal or electron impact techniques. No new stable isotopes were discovered. Relative abundances of all naturally occurring nuclides (about 300) are given in Appendix I.

2. Determination of Relative Abundances

In his pioneering work on the determination of isotopic abundances Aston employed photographic detection. For many years, however, electrical detectors have been used almost exclusively for abundance measurements. Isotope ratios can be determined by alternately focusing the particular ion beams onto the collector and measuring individual ion currents. Such measurements are best made by changing the magnetic field strength and keeping the acceleration voltage and all other instrumental parameters constant. A more advanced technique, one that eliminates many problems connected to instrumental instability, is based upon the use of two collectors, as described in Section 5-I-A. Here the ions describe different paths and the current-measuring devices are different; consequently the results are only relative. In many cases, however, the knowledge of absolute composition is not necessary, and an intercomparison of samples with small differences in composition is adequate (e.g., mechanism studies in chemistry and biology).

Depending on the type of sample, both electron impact and thermal ionization sources may be used; in general, the former yield somewhat better precision. The most favorable case is, naturally, when the isotope ratio is $1:1$, and a precision of $\pm 0.001\%$ of the ratio is normally obtainable. Precision becomes poorer as the ratio increases and smaller and smaller ion currents must be measured. Still, ratios as high as $1:10^6$ may be measured with a precision of a few percent.

Errors in abundance measurements are caused by discrimination effects in the sample introduction system, in the process of ionization, during mass separation in the analyzer, and in the course of detection, particularly when electron multipliers are employed. The most important discrimination effects have already been mentioned in Part I in the description of the main components of mass spectrometer systems; a detailed discussion is given by Bainbridge (3).

3. Determination of "Absolute" Abundances

With the availability of separated isotopes, calibration of the mass spectrometer with essentially pure components and accurately known synthetic mixtures became possible, thereby eliminating many inherent errors in abundance measurements. The first absolute abundance measurements were reported by Nier (4), who achieved an accuracy of 0.1% for a number of inorganic gases, including argon, oxygen, and xenon. This work also showed that earlier measurements were accurate within the 1% accuracy formerly claimed.

The first step in absolute isotopic abundance measurements is to obtain a "working standard" often called a "secondary standard." It is used so that the reference material need not be the prepared isotopic standard. This is a pure sample in the desired chemical form prepared from a carefully selected source (there is a natural variation in isotopic abundances, see later). The isotopic ratio in this "working standard" is first determined by a relative method (dual collector) to establish the constancy of the sample composition. The second step involves the preparation of a standard of approximately the same composition as the "working standard" from separated isotopes of $>99\%$ purity; normally a gravimetric technique is employed. Often more than one standard is made to bracket the natural abundances of the principal isotopes of the element. Finally, repetitive measurements are made with both the unknown and the artificially prepared standards and a value for the instrumental bias is obtained which can be used to correct the results for the "working standard."

The importance of absolute isotopic determinations lies in the fact that they serve as a basis for calculating atomic weights. Absolute isotopic composition has been established for only a few elements and the field is currently a rather active one in several laboratories. Details of the determination of absolute isotopic abundance ratios and the atomic weight of a reference sample of chromium are described by Shields et al. (5). This paper describes current experimental techniques and details how the limits of possible errors are determined. The resulting absolute values are $^{50}Cr/^{52}Cr = 0.051859 \pm 0.000100$, $^{53}Cr/^{52}Cr = 0.113386 \pm 0.000145$, $^{54}Cr/^{52}Cr = 0.028222 \pm 0.000059$, yielding an atomic weight of 51.99612 ± 0.00033 for chromium on the ^{12}C scale.

4. Natural Variations in Isotopic Abundance

Variations in the natural abundance of isotopes have been reported since the early days of mass spectrometry. These variations are caused by nuclear transformations and such natural physicochemical processes as evaporation, diffusion, and chemical exchange. These effects are naturally more pronounced for the lighter elements. Extensive studies have been made on the variations in the isotopic constitution of hydrogen, carbon, oxygen, and sulfur. The study of these variations, a distinct branch of geochemistry, is an important tool in obtaining a broad understanding of the long-range changes in the natural environment.

The natural abundance of ^{13}C varies about 4%, as shown in Figure 13-1 (6). The $^{13}C/^{12}C$ ratio is thus 0.01118 for atmospheric carbon dioxide

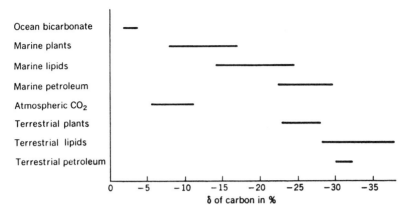

Figure 13-1. Variations in the natural abundance of carbon-13 (6). The unit of the abscissa is $\delta = (^{13}C/^{12}C \text{ of sample} - {}^{13}C/^{12}C \text{ of standard})/(^{13}C/^{12}C \text{ of standard})$.

(1.1056% ^{13}C) while it is 0.01082 (1.0704% ^{13}C) in the lower extreme of terrestrial plants. These variations, although not fully understood, are not surprising in the light of continuous biochemical and geochemical processes and exchanges involving carbon dioxide and carbonates.

It is known that the ^{18}O content of atmospheric oxygen is about 3% greater than that of fresh water. Also, carbon dioxide and ocean water are enriched in ^{18}O with respect to fresh water by 3.5% and 0.6%, respectively. Since the constitution of meteoric sulfur is constant, but terrestial sulfates are enriched and sulfides are depleted in ^{34}S, it has been suggested that there is a *sulfur cycle* in the sea. For details the reader is referred to texts on geochemistry.

5. *Isotopic Abundances and Relative Positive Ion Currents*

In isotopic abundance measurements the information desired is abundance in atomic per cent, while the information obtained is in the form of current ratios. The conversion is illustrated for carbon dioxide:

$$R = \frac{[^{12}CO_2]}{[^{13}CO_2]} \tag{13-1}$$

where R is the ratio obtained in the actual measurement. The abundance of ^{13}C in atomic per cent is

$$\alpha = \frac{[^{13}CO_2]}{[^{12}CO_2] + [^{13}CO_2]} \times 100 = \frac{[^{13}CO_2]}{R} \times 100 \tag{13-2}$$

where the brackets denote molecular concentrations. Dividing the first equation by $[^{13}CO_2]$ and combining with the second equation yields

$$\alpha = 100/(R + 1) \qquad (13\text{-}3)$$

In the case of nitrogen one has to consider the equilibrium

$$^{14}N^{14}N + {}^{15}N^{15}N = {}^{14}N^{15}N + {}^{14}N^{15}N \qquad (13\text{-}4)$$

for which the equilibrium constant is 4. The ^{15}N atomic per cent is obtained as

$$\alpha = 100/(2R + 1) \qquad (13\text{-}5)$$

where R is the $^{14}N/^{15}N$ ratio.

B. Atomic Weights

1. Calculation of Atomic Weights

The term atomic weight (a misnomer) refers to the "average relative atomic mass of the naturally occurring element." As discussed in Section 1-I-B, the mass scale currently in use is based on carbon-12 whose mass is taken exactly 12.000000 amu. Historically, atomic weights were determined by gravimetric ratios of a halide of the element to silver chloride or bromide. Transformation into the oxygen-based system was made through the silver–silver nitrate system. In recent years atomic weights have been increasingly determined by mass spectrometry, and in the latest compilation of atomic weights (1961) the majority of the values are based on mass spectrometric determinations.

Mass spectrometry provides accurate values for both the nuclidic masses and the isotopic composition of the elements, making the calculation of relative atomic weights a comparatively easy task. The atomic weight is calculated by multiplying the atomic fractions of the isotopes by their respective nuclidic masses and adding the results. Masses are measured by large-radius mass spectrometers specially designed for precise mass measurements or by mass spectrographs using the doublet method.

There are 21 elements which exist in nature in only a single stable atomic species; their atomic masses are known within a few micromass units. There are a few elements, such as H, N, B, O, and U, where variations in natural isotopic composition limits the accuracy of the atomic weights. For some other elements, such as Ti, Mo, Hg, and Pb, atomic weights are still based on gravimetric methods. It is suggested in a review of the field (7) that physical methods such as mass spectrometry will be more and more relied upon in atomic weight determinations. The accuracy of present mass measurements is more than adequate, but improvement is needed in isotopic composition measurements.

2. Atomic Masses and Nuclear Stability

Nuclear masses are not equal to the sum of the masses of the constituent protons, neutrons, and electrons. Consideration of exact atomic weights shows that mass is not conserved in nuclear reactions; however, mass and energy, taken together, are conserved according to the Einstein equation. The *binding energy*, E_b, of a nucleus is defined as the difference between the sum of the masses of the Z protons and N neutrons in their free state and the mass of the nucleus containing the $A = Z + N$ nucleons,

$$E_b \text{ (MeV)} = 931 \left[ZM_H + (A - Z)M_n - M \right] \tag{13-6}$$

where M_H is the atomic mass of hydrogen of mass number 1, M is the atomic mass of the isotope of atomic number Z and mass number A, and M_n is the mass of the neutron. The mass of the neutral atom of $_1^1H$ is used instead of the mass of the proton, and the mass of the appropriate neutral atom of mass number A is employed instead of the mass of the nucleus, since the masses of the electrons cancel out. Figure 13-2 shows the binding energy per nucleon, E_b/A, plotted against the mass number A. It is seen that variations in E_b/A are considerable between mass numbers 1 and 20. Next there is a slow increase from 8 MeV to 8.5 MeV ($A \sim 60$), followed by a slow decrease for the heavier elements.

Another widely used concept for the presentation of the data obtained in precise atomic mass determinations is the *mass defect*, Δ, defined as the difference between the atomic mass M of an isotope and its mass number A,

$$\Delta = M - A \tag{13-7}$$

A more exact treatment is to consider the difference between the isotopic

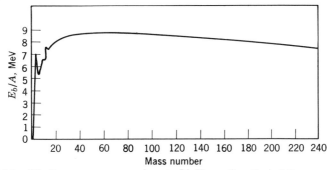

Figure 13-2.   Binding energy per nucleon in MeV as a function of the mass number A.

weight and the total weight of the individual electrons, protons, and neutrons which make up the atom. The formula thus obtained for the "true mass defect" is identical to that shown in equation 13-6. This quantity corresponds to the loss of mass or, more correctly, the mass which would be converted into energy, if a particular atom were to be assembled from the requisite numbers of electrons, protons, and neutrons. The same amount of energy would be required to break up the atom into its constituent particles (binding energy). The mass defect of the whole atom per nucleon is called the *packing fraction*, F, a term introduced by Aston,

$$F = \frac{\Delta}{A} = \frac{M - A}{A} \tag{13-8}$$

Since F is a very small number, Aston multiplied results by 10^4 so as to obtain figures which were easier to record. A packing fraction curve (F versus A) is shown in Figure 13-3. On the original packing fraction curves F was taken as zero for the oxygen-16 isotope. The binding energy and packing fraction curves are essentially mirror images, the maximum on the E_b/A curve occurring at the same mass number as the minimum on the F curve. Most of the data in these curves are obtained in mass spectrometric measurements. The breaks in the curves

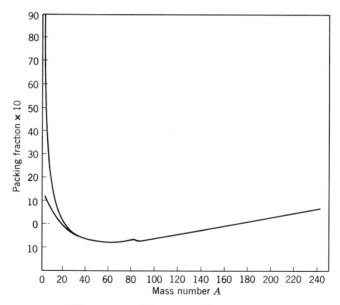

Figure 13-3.   The packing fraction curve.

(such as around mass number 90) are explained on the basis of nuclear shell effects. The E_b and E_b/A values and their variations provide important information for the study of the role of various forces in the stability of the nuclei and other nuclear properties. For an illustration of modern mass spectrometric measurements of atomic masses and their uses in nuclear systematics, reference is made to two papers by Ries et al. (8).

C. Determination of Half-Lives and Neutron Capture Cross Sections

There are three techniques of determining the half-life of a radioactive material: decay method, daughter-growth method, and specific activity method. In the *decay* technique, which is rather straightforward, the change in activity in successive intervals of time is measured and the half-life, $T_{1/2}$, is given by

$$T_{1/2} = 0.693 \, t/\ln \, (N/N_0) \tag{13-9}$$

where N/N_0 denotes that fraction of the atoms remaining after a decay time of t. Mass spectrometric determinations are useful in the few days to 50 years half-life range. The sensitivity of the instrument for the particular element has to be carefully considered. The best known application of this technique is the determination of the half-life of ^{85}Kr as 10.27 ± 0.18 years by Thode et al. (9), who measured for some seven years the abundance of ^{85}Kr relative to the stable isotopes of krypton. Half-lives of 30.35 ± 0.38 years for ^{137}Cs and 2.046 ± 0.004 years for ^{134}Cs have been determined by Dietz et al. (10) using a decay method combined with an internal standard technique.

In the *daughter-growth* method the accumulation of the daughter product is followed as a function of time, and the half-life is determined from

$$N_d = N_0 \left(1 - \exp \frac{0.693t}{T_{1/2}} \right) \tag{13-10}$$

where N_d is the number of daughter atoms and N_0 is the number of parent atoms at $t = 0$. The technique is particularly advantageous for long half-lives. A classic example is the determination the half-life of ^{235}U (7.13×10^8 years) in 1939 by Nier (11), who measured the quantity of ^{207}Pb, the end product of the natural disintegration series which begins with ^{235}U. Several uranium minerals of different ages were analyzed.

The *specific activity* technique employs mass spectrometry to determine the number of atoms N in the differential decay equation

$$T_{1/2} = -\frac{0.693}{dN/dt} \times 100 \tag{13-11}$$

while the disintegration rate, dN/dt, is measured by absolute counting. A variation of this method was used by Wiles and Tomlinson (12) and by Brown et al. (13) for the determination of the half-life of ^{137}Cs. Comparison of their work with that of Dietz et al. (10) provides a good insight into the role of mass spectrometry in such measurements.

When elements are irradiated with slow neutrons, the capture of a neutron results in an increase of one mass unit in the mass of the particular isotope involved, causing a change in the isotopic composition. When the end product of such a (n,γ) reaction is a stable isotope, only mass spectrometry can provide information concerning the capture cross section, σ_c. The number of target nuclei remaining at the end of the irradiation (i.e., after irradiation time t), N, is given by

$$N = N_0 \exp\left(-\sigma_c \phi t\right) \tag{13-12}$$

where N_0 is the number of target nuclei at the beginning and ϕ is the neutron flux. From this equation absolute cross section may be evaluated when N/N_0 is measured by mass spectrometry and the integrated neutron flux, ϕt, is determined by the simultaneous irradiation of a flux monitor (e.g., BF_3) with the sample. The numerical magnitude of cross sections is illustrated by a few typical values: for ^{113}Cd (stable) the cross section is 20,000 barns, while for Cd with mass numbers 110, 111, 112, 114, and 116 it is less than 500 barns (14); the neutron capture cross section for ^{83}Kr (half-life 83 days) is 216 $\pm$ 43 barns, for ^{131}I (half-life 8 days) it is 51 $\pm$ 7 barns (1 barn $= 10^{-24}$ cm^2)(15).

D. Isotope Separators

Electromagnetic separators are large- radius (e.g., 24 in.), 180° deflection mass spectrometers. Ions are normally obtained in low voltage (100–150 V) arc discharge sources. The arc current is about 3–5 A, while the ion current obtained is of the order of a few hundred mA. Ionization efficiency is usually poor, of the order of 5%. Gaseous materials, often in the form of chlorides (e.g., $SnCl_4$, $SiCl_4$) are continuously fed into the arc chamber through a controlled leak. The pressure inside the arc chamber must be kept constant to obtain a steady ion beam. Compounds which can be volatilized in the 100–600°C temperature range (chlorides, bromides of many elements) are introduced through heated stainless steel "charge bottles." Temperatures as high as 1200°C can be obtained with graphite "charge bottles."

The pumping systems of large separators are impressive indeed. The main requirement is pumping capacity rather than ultimate vacuum. In a large instrument, total pumping capacity as high as 20,000 liters/sec is employed to maintain a vacuum of 10^{-5} torr.

The collection of the separated isotopes presents many problems. First of all, considerable cooling is necessary. When an ion beam of 100 mA, accelerated to 35 kV, impinges on a metallic receiver, the heat generated is of the order of 3.5 kW. Other problems which must be considered in the design of a collector system are erosion, sputtering, efficiency, and technique of isotope removal. Graphite and copper are the most frequently used collector materials. The collected material is often removed from graphite by simple scraping or by the burning of the entire pocket in oxygen. Dissolution in nitric acid is a frequent removal method with copper pockets. Mercury isotopes may be collected in silver pockets from which removal is easily accomplished by heating in vacuum.

Enrichment factor and production rate are two important performance parameters in electromagnetic separators. The enrichment factor is the ratio of the ratios of the isotopic concentrations in the product and the starting material, i.e., A/B in the product over A/B in the feed gas, where A is the "wanted" isotope. Enrichment factors are normally in the 30–50 range; in special applications enrichment may be as high as 1000. The production rate of the separators greatly depends on the throughput. A current of 1 mA corresponds to 6.3×10^{15} unit charges per second. A current of 100 mA will produce about 10 g of a nuclide of mass 100 in a full day's operation; about one-half of this amount may be recovered from the collector. Since the efficiency of ionization, as mentioned, is below 10%, at least 100 g of charge material is required. When the "wanted" isotope is a rare nuclide, production rate may be very poor; for example, ^{40}K (Appendix I) can only be produced at a rate of 0.5 mg/day. In cases like this, other separation methods such as chemical exchange, thermal diffusion, and distillation should be used if at all possible.

In the early 1940's most isotopes were produced at the milligram level and efforts were mainly concentrated on the separation of ^{235}U from ^{238}U. By 1950 many isotopes were produced in gram quantities and today just about any isotope may be obtained in a variety of enrichments and in many convenient chemical forms. The price of most enriched isotopes decreased considerably in recent years and the use of isotopes in many areas of physical and chemical research is now becoming commonplace.

II. GEOCHEMISTRY AND COSMOCHEMISTRY

The availability of high precision mass spectrometers has resulted in the development of a new general field known as *isotope geology* which deals with the determination of isotopic abundances and their deviations from their "normal mean" values of terrestrial elements and minerals, and of meteoritic samples. From such information theories can be developed and calculations made on the age and history of the earth's crust and, eventually, of the planetary system. So far, meteorites are the only samples of solid material available from extraterrestrial space. However, it is probably only a matter of a few years until samples from the moon's surface and from other planets will become available for mass spectrometric analysis. In the meantime, the composition of the upper atmosphere is already being actively studied. Recent results will be reviewed in the second part of this section.

A. Isotope Geology

1. Geochemical Prospecting

Geochemical prospecting is the use of geochemical data in the search for valuable mineral deposits. Here mass spectrometry is mainly used as a tool for trace element analysis, searching for uranium, thorium, etc. The major requirement is high absolute sensitivity and the technique most frequently employed is isotope dilution. To compensate for possible variations of isotopic abundances from one deposit to another, due to fractionation by physical, chemical, or biological processes, three samples are usually analyzed: a straight extract of the mineral, a known mixture of the extract and an isotopically enriched tracer, and a pure sample of the enriched isotope. This technique also minimizes problems arising from contamination, standardization, etc. (Section 11-III-A). Detection limits for the various elements, as already discussed, are determined by chemical rather than mass spectrometric considerations. Concentrational sensitivity may be as low as 1 part in 10^{12}, while absolute sensitivities are between 10^{-13} and 10^{-15} g for many materials.

2. Geochronometry

Geochronometry is the study of the absolute age of the rocks and of the earth. Geologic age determinations are based on the fact that there are minerals which contain radioactive isotopes having a half-life comparable with the duration of geological times, and stable daughter (end) products with which they are in equilibrium. The general age relationship is

$$t = (1/\lambda) \ln (B/A + 1) \tag{13-13}$$

where t is the age in units of 10^6 years, A is the number of radioactive parent atoms, B is that of the stable daughter atoms, and λ is the constant of disintegration related to the half-life by the expression $t_{1/2} = 0.693/\lambda$. There are four basic assumptions involved: (*a*) the parent atoms were present when the mineral crystallized from the silicate melt to form an igneous rock, (*b*) the concentration of daughter atoms was zero (or known) at the beginning, (*c*) the value of $t_{1/2}$ for the radioactive decay involved is of the same order of magnitude as the time span to be determined, and (*d*) the system remained chemically closed during its history or correction can be made for leaching. The most important methods of quantitative geochronometry are summarized in Table 13-1.

Table 13-1.   Methods of Geochronometry

Process[a]	Comparison nuclide	$t_{1/2}$, yr	Effective range, yr	Minerals
$^{235}U \xrightarrow{\alpha} {}^{207}Pb$	^{204}Pb	7.2×10^8	T_0-10^7	Uraninite, zircon monazite
$^{238}U \xrightarrow{\alpha} {}^{206}Pb$	^{204}Pb	4.5×10^9	T_0-10^7	Uraninite, zircon monazite
$^{232}Th \xrightarrow{\alpha} {}^{208}Pb$	^{204}Pb	1.4×10^{10}	T_0-10^7	Uraninite, zircon monazite
$^{40}K \xrightarrow[\text{capture}]{\text{electron}} {}^{40}A$	$^{36}Ar, {}^{38}Ar$	1.3×10^9	T_0-10^4	Muscovite, biotite, glanconite
$^{87}Rb \xrightarrow{\beta} {}^{87}Sr$	$^{86}Sr, {}^{88}Sr$	5.0×10^{10}	T_0-10^8	Muscovite, biotite, glanconite
		5.6×10^3	5×10^4 to present	Carbon materials once in contact with atmosphere-biosphere system.

[a] First element: radioactive nuclide, second element: daughter nuclide.

Of the four stable isotopes of lead, three are end products of the naturally occurring radioactive decay of uranium and thorium. Mass spectrometric measurements by Aston have shown, as early as 1929, that the relative abundances of the lead isotopes are significantly different in radiogenic and "normal" (i.e., very low U and Th content) minerals. It has been said that uranium (thorium) and lead in a mineral play the part of sand in an hourglass: uranium corresponds to the sand at the top, the lead to that at the bottom. In the *uranium–lead* and *thorium–lead* methods two determinations are needed: The U (Th) content must be determined by chemical methods and the isotopic analysis of the radiogenic lead is done by mass spectrometry. With the knowledge of the decay constant involved, the age of the mineral can be calculated. Correction for possible "ordinary" lead contamination can be made on the basis of the ^{204}Pb which, fortunately, is nonradiogenic.

A more reliable technique is the so-called *lead–lead* method developed by Nier (16). It is based on the fact that the ^{238}U → ^{206}Pb and ^{235}U → ^{207}Pb decay processes have half-lives differing by a factor of about 6. This means that the ^{207}Pb/^{206}Pb ratio increases systematically with the age of the mineral and age may be determined without the knowledge of the uranium content, thus avoiding the generally less accurate chemical analyses. Four measurements are needed in this method and three of them involve mass spectrometry. First, a determination of the present abundance ratio of the two uranium isotopes must be made. Accurate measurements of the ^{235}U/^{238}U ratio on many samples from various sources yielded 1/139, which is now considered as a constant. The next step is the measurement of the relative abundances of the two lead isotopes by mass spectrometry. A correction is needed when nonradiogenic lead is also present in the mineral investigated. This calls for a third analysis, searching for the nonradiogenic ^{204}Pb isotope. The correction is made on the basis of the "ordinary" lead isotope composition, as long as there is only a small contamination. Finally, the decay constants of the two uranium isotopes must be known. The age is determined from the formula

$$\frac{^{207}\text{Pb}}{^{206}\text{Pb}} = \frac{1}{139}\frac{\exp{(\lambda_{235}t - 1)}}{\exp{(\lambda_{238}t - 1)}} \tag{13-14}$$

where the λ's refer to the uranium isotopes. Lead samples are usually introduced into the mass spectrometer in the form of halides (i.e., iodide) and electron impact is used for ionization. For microgram quantities, surface ionization with secondary electron multiplier detection is often employed.

Figure 13-4 is a block diagram illustrating the isotopic composition of thorium, uranium, and common lead. To explain the somewhat different composition of "ordinary" lead from various sources, Nier distinguishes (17) between "primeval" and "common" lead: The first term refers to lead as it existed at the time of the formation of the earth's crust, the second describes lead as it is known today, consisting of primordial lead (solid bars) which has been contaminated by approximately equal amounts of thorium lead (horizontally striped bars) and uranium lead (vertically striped bars). Age measurements for a few uranium minerals are shown in Table 13-2 (18). It is seen that the lead–lead method provides the most consistent data for minerals from the same locality. The age of the earth is defined as the time it has existed with its present mass and density. A minimum age is set by

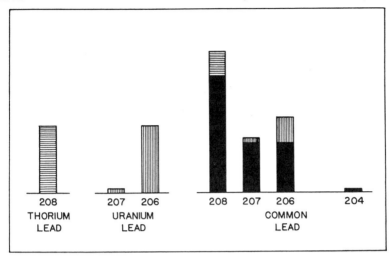

Figure 13-4. Isotopic composition of lead from various sources. (From A. O. Nier, *Am. Scientist*, **54**, 359 (1966).)

the age of the oldest rocks on earth, while the maximum is limited by the cosmogenic origin of the elements. The range, based on uranium measurements is 3×10^9 to 6×10^9 years. The age of meteorites has been estimated by Anders (19) to be 4.6×10^9 years.

The *potassium–argon* method of age determination utilizes the decay of ^{40}K to ^{40}Ar. It is believed that atmospheric ^{40}Ar originated during the

Table 13-2. **Age Measurements for Uranium Minerals (Millions of Years)**

		$^{207}Pb/^{206}Pb$	$^{238}U/^{206}Pb$	$^{235}U/^{207}Pb$	$^{232}Th/^{208}Pb$
Huron Claim,	Uraninite	2490	1370	1980	1270
Manitoba		2565	—	—	—
		2550	—	—	—
	Monazite	2600	3200	2850	1860
Conger	Uraninite	1030	1010	1015	1010
Township,		990	994	993	897
Ontario					
Beaverlodge	Pitchblende	940	190	260	—
Lake, Sask-		930	822	850	—
atchewan		920	592	664	—

 ^a Data from reference 18 and J. T. Wilson, R. D. Russell, R. M. Farquhar, in *Encyclopedia of Physics*, S. Flügge, Ed., Springer Verlag, Berlin, 1956, Part X, Chapter 14.

period the earth's crust was in a molten state, and therefore the ratio $^{40}Ar/^{36}Ar$ should be higher in old rocks than in atmospheric argon. This has been experimentally verified (20). The widespread occurrence of potassium in minerals and the relative paucity of argon make this method attractive. Analytical precision requirements, in view of geological uncertainties, are not severe; a precision of 1% is adequate. Experimental problems center around sensitivity. The total quantity of argon available might be as small as 2×10^{-4} std cc per gram of mineral. Since the presence of the ^{36}Ar isotope is taken as evidence of atmospheric argon contamination for which correction must be made, and since this isotope has an abundance in normal argon of only 0.3%, gas quantities of the order of 10^{-6} cc must be detected and estimated. The best instrument for the purpose is the one designed by Reynolds (Section 8-III). It is a static instrument (21), i.e., the ion source is completely closed off from the pumps after sample introduction. The detection of as few as 1.4×10^7 atoms of ^{36}Ar has been reported. Values for the age of the earth, and limits for the age of the universe (meteorite measurements) by the potassium–argon technique are generally in acceptable agreement with those resulting from lead measurements.

The *rubidium–strontium* technique involves the determination of the total Rb and Sr content by isotope dilution (few per cent accuracy), and the isotope composition of Sr in a separate measurement. As in previous methods, age is calculated on the basis of the decay law. Correction for ordinary strontium is based on the ^{86}Sr isotope. The main experimental difficulty lies in the required complete chemical separation of Rb and Sr (same mass). The ratio analysis is normally carried out on a single-filament emission source; sensitivity and background usually present no problems.

Radioactive carbon dating does not involve mass spectrometry since the amount of ^{14}C to be measured $(1:10^{12})$ is below sensitivity limits. The technique is based on the radioactive decay of ^{14}C, which is an isotope produced by cosmic rays. Primary cosmic rays produce neutrons in the atmosphere which, in turn, react with ^{14}N nuclei, converting them to ^{14}C, after which they enter the biogeochemical cycle (22).

A relatively recent technique, known as *xenology*, is another powerful tool in the study of the age and history of meteorites and the solar system. Developed by Reynolds (23), the method is based on the fact that the radioactive isotope ^{129}I decays with a half-life of approximately 17 million years to form ^{129}Xe. Figure 13-5 shows the mass spectrum of a xenon sample of 7×10^{-9} cc STP, extracted from the Richardton stone meteorite and analyzed in a static manner (21). The horizontal lines show the

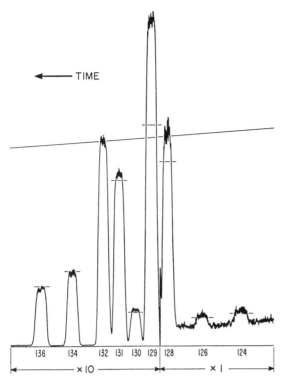

Figure 13-5.   Mass spectrum of xenon extracted from Richardton stone meteorite (23). Horizontal lines show the comparison of terrestrial xenon. The unexpectedly high ^{129}Xe peak is called the "special xenon anomaly" while the other, less striking, departures from normality in the isotopic composition of Richardton xenon are known as "general anomalies."

comparison spectrum of terrestrial xenon and the excess ^{129}Xe isotope is clearly seen. This effect has been verified and shown to be a general property of the chondritic stone meterorites. Calculations revealed that this sample might be as old as the universe. The theoretical implications of the presently available results of xenology are still being strongly debated. The importance of the technique, however, is being increasingly recognized.

3. Geologic Thermometry

Geologic thermometry deals with the determination of temperatures at which geologic processes take place. In addition to direct temperature measurements in lavas, mines, hot springs, etc., and to indirect methods based on transformations, decompositions, etc., the *isotopic* methods have proven to be extremely successful. Isotopic thermometry is based on the fact that the distribution of the isotopes of an element between pairs of compounds in equilibrium is slightly different at different temperatures. The classical example is the calcium carbonate and water system, studied by Urey (24),

$$\tfrac{1}{3}C^{16}O_3{}^{2-} + H_2{}^{18}O \rightleftharpoons \tfrac{1}{3}C^{18}O_3{}^{2-} + H_2{}^{16}O \qquad (13\text{-}15)$$

Since the equilibrium constant for this reaction is slightly greater than unity, marine fossils must exhibit a slight (about 3%) enrichment in ^{18}O over that of seawater, the exact extent depending on the temperature of the water at the time of deposition. Thus, from the measurement of $^{18}O/^{16}O$ ratios, ocean paleo-temperatures may be determined.

The isotope ratios must be measured accurately since the temperature coefficients of isotopic effects are small. Shackleton (25) discusses the problems of the high precision isotopic analysis of oxygen and carbon in CO_2 and describes techniques for achieving precision of $\pm 0.1\%$ with the use of only 0.1 ml size samples. This enables one to determine ocean paleo-temperatures to $\pm 0.5°C$, a truly amazing accuracy. The oxygen isotopic ratio is usually expressed in units of parts per thousand with respect to an arbitrary standard,

$$\delta = \frac{{}^{18}O/^{16}O \text{ (sample)} - {}^{18}O/^{16}O \text{ (standard)}}{{}^{18}O/^{16}O \text{ (standard)}} \qquad (13\text{-}16)$$

The relationship between the δ value and the temperature at which the calcium carbonate sample was deposited is given (26) as

$$T = (16.5 - 4.3\delta + 0.14\delta^2) \ °C \qquad (13\text{-}17)$$

This experimental equation may be modified if necessary for variations in the isotopic composition of the ocean. Figure 13-6 shows the seasonal temperature variations indicated by the oxygen-18 content of the growth rings of Jurassic belemnite. This work by Epstein (27) is considered by many as one of the most extraordinary achievements of mass spectrometry.

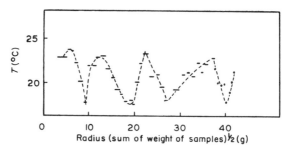

Figure 13-6. Seasonal temperature variations indicated by the ^{18}O content of the growth rings of Jurassic belemnite (27).

B. Upper Atmosphere and Space Research

1. General

Perhaps the most glamorous application of mass spectrometry in recent years has been the study of the composition of the upper atmosphere. Instruments are now being designed and built for the analysis of the lunar atmosphere, the Martian dust, the "composition" of deep space, etc. Many of these instruments will be on their long journeys in outer space within a few years and the data telemetered back to earth will provide answers to age-old questions. It is likely that the first indication of real life, if any, in space will be a mass spectrometric peak.

In the meantime, mass spectrometers are now actively used in the study of the upper atmosphere. Many mass spectrometers have been sent up in sounding rockets (research rockets) during the last decade, and a considerable amount of information is now available about the neutral as well as the ionic composition of the atmosphere up to about 200 km. The knowledge of the composition, both neutral and ionic, of the upper atmosphere is one of the central goals of astrophysics. Since the mean free paths of molecules or charged particles are comparable to, or even considerably greater than, the characteristic dimensions of any measuring equipment, and also since the composition continuously changes due to various ionization and photochemical processes, measurement of almost every atmospheric parameter (pressure, temperature, density, etc.) is meaningless without knowledge of composition. Problems such as the location of the gravitational separation level for gases (i.e., above which gaseous composition regularly varies with altitude), the dissipation of the atmosphere into outer space (helium problem), and the nature of the various ion layers (radio communication) cannot be studied without a good knowledge of the composition of the particular atmospheric region. The composition of neutral molecules may be studied by several methods and many sampling techniques have been developed with balloon ascents into the stratosphere, but the ionic composition can only be investigated by mass spectrometers.

Radiofrequency mass spectrometers, omegatrons, and quadrupole instruments, as well as magnetic type machines have been employed in these studies. Each type has its merits and shortcomings when the special requirements of space research are considered. In rocket-borne mass spectrometers, for example, a significant amount of outgassing occurs as soon as the rarefied layers of the atmosphere are reached and there is usually no way to get rid of this contamination. The surface of satellite-borne instruments, on the other hand, will become thoroughly outgassed during their prolonged stay in space. Size, weight, and data form, as well as resolution, sensitivity, and useful mass range must be considered in the selection of the instrument for a particular problem. A detailed discussion of mass spectrometers from the point of view of space research is given by Jayaram (28), who also discusses many applications. Both American and Russian achievements in the analysis of the upper atmosphere prior to 1960 are described by Mirtov (29).

2. Ion Composition of the D Region

The "D region" is that part of the ionosphere which extends from about 50 to 90 km. It is too high for balloons and aircraft and too low for satellites. The pressure in this region is in the range of 5×10^{-1} to 3×10^{-3} torr, and so the rocket must carry along with the mass spectrometer a pumping system which must operate at zero gravity. Special titanium getter-ion pumps and molecular sieve pumps containing zeolite cooled with liquid nitrogen have been developed for this purpose.

The first positive ion composition measurements in the 64–112 km range were made by Narcisi and Bailey (30), who on October 31, 1963 flew a quadrupole mass spectrometer system aboard a Nike Cajun rocket. Figure 13-7 shows a cross section of the front end of the payload in the sampling configuration after nose tip and vacuum cup ejection. To maintain maximum available sensitivity, resolution was set at 16 (10% valley). Pressure was maintained around 1 μ by a liquid nitrogen cooled zeolite pump. Scanning was accomplished by varying the rf and dc voltages as an exponential function of time (constant dc/rf ratio). After the nose tip and vacuum cup were ejected at 63.4 km, about 100 spectra were obtained during ascent and descent.

Results are shown in Figures 13-8 and 13-9. Ions at m/e 19, 30, and 37 were found predominant in the D region with the peak at mass 32 rapidly increasing above 75 km. At an altitude of 82.5 km the m/e 19 and 37 peaks disappeared rather sharply with the simultaneous appearance of $^{23}Na^+$, $^{24}Mg^+$, $^{25}Mg^+$, $^{26}Mg^+$, and $^{40}Ca^+$ ions, each exhibiting an essentially similar altitude profile. The origin of these metallic ions is not definitely known, although they may be produced by the vaporization of meteors. Above 82 km ions of mass 30 and 32 appeared to be the most predominant. Many other ions were also detected in smaller quantities and, in a sense, the experiment created more questions than it answered. Several more flights are being planned.

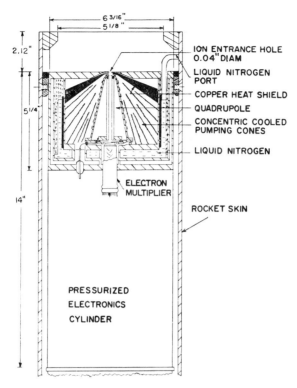

Figure 13-7. Cross section of quadrupole mass spectrometer in rocket, sampling configuration (30).

3. *Neutral Composition in the 100–200 km Range*

Up to about 100 km there is a more or less complete mixing of the gases, and the composition is essentially the same as on the ground; pressure decreases to the order of 10^{-4} torr. At an altitude of 100 km, molecular oxygen (and to some extent nitrogen) begins to dissociate into monoatomic forms as the sun's ultraviolet radiation (170–1027 Å) becomes more intense,

$$O + h\nu \rightarrow O^+ + e^-$$
$$O_2 + h\nu \rightarrow O_2^+ + e^- \tag{13-18}$$
$$O^+ + O_2 \rightarrow O + O_2^+$$

Also diffusive separation of the gases begins at this altitude, making the 100–200 km range extremely important.

The neutral composition of the atmosphere in the 100–200 km range has been investigated by Nier et al. (31), who made two Aerobee rocket

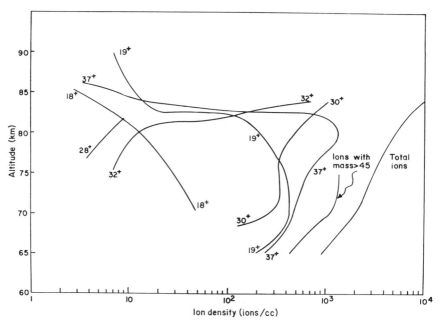

Figure 13-8. Concentration of positive ions detected within altitude range 64–83 km (30).

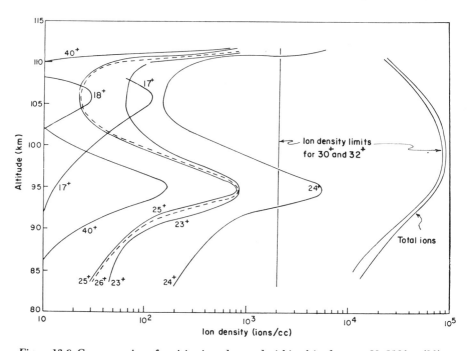

Figure 13-9. Concentration of positive ions detected within altitude range 83–112 km (30).

launchings (June 6, 1963 and April 15, 1965). The first rocket contained a 90° single-focusing magnetic mass spectrometer and also a double-focusing machine (which failed to operate), while the second rocket contained three 90° magnetic mass spectrometers of more or less conventional design. Figure 13-10 shows the intensity of the mass 28 peak as a function of altitude in terms of equivalent telemeter chart cm versus altitude. The rocket spun at a rate of one revolution in 4 sec and this is seen to cause violent variations in intensity. The "beat effect" is caused by the fact that The period of rocket spin was an integral multiple of spectral scan rate. (2 sec). The intensity modulation results from the fact that sometimes the ion source is "looking" in the forward direction, while other times the opening in the ion source is facing toward the "back," as if it was "running away" from the molecules. In the latter case only the fast tail of the Maxwellian distribution of speeds of the molecules is collected by the source. The curve through the nodal points (source opening at right angles to the direction of motion) offers a means to calculate the density of the atmos-

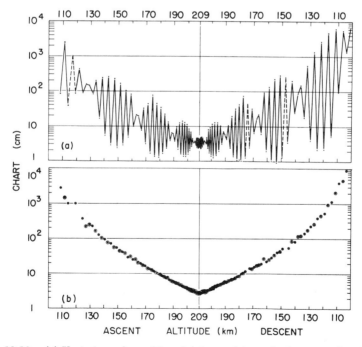

Figure 13-10. (a) Variations of mass 28 peak (nitrogen) intensity in terms of equivalent telemeter chart cm versus altitude, 100–200 km zone; (b) appearance of curve after correction for velocity and spin of rocket and temperature variations in the atmosphere (31).

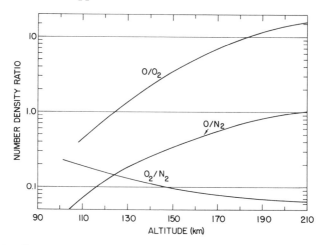

Figure 13-11. Variation of O/O_2, O_2/N_2, and O/N_2 ratios with altitude in the 100–200 km range (31).

phere. Similar curves can be obtained for other components, including oxygen. The effects of diffusive separation are shown in Figure 13-11, which depicts the variation of the O/O_2, O_2/N_2, and O/N_2 ratios with altitude. Near 100 km the O/N_2 ratio approaches zero since atomic oxygen recombines quickly after its formation. At 100 km altitude the composition appears similar to that on ground, but at 200 km molecular nitrogen and atomic oxygen are present in a 1:1 ratio. At an altitude of 600 km helium becomes the main constituent (at ground level He has an abundance of only 1/200,000). Since the modulation effect depends upon the ratio of average molecular velocity to the component of rocket velocity in the direction of the axis of the ion source, gas temperatures and other properties may be estimated (31).

4. Studies in the Thermosphere

The thermosphere begins at about 85 km and extends to about 360 km during periods of minimum solar activity and up to 700 km during maximum activity. In the lower thermosphere temperature increases rapidly with increasing altitude. Within the 250–400 km range, however, temperature is quite constant at any given time. In this isothermal region the temperature is about 700°K at night, while during the period of maximum solar activity (daytime) it is as high as 2000°K. The temperature increase is due to absorption of solar radiation which causes dissociation and ionization of particles with considerable amounts of heat released. Molecular oxygen at the lower level and both atomic and molecular nitrogen and oxygen at the upper levels constitute the atmosphere. Of course, high temperature here does not mean conventional "hotness," but is based on the kinetic energies of the individual particles in this highly rarefied region. Electron and molecular nitrogen temperature and density measurements in the thermosphere have been carried out by Spencer et al. (32) employing a tuned (to nitrogen) omegatron mass spectrometer (33). Results showed good agreement with theoretical expectations, but, the experiments resulted in a series of new questions which have to be investigated in future experiments. This pattern of raising new problems and questions in every launching appears general, and quite understandable, in modern rocket experiments. The field is an extremely fast moving one, and the pioneering investigations described above are soon likely to become "classical" experiments.

III. BIOCHEMISTRY

Recent developments in instrumentation (greater mass range, highly increased resolution and sensitivity, the direct sample insertion probe) and analytical techniques (GC–MS combination, chemical pretreatment to increase volatility) have made the mass spectrometric studies of complex organic molecules and natural products of biochemical interest both possible and common. With the rapidly growing number of mass spectrometers in biochemistry departments, medical schools, and hospital research laboratories, the number of publications on biochemical applications has been increasing exponentially during the last five years.

The general mass spectrometric techniques employed in studying organic compounds of biochemical interest are essentially the same as described in Chapter 10, and need not be repeated here. In this section a few special analytical techniques are discussed and representative applications described. The length of this section should not be taken as indicative of the importance of the subject. On the contrary, it is the belief of this author that the application of mass spectrometry to problems in biochemistry is the most important area within the discipline today, and the mass spectrometer is to become the most cherished tool of biochemists. Among the books on the mass spectrometry of organic compounds (Section 14-II-B), the texts by Budzikiewicz et al. are perhaps the best for biochemists.

Mass spectrometers have long been employed for the simultaneous analysis of *respiratory gases* (34,35). More recently commercial instruments (Fig. 8-5) have become available for continuous gas analysis (36). In an interesting study Adamczyk et al. (37) measured the excretion of CO_2, He, and Ar by the human skin at several places on the body during and after respiration of ordinary air and various gas mixtures. Maximum excretion of carbon dioxide (when breathing air) occurred at the axillae and forehead. Helium excreted quite strongly also from the chest and neck. Figure 13-12 shows the excretion of helium during continuous respiration of a 80% He, 4% N_2, and 4% O_2 mixture: helium excretion started almost immediately and a maximum was reached in about 15 min. When respiration with air commenced, the excretion of helium decreased exponentially. Similar results were obtained with argon, and the rates of excretion for these two gases appear to be related to their diffusion coefficients. Further studies may lead to important applications in the selection of breathing atmospheres for space flights and oceanography, and in blood circulation measurements.

A clinical technique for the determination of small amounts of diethyl ether in blood was reported many years ago by Jones et al. (38). The ether is partially extracted from the blood by equilibration with a gas containing a known amount of argon, followed by the measurement of the argon/ether ratio with a mass spectrometer. In more recent studies, mass spectrometric measurements are utilized to identify impurities in inhalant anesthetic compounds (39).

The use of radioactive labeling for the study of biochemical processes is a well-established and much publicized technique. The availability of enriched stable isotopes presents another technique, one which offers certain

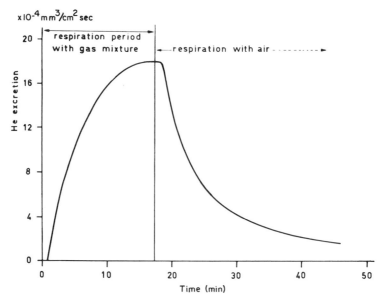

Figure 13-12.   Excretion of helium under continuous respiration of a mixture of 80% He, 4% N_2, and 16% O_2 (37).

important advantages. Biochemists are mainly interested in the elements hydrogen, carbon, nitrogen, oxygen, and sulfur, of which only ^3H, ^{14}C, and ^{35}S are available in radioactive form with experimentally convenient half-lives. It may be questionable if these isotopes which do not occur in nature could be employed with faith to study the behavior of their isotopically normal counterparts. Moreover, radioactivity is often objectionable in research on humans. All five of the above elements are available in enriched stable isotope form and prices have recently been reduced to such an extent that widespread use has become possible.

The most popular stable isotope is deuterium. Deuterium may be incorporated into a molecule by exchange with a labile hydrogen atom, by replacement of a functional group with deuterium, or by the saturation of multiple bonds with deuterium. Details and experimental techniques are available in many books on organic mass spectrometry. Since deuteration can be carried out easily even on milligram quantities of material, the technique has found a wide variety of uses.

Carbon-13 is more expensive and the substitution is more difficult than in the case of deuterium. There are no major problems in the use of the enriched isotopes of nitrogen, oxygen, and sulfur.

The first use of mass spectrometry in studying biochemical processes, almost 30 years ago, was reported by Schoenheimer and Rittenberg (40), who used deuterium labeling to investigate the metabolism of amino acids and fatty acids. They developed the concept of "dynamic state of body constituents" according to which the apparently static state of many constituents results from rapid opposing reactions at nearly equal rates. The first problem, one that may be quite difficult indeed, is the preparation (or synthesis) of the required compounds containing the enriched isotope (41). The role of mass spectrometry in such experiments is relatively simple. Isotope ratios are measured in gases or gaseous degradation products of organic molecules using more or less standard techniques. Often the final product is carbon dioxide and ^{18}O must be determined in very small gas samples (42). The isotope dilution technique is also frequently employed. Stable isotopes have been used in problems ranging from the determination of the "life span" of red blood cells and water in the human body (127 and 14 days, respectively) to the investigation of the mechanism of complex biosyntheses. Several examples of studies on fragmentation mechanisms of compounds of biological interest have been described in a review article by Stenhagen (43). A new technique for studying mechanisms of mass sprectral reactions using ^{14}C-labeled compounds has recently been described by Knöppel and Beyrich (44).

The combination of mass spectrometry with gas chromatography is likely to become one of the most important tools in biochemical and biomedical research. The field is very new but has shown a spectacular rise during the last few years. For an illustration, reference is made to the work of Ryhage and his co-workers (45), who recently reported the identification of mono- and dihydroxy bile acids in human feces by GC-MS technique in the course of their investigations on the influence of diet on bile acid excretion. It is known that the mixture of primary bile acids (cholic and chonedeoxycholic acids) synthesized by the liver undergoes bacterial transformation in the intestine and the excretion in the feces contains a rather complex mixture of primary and secondary bile acids. After chemical preparation and isolation, 10-μg samples were analyzed as trifluoroacetates in the GC-MS combination shown in Figure 10-16*b*; spectra were recorded to mass 500 in 6 sec. Figure 13-13 shows a comparison between the mass spectra of an unknown component from feces and that of authentic artificially prepared methyl 3β,7α-di(trifluoroacetoxy)-5β-cholanoate used for identification. With the combined data from gas–liquid chromatography and mass spectrometry, several methyl cholanoates isolated from biological materials have been identified using comparisons with synthetic compounds to confirm proposed structures. Several other examples are described by Stenhagen (43).

Chemical pretreatment of the samples is an important and frequently employed technique in connection with biochemical applications. The primary use of converting the compound of interest into a derivative is to increase volatility. A derivative may also have a characteristic and known fragmentation mechanism which will aid in the interpretation of the spectrum.

The use of trimethylsilyl (TMS) derivatives in mass spectrometry was first introduced in 1957 by Sharkey et al. (46), primarily for the analysis of mixtures of alcohols. The characteristic fragmentation of the trimethylsilyl group can be utilized in many applications. A large peak due to the loss of the methyl group from the highly branched silicon atom is an important help in determining molecular weights. In recent years TMS derivatives have been widely used in gas–liquid chromatography, in GC–MS combination analysis, and in the high-resolution mass spectrometry of natural products, on account of their ease of preparation, thermal stability, and volatility. The use of deuterium-labeled trimethylsilyl derivatives in mass spectrometry is described by McCloskey et al. (47); the reactions of labile TMS derivatives with fluorocarbons in GC–MS systems is discussed by Foltz et al. (48). The use of fluoroalcohol esters as derivatives has been

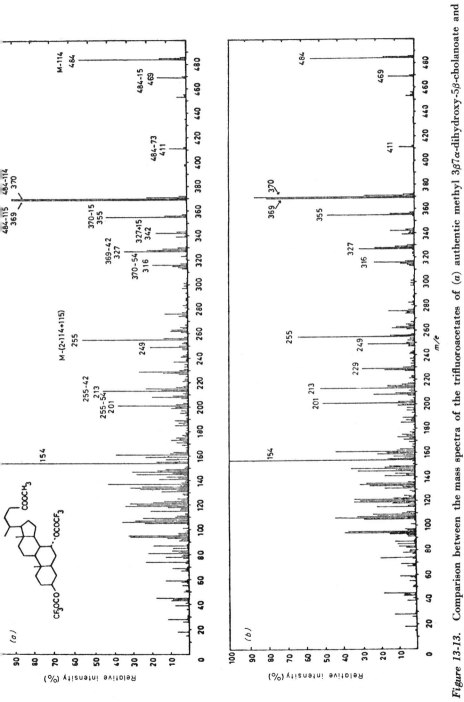

Figure 13-13. Comparison between the mass spectra of the trifluoroacetates of (*a*) authentic methyl 3β7α-dihydroxy-5β-cholanoate and (*b*) an isolated compound from feces (45).

suggested by Teeter (49). The application of the TMS derivative technique to biochemistry is illustrated in the studies of vitamins A, B_6, C, K_1, and K_2 by Vetter and his co-workers (50), who made quantitative determinations on separated isomers and identified many fragments.

Another area where rather high-flown adjectives must be used to describe future possibilities is the determination of organic structures by high resolution mass spectrometry. Perhaps an entirely new area of the study of *peptides* started with the simultaneous publications (in 1966) by Biemann and his co-workers (51) and McLafferty and his co-workers (52) of computerized techniques for the determination of amino acid sequences in oligopeptides from high resolution spectra. Both methods are basically similar and start with the realization that, barring rearrangements, the structure of a linear molecule is determined unequivocally by using only the possible fragments which contain one end of the chain. The possibility to determine the exact mass of hundreds of peaks in a short time permits the development of sequence determination utilizing measurements on N-terminal groups. The selection and preparation of suitable compounds and derivatives is a problem in organic chemistry. The high resolution mass spectrometric analysis of the samples (which may be of microgram size) is performed using techniques described in Section 10-III. Computer interpretation of the spectra is achieved by the development of a logical programming system incorporating the rules and boundary conditions provided by the basic assumptions. Mass spectrometric studies of peptides is an increasingly active field as evidenced by the large number of recent references quoted by Biemann and McLafferty in their papers (51,52). With further developments in computerized spectrum interpretation, an even faster increase in application for protein studies is expected.

Carbohydrates, steroidal alkaloids, estrogens, and practically all classes of natural products have been investigated by mass spectrometry during the last few years. The volumes of Budzikiewicz, Djerassi, and Williams on the structure elucidation of natural products by mass spectrometry review hundreds of applications (53).

There is a growing awareness of the importance of *trace elements* in biological processes. The discovery of the toxicity, and later the essentiality, of selenium is a good example of the dependence of biochemists upon trace analytical chemistry. Nearly every element may at times appear in living material, and for some of the elements little or nothing is known of their function or mechanisms of transport into and out of living materials.

Spark source mass spectrography offers a possibility for the simultaneous determination of many elements with high sensitivity. This is a new field, and few applications have been reported. Wolstenholme (54) reported the determination of trace elements in dried blood plasma, and Hardwick and Martin (55) developed a technique for the estimation of trace elements in dental tissues. A method for the survey analysis of biological materials by spark source mass spectrometry has recently been published by Evans and Morrison (56). As many as 50 trace elements in the 100 ppm to few ppb concentration range were determined in blood serum, kidney tumor, lung tissue, bone, and plant leaves.

IV. VACUUM TECHNOLOGY

Recent utilization of high vacuum ranges from test chambers simulating space conditions for astronaut training to freeze-drying of foods. Very often it is essential to know the qualitative, and also semiquantitative, composition of the residual gas. This applies to production line processes as well as laboratory research studies, especially in dynamic systems. In the semiconductor field, for example, partial pressures of individual components in the residual gas in a vacuum evaporator must be known to control conditions under which impurities might be incorporated into the films giving rise to undesired electrical, magnetic, etc. properties. The selection of proper pumping equipment for a particular problem requires information on the partial pressures of expected gases since the efficiency of most pumps depends on the composition of the gas to be removed. Similarly, a search for the cause of inadequate pumping in a system is greatly aided by partial pressure analysis. Another important field of application is the study of adsorption, condensation, and desorption properties of construction materials. It should be noted here that most pressure- and density-measuring gauges show different sensitivities for various gases below 10^{-4} torr and the total pressure often must be given in terms of "nitrogen equivalent pressure"; the error in extreme cases may be as much as a factor of 10.

In principle, any mass spectrometer can be used as a residual gas analyzer. For the analysis of very low pressures, however, certain special features are required and many instruments have been designed specifically for residual gas analysis. A revealing discussion of partial pressure analysis, including instrumentation and applications, is given by Huber (57).

An important, but often overlooked, consideration in selecting a partial pressure analyzer is the requirement that the gauge must become an *integral part* of the vacuum system under study. This calls for simple design and provision for baking. Connection to the vacuum system must be made through a tubulation of high conductance, the ideal arrangement being the "nude" source which is completely immersed in the vacuum chamber itself. Analyzers requiring a magnet are disadvantageous from this point of view, although proper mechanical design can often eliminate many problems caused by the presence of the magnet. To reduce the level of outgassing, total surface area must be kept small; the very small overall dimensions of the measuring element in omegatrons have much merit in this respect. Attention must be paid to matching the construction materials of the analyzer with those of the vacuum system; e.g., omegatrons are quite conveniently sealed to all-glass systems, while the bulkier magnetic deflection instruments usually require metal flanges and are best used with all-metal systems.

Among instrument performance parameters, *partial pressure sensitivity* (i.e., smallest detectable portion of a gas component present in the gas mixture) is perhaps the most important. The ultimate sensitivities of both magnetic deflection and resonance type instruments are approximately the same—in the 10^{-10} to 10^{-12} torr range. It is important, however, to consider *how* the high sensitivity is achieved. Electron multipliers can usually be employed and they provide high sensitivity and fast response, but are somewhat more difficult to handle and outgas than electrometer detectors. Sector field machines can, of course, easily be connected to multipliers and such arrangements provide the most sensitive analyzers. The best available sensitivity at this time is about 10^{-14} torr. Amplification with electron multipliers is impossible for omegatrons. The relatively high sensitivity of omegatrons is partially due to the high instrument transmission ratio (i.e., ratio of ions collected after separation to total number of ions produced). A typical omegatron sensitivity is 10^{-14} A/10^{-9} torr pressure at an emission current of 1 μA. These three types of gauges are used in ultrahigh vacuum studies most of the time. Farvitrons, topatrons, and TOF instruments have lower sensitivities but offer other advantages. The increasingly popular mass filters are available in both high and lower sensitivity versions. The upper limit of the useful pressure range is mainly dependent upon the path lengths of the ions in the analyzer. Magnetic deflection instruments (upper limit 10^{-4} torr) are somewhat advantageous; for omegatrons the limit is 10^{-5} torr due to the long ion paths of the resonant ions.

As far as resolution (and useful mass range) is concerned, magnetic deflection instruments are considerably more versatile than resonance types. The number of gases to be identified and measured in ultrahigh vacuum systems is relatively small: H_2, CO, N_2, Ar, CO_2, traces of methane, and higher hydrocarbons (water, oxygen, etc. usually disappear during bakeout). An important observation by Craig (58) is that the m/e 43 peak ($C_3H_7^+$) is so characteristic of all types of hydrocarbons with more than 4 carbon atoms that the total hydrocarbon content present can be estimated within a factor of 2 by taking the partial pressure based on an assumed m/e sensitivity of 1.4 relative to nitrogen. This means that a resolution of 50 which permits a distinction between carbon dioxide and the m/e 43 peak enables one to estimate total hydrocarbon partial pressure. For the identification of peaks from silicone oils and polyphenyl ether oils, a resolution of up to 150 is required. Resolution with magnetic deflection instruments is normally about 50 (1% valley); however, it may often be increased (at reduced sensitivity) to 100 or more. Resolution of omegatrons in about 40. It is recalled (Section 3-III-B) that the resolving power of omegatrons depends on the mass, resulting in a sharp drop in resolving power towards the higher masses. The resolution of a farvitron (electrostatic mass spectrometer) is about 20. For mass filters the resolution is the same over the entire mass range ($\Delta M = 1$). The resolution of topatrons can be adjusted (depending on sensitivity) in the 20–30 range. The time of response of residual analyzers is determined by the time constant of the ion current amplifier in all cases, except for the farvitron where it is 50 cps. Speed is normally not a major problem; however, for very rapid processes (e.g., adsorption) time-of-flight, quadrupole, or monopole residual analyzers may be employed.

It appears from the above discussion that the magnetic deflection type analyzers are best suited for residual gas analysis, although many resonance type instruments are currently being used for various applications, particularly as indicator gauges in "medium" high vacuum applications. Farvitrons, for example, are quite simple and inexpensive, and few adjustments are needed. However, no quantitative determinations and no total pressure measurements are possible, and components below 3% contribution cannot be detected. Mass filters also require relatively few adjustments and here there is no need to inject the ions with constant energy, and therefore high yield is achievable. A disadvantage is the needed high consistency of the high frequency voltage in relation to the dc voltage applied.

The influence of contaminating background gases on the behavior of vacuum-deposited thin films has been widely investigated. Caswell, for example, studied the effects of residual gases on indium and tin films (59); he also reviewed all work prior to 1963 (60). An ultrahigh vacuum

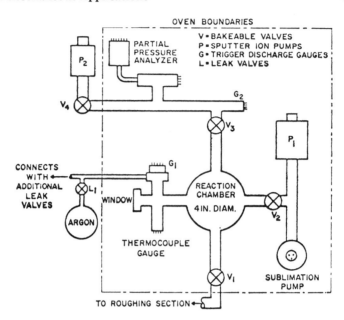

Figure 13-14. Schematic drawing of ultrahigh vacuum system for thin film and residual gas analysis studies (61).

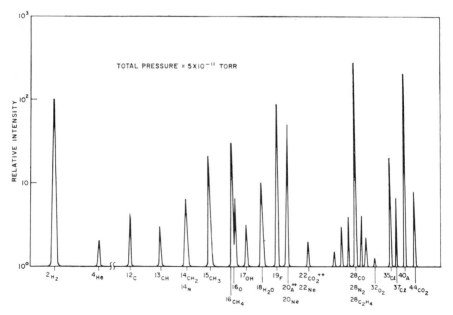

Figure 13-15. Partial pressure analyzer spectrum of the vacuum system shown in Figure 13-14 (61).

station specifically designed for thin film and residual gas analysis studies has recently been described by Rozgonyi et al. (61). A schematic drawing of their system is shown in Figure 13-14. The system is bakable to 450°C, ultimate vacuum is 10^{-10} torr, and the lowest measurable partial pressure is 10^{-14} torr. Figure 13-15 shows a partial pressure analyzer spectrum at a total pressure of 5×10^{-11} torr. A system of this type may be used to obtain both qualitative and quantitative correlations between residual gases and properties of deposited films, and to study the influence of gauge operation on background gases during high vacuum mass spectrometry (62).

Other residual gas studies include the investigation of pump systems, adsorption–desorption measurements, and the study of permeation and diffusion of structural materials. Many additional references are given by Huber (57).

References

1. White, F. A., T. L. Collins, and F. M. Rourke, *Phys. Rev.*, **97**, 566 (1955); **101**, 1786 (1956).
2. Leipziger, F. D., *Appl. Spectry.*, **17**, 158 (1963).
*3. Bainbridge, K. T., in *Experimental Nuclear Physics*, E. Segre, Ed., Vol. I, Wiley, New York, 1953, Part V.
*4. Nier, A. O., *Phys. Rev.*, **77**, 789 (1950); **79**, 450 (1950); also Halstead, R. E., and A. O. Nier, *Rev. Sci. Instr.*, **21**, 1019 (1950).
5. Shields, W. R., T. J. Murphy, E. J. Catanzaro, and E. L. Garner, *J. Res. Natl. Bur. Std.*, **70A**, 193 (1966).
6. Park, R., and H. N. Dunning, *Geochim. Cosmochim. Acta*, **22**, 99 (1961).
*7. Cameron, A. E., *Anal. Chem.*, **35**, 23A (1963).
8. Ries, R. R., R. A. Damerow, and W. H. Johnson, *Phys. Rev.*, **132**, 1662, 1673 (1963).
9. Thode, H. G., and R. L. Graham, *Can. J. Res.*, **A25**, 1 (1947); also Wanless, R. K., and H. G. Thode, *ibid.*, **31**, 517 (1953).
*10. Dietz, L. A., C. F. Pachucki, and G. A. Land, *Anal. Chem.*, **34**, 709 (1962); **35**, 797 (1963).
11. Nier, A. O., *J. Am. Chem. Soc.*, **60**, 1571 (1938); *Phys. Rev.*, **55**, 153 (1939).
*12. Wiles, D. M., and R. H. Tomlinson, *Phys. Rev.*, **99**, 188 (1955).
13. Brown, F., G. R. Hall, and A. J. Walter, *J. Inorg. Nucl. Chem.*, **1**, 241 (1955).
14. Dempster, A. J., *Phys. Rev.*, **71**, 829 (1947).
15. Kenneth, T. J., and H. G. J. Thode, *J. Inorg. Nucl. Chem.*, **5**, 253 (1958).
16. Nier, A. O., *Phys. Rev.*, **55**, 150 (1939).
*17. Nier, A. O., *J. Am. Chem. Soc.*, **60**, 1571 (1938).
18. Collins, P. J., R. D. Russel, and R. M. Farquhar, *Can. J. Phys.*, **31**, 402 (1953).

19. Anders, E., *Rev. Mod. Phys.*, **34**, 287 (1962).
20. Aldrich, L. T., and A. O. Nier, *Phys. Rev.*, **74**, 876 (1948).
*21. Reynolds, J. H., *Rev. Sci. Instr.*, **27**, 928 (1956).
22. Libby, W. F., *Radiocarbon Dating*, 2nd ed., University of Chicago Press, 1955.
23. Reynolds, J. H., *Phys. Rev. Letters*, **4**, 8 (1960); also *J. Geophys Res.*, **68**, 2939 (1963).
*24. Urey, H. C., *J. Am. Chem. Soc.*, **69**, 562 (1947).
25. Shackleton, N. J., *J. Sci. Instr.*, **42**, 689 (1965).
26. Epstein, S., R. Buchsbaum, H. Lowenstam, and H. C. Urey, *Bull. Geol. Soc. Am.*, **62**, 417 (1953).
27. Epstein, S., "Mass Spectroscopy in Physics Research," *Natl. Bur. Std. Circ.* **522**, 133 (1953).
28. Jayaram, R., *Mass Spectrometry*, Plenum Press, New York, 1966.
29. Mirtov, B. A., *Gaseous Composition of the Atmosphere and Its Analysis* (translated from the Russian), Israel Program for Scientific Translations, Jerusalem, 1964.
*30. Narcisi, R. S., and A. D. Bailey, *J. Geophys. Res.*, **70**, 3687 (1965).
*31. Nier, A. O., J. H. Hoffman, C. Y. Johnson, and J. C. Holmes, *J. Geophys. Res.*, **69**, 979, 4629 (1964); Hedin, A. E., C. P. Avery, and C. D. Tschetter, *ibid.*, **69**, 4637 (1964); Hedin, A. E., and A. O. Nier, *ibid.*, **70**, 1273 (1965).
32. Spencer, N. W., L. H. Brace, G. R. Carignan, D. R. Taeusch, and H. B. Niemann, *J. Geophys. Res.*, **70**, 2665 (1965).
33. Niemann, H. B., and B. C. Kennedy, *Rev. Sci. Instr.*, **37**, 722 (1966).
34. Hunter, J. A., R. W. Stacy, and F. A. Hitchock, *Rev. Sci. Instr.*, **20**, 333 (1949).
35. Fowler, K. T., and P. Hugh-Jones, *Brit. Med. J.*, **1**, 1205 (1957).
36. Brunnèe, C., and C. Delgmann, *Z. Anal. Chem.*, **197**, 51 (1963).
*37. Adamczyk, B., A. J. H. Boerboom, and J. Kistemaker, *J. Appl. Physiol.*, **21**, 1903 (1966).
38. Jones, S. C., J. M. Saari, R. A. Devloo, A. Faulconer, and E. J. Baldes, *Anesthesiology*, **14**, 490 (1953).
39. Chapman, J., R. Hill, J. Muir, C. W. Suckling, and D. J. Viney, *J. Pharm. Pharmacol.*, **19**, 231 (1967).
40. Schoenheimer, R., and D. Rittenberg, *J. Biol. Chem.*, **127**, 285 (1939).
41. Arnstein, H. R. V., and R. Bentley, *Quart. Rev. Chem. Soc.*, **4**, 172 (1950).
42. Han, I., and G. J. Fritz, *Anal. Chem.*, **37**, 1442 (1965).
*43. Stenhagen, E., *Chimia*, **20**, 346 (1966).
44. Knöppel, H., and W. Beyrich, *Tetrahedron Letters*, **3**, 291 (1968).
*45. Eneroth, P., B. Gordon, R. Ryhage, and J. Sjövall, *J. Lipid Res.*, **7**, 511 (1966).
46. Sharkey, A. G., R. A. Friedel, and S. H. Langer, *Anal. Chem.*, **29**, 770 (1957).
47. McCloskey, J. A., R. N. Stillwell, and A. M. Lawson, *Anal. Chem.*, **40**, 233 (1968).
48. Foltz, R. L., M. B. Neher, and E. R. Hinnenkamp, *Anal. Chem.*, **40**, 1339 (1968).

49. Teeter, R. M., *Anal. Chem.*, **39**, 1742 (1967).

50. Richter, W., M. Vecchi, W. Vetter, and W. Walther, *Helv. Chim. Acta*, **50**, 364 (1967); M. Vecchi, W. Vetter, and W. Walther, *ibid.*, **50**, 1243 (1967); W. Vetter, M. Vecchi, K. Gutmann, R. Ruegg, W. Walther, and P. Meyer, *ibid.*, **50**, 1866 (1967); M. Vecchi and K. Kaiser, *J. Chromatog.*, **26**, 22 (1967).

*51. Biemann, K., C. Cone, B. R. Webster, and G. P. Arsenault, *J. Am. Chem. Soc.*, **88**, 5598 (1966).

*52. Senn, M., R. Venkataraghavan, and F. W. McLafferty, *J. Am. Chem. Soc.*, **88**, 5593 (1966).

53. Budzikiewitz, H., C. Djerassi, D. H. Williams, *Structural Elucidation of Natural Products by Mass Spectrometry*, Vols. I and II, Holden-Day, San Francisco, 1964.

54. Wolstenholme, W. A., *Nature*, **203**, 1284 (1964).

55. Hardwick, J. L., and C. I. Martin, *Helv. Odontol. Acta*, **11**, 1, 62 (1967).

*56. Evans, C. A., and G. H. Morrison, *Anal. Chem.*, **40**, 869 (1968).

*57. Huber, W. K., *Vacuum*, **13**, 399, 469 (1964).

58. Craig, R. D., and E. H. Harden, *Vacuum*, **16**, 67 (1966).

59. Caswell, H. L., *J. Appl. Phys.*, **32**, 105, 2641 (1961).

60. Caswell, H. L., in *Physics of Thin Films*, Vol. 1, G. Hass, Ed., Academic Press, New York, 1963, p. 1.

61. Rozgonyi, G. A., W. J. Polito, and B. Schwartz, *Vacuum*, **16**, 121 (1966).

62. Rozgonyi, G. A., and J. Sosniak, *Vacuum*, **18**, 1 (1968).

Chapter Fourteen

Information and Data

I. CLASSICAL PUBLICATIONS

In 1886, E. Goldstein, a physicist at the Berlin observatory, published a paper reporting the discovery of a luminous ray emerging in straight lines from the holes of a perforated metal disc employed as a cathode in a discharge tube. He called the rays *Kanalstrahlen* (canal rays). This paper [*Berl. Ber.*, **39**, 691 (1886)] may be considered the first paper in mass spectrometry although the term mass spectrum was suggested by Aston only some 35 years later. Perrin's suggestion (1895) that the rays were associated with a positive charge was confirmed by W. Wien, a professor at Munich and Nobel prize winner (1911) by studying their deflection in electric and magnetic fields [*Verh. Phys. Ges.*, **17**, 1898 (1898); *Ann. Phys. Leipzig*, **8**, 224 (1902)].

Based upon these foundations, mass spectrometry developed as a distinct field during the period 1911–1925 mainly as a result of the works of the three founding fathers, Thomson, Aston, and Dempster. Joseph John Thomson (1856–1940), known to many as "J.J.," was a professor of Physics at Cambridge. He was one of the few true geniuses of the history of science. The Nobel prize which he received in 1906 was indeed only a token reward for his tremendous contribution to the knowledge of mankind. A recent book entitled *J. J. Thomson and the Cavendish Laboratory in His Day*, written by his son G. P. Thomson (Doubleday & Co., New York, 1965) is of basic historical importance and is fascinating reading.

Thomson's work was continued by his assistant Francis Williams Aston (1877–1945), an English chemist whose name has become almost synonymous with mass spectrometry. He discovered over 200 naturally occurring isotopes and paved the way for the rapid development of modern nuclear theory and practice. His paper entitled "The Constitution of Atmospheric Neon" [*Phil. Mag. 6th Ser.*, **39**, 449 (1920)] was probably the first paper in applied mass spectrometry. In 1922 he was awarded the Nobel prize in chemistry. In his address accepting the Prize, Aston clearly forecast both

J. J. Thompson   *F. W. Aston*
        Courtesy of V. H. Dibeler   Courtesy of E. Wichers

A. J. Dempster
    Courtesy of the University of Chicago

nuclear bombs and power plants, based upon the future use of the energy bound in the atomic nucleus.

Aston's vision was brought a long and hard step closer to reality by the third founding father of mass spectrometry, the Canadian-American physics professor at the University of Chicago, Arthur Jeffrey Dempster (1886–1950), who in 1935 provided mass spectrometric evidence that 7 out of 1000 uranium atoms were uranium-235 isotopes. Dempster started to build mass spectrometers at about the same time Aston's first instrument appeared. While Aston applied velocity focusing, Dempster utilized ions of constant energy in directional focusing analyzers. At that time mass spectrographs and mass spectrometers were distinguished quite sharply. By 1921 Dempster was carrying out accurate abundance measurements employing both electron impact and thermal evaporation ion sources in basically the same manner as they are being used today. He also discovered the spark-source and vacuum vibrator techniques for the analysis of solids. Eight years later he proposed (together with W. Bartky) the use of combined direction and velocity focusing. The first double-focusing instruments, however, were built only around 1934 when the Dempster, Mattauch-Herzog, and Bainbridge-Jordan geometries appeared at about the same time.

There are three books in the literature of mass spectrometry that may be termed classics:

Aston, F. W., *Isotopes*, E. Arnold, London, 1923.
Aston, F. W., *Mass Spectra and Isotopes*, 2nd ed., E. Arnold, London, 1942.
Thomson, J. J., *Rays of Positive Electricity and Their Application to Chemical Analysis*, Longmans, Green, London, 1913.

Aston's two books should indeed be read by those wishing to choose mass spectrometry as their profession. His somewhat personalized description of instruments, experiments, and evaluations of results leaves one with the impression that Aston knew everything that can be known about mass spectrometry.

For those interested in learning about mass spectrometry by tracing its history through original references, 20 selected papers are listed in the following. All are written in English and appeared in journals generally available in university libraries.

1. Rays of Positive Electricity.
   Thomson, J. J., *Phil. Mag.*, **21**, 225 (1911).
2. A New Method of Positive Ray Analysis.
   Dempster, A. J., *Phys. Rev.*, **11**, 316 (1918).

3. A Positive Ray Spectrograph.
   Aston, F. W., *Phil. Mag.*, **38**, 707 (1919).
4. The Mass Spectra of Chemical Elements.
   Aston, F. W., *Phil. Mag.*, **39**, 611 (1920).
5. The Application of Anode Rays to the Investigation of Isotopes.
   Thomson, J. J. *Phil. Mag.*, **42**, 857 (1921).
6. Positive Ray Analysis of Potassium, Calcium, and Zinc.
   Dempster, A. J., *Phys. Rev.*, **20**, 631 (1922).
7. The Mass Spectra of Chemical Elements.
   Aston, F. W., *Phil. Mag.*, **45**, 934 (1923); **47**, 385 (1924).
8. Photographic Plates for the Detection of Mass Rays.
   Aston, F. W., *Cambridge Phil. Soc. Proc.*, **22**, 548 (1925).
9. Thermionic Effects Caused by Vapours of Alkali Metals.
   Langmuir, I., and Kingdon, K. H., *Proc. Roy. Soc. (London), Ser. A*, **107**, 61 (1925).
10. Primary and Secondary Products of Ionization in Hydrogen.
    Smyth, H. D., *Phys. Rev.*, **25**, 452 (1925).
11. A Velocity Filter for Electrons and Ions.
    Smythe, W. R., *Phys. Rev.*, **28**, 1275 (1926).
12. A New Mass Spectrograph and the Whole Number Rule.
    Aston, F. W., *Proc. Roy. Soc. (London)*, A**115**, 487 (1927).
13. A New Method of Positive Ray Analysis and Its Application to the Measurement of Ionization Potentials in Mercury Vapor.
    Bleakney, W., *Phys. Rev.*, **34**, 157 (1929).
14. Paths of Charged Particles in Electric and Magnetic Fields.
    Bartky, W., and Dempster, A. J., *Phys. Rev.*, **33**, 1019 (1929).
15. Atomic Weight of Cesium. Use of the Word "Mass Spectrograph."
    Aston, F. W., *Nature*, **127**, 813 (1931).
16. The Blackening of Photographic Plates by Positive Ions of the Alkali Metals.
    Bainbridge, K. T., *J. Franklin. Inst.*, **212**, 489 (1931).
17. The Isotopic Constitution of Zinc.
    Bainbridge, K. T., *Phys. Rev.*, **39**, 847 (1932).
18. The Ionization Potential of Molecular Hydrogen.
    Bleakney, W., *Phys. Rev.*, **40**, 496 (1932).
19. A New Mass Spectrometer.
    Smythe, W. R., and Mattauch, J., *Phys. Rev.*, **40**, 429 (1932).
20. The Emission of Ions and Electrons from Heated Sources.
    Dempster, A. J., *Phys. Rev.*, **46**, 165 (1934).

II. BOOKS AND REVIEWS

A. Bibliography: General

The following bibliography, although believed to be fairly complete, is not meant to be exhaustive. The subdivision into "General" and "Applications" is somewhat arbitrary because many books, such as Beynon's, feature rather detailed sections on general principles and instrumentation followed by applications in specialized fields.

Barnard, G. P., *Modern Mass Spectrometry*, The Institute of Physics, London, 1953.

Beynon, J. H., *Mass Spectrometry and Its Applications to Organic Chemistry*, Elsevier, Amsterdam, 1960.

Blauth, E. W., *Dynamic Mass Spectrometers* (Translated from German), Elsevier, Amsterdam, 1966.

Brunnee, C., and H. Voshage, *Massenspektrometrie*, K. Thiemig, Munich, 1964.

Duckworth, H. E., *Mass Spectroscopy*, Cambridge University Press, London, 1958.

Ewald, H., and H. Hintenberger, *Methoden und Anwendungen der Massenspektroskopie*, Verlag Chemie, Weinheim, 1952; English translation by USAEC, Translation Series AEC-tr-5080, Office of Technical Services, Washington, D.C., 1962.

Inghram, M. G., and R. J. Hayden, *A Handbook on Mass Spectroscopy*, Nuclear Science Series Report No. 14, National Academy of Sciences, National Research Council Publication 311, Washington, D.C., 1954.

Jayaram, R., *Mass Spectrometry, Theory and Applications*, Plenum Press, New York, 1966.

Kienitz, H., *Massenspektrometrie*, Verlag Chemie, Weinheim, 1968.

Kiser, R. W., *Introduction to Mass Spectrometry and Its Applications*, Prentice-Hall, Englewood Cliffs, N.J., 1965.

"Mass Spectrometry in Physics Research," National Bureau of Standards, Circular 522, U.S. Government Printing Office, Washington, D.C., 1953.

McDowell, C. A., Ed., *Mass Spectrometry*, McGraw-Hill, New York, 1963.

Reed, R. I., Ed., *Mass Spectrometry*, NATO Advanced Study Institute Publication on Theory, Design, and Applications, Academic Press, London, 1965.

Reed, R. I., Ed., *Modern Aspects of Mass Spectrometry*, NATO Advanced Study Institute, Plenum Press, New York, 1968.

Rieck, G. R., *Einführung in die Massenspektroskopie* (translated from Russian), VEB Deutscher Verlag der Wissenschaften, Berlin, 1956.

Robertson, A. J. B., *Mass Spectrometry*, Methuen, London, 1954.

White, F. A., *Mass Spectrometry in Science and Technology*, Wiley, New York, 1968.

Barnard's book was the first of the modern books on mass spectrometry. Today it is outdated in many respects, particularly because it does not deal with high resolution instruments; however, it provides a good background on fundamental applications. The book by Ewald and Hintenberger, published also in 1953, is an excellent help in understanding basic ion optics. The Inghram-Hayden handbook (copies are difficult to obtain) has become a classic in mass spectrometry. A small booklet of only 50 pages, it gives a remarkably clear introduction to ion optics and the most salient features of sources, detectors, and inlet systems. Another book (also difficult to come by) one which may be considered the perfect complement to Inghram's book, is the NBS *Circular* 522 on the applications of mass spectrometry in physics research. This volume contains the proceedings of the Bureau's semicentennial symposium on mass spectrometry. The contributors were some of the best known people in the field and many truly original papers were presented. The discussions following the papers are quite revealing.

Of all the books listed in this review, Beynon's book is the most recommended. Although primarily intended for the organic chemist, it is considered indispensable to almost everybody in the field of applied mass spectrometry. General principles and instrumentation are considered from an applications rather than design point of view. Many small but important details are discussed and countless practical suggestions, all highly useful, are made.

The book edited by McDowell is a collection of chapters on several major segments of mass spectrometry, written by experts. Detailed discussions of ion optics and high resolution mass spectrometry are primarily directed toward those interested in instrument design. The sections on applications, all excellent, may be somewhat advanced and detailed for the novice. The NATO books, edited by Reed, also consist of individual articles on various aspects of mass spectrometry (both instrumentation and application). These up-to-date books are recommended for beginners.

The books by Brunnee-Voshagen (in German) and Kiser are modern general introductory texts, similar in nature to the present one. Jayaram's book deals mainly with instrumentation and technique for ionospheric research.

B. Bibliography: Applications

ORGANIC CHEMISTRY

Beynon, J. H., *Mass Spectrometry and Its Applications to Organic Chemistry*, Elsevier, Amsterdam, 1960.

Beynon, J. H., R. A. Saunders, and A. E. Williams, *The Mass Spectra of Organic Molecules*, American Elsevier, New York, 1968.

Biemann, K., *Mass Spectrometry: Applications to Organic Chemistry*, McGraw-Hill, New York, 1962.

Budzikiewicz, H., C. Djerassi, and D. H. Williams, *Interpretation of Mass Spectra of Organic Compounds*, Holden-Day, San Francisco, 1964.

Budzikiewicz, H., C. Djerassi, and D. H. Williams, *Structure Elucidation of Natural Products by Mass Spectrometry;* Volume I: Alkaloids; Volume II: Steroids, Terpenoids, Sugars, and Miscellaneous Classes, Holden-Day, San Francisco, 1964.

Budzikiewicz, H., C. Djerassi, and D. H. Williams, *Mass Spectrometry of Organic Compounds*, Holden-Day, San Francisco, 1967.

Hill, H. C., *Introduction to Mass Spectrometry*, Heyden, London, 1966.

McLafferty, F. W., "Mass Spectrometry," in F. C. Nachod and W. D. Phillips, Eds., *Determination of Organic Structures by Physical Methods*, Academic Press, New York, 1962.

McLafferty, F. W., Ed., *Mass Spectrometry of Organic Ions*, Academic Press, New York, 1963.

McLafferty, F. W., *Interpretation of Mass Spectra*, Benjamin, New York, 1966.

Quayle, A., and R. I. Reed, in D. W. Mathieson, Ed., *Interpretation of Organic Spectra*, Academic Press, New York, 1965.

Reed, R. I., *Applications of Mass Spectrometry to Organic Chemistry*, Academic Press, New York, 1966.

Silverstein, R. M., and G. C. Bassler, *Spectrometric Identification of Organic Compounds*, 2nd ed., Wiley, New York, 1967.

Spiteller, G., *Massenspektrometrische Strukturanalyse Organischer Verbindungen*, Verlag Chemie, Weinhein, 1966.

ELECTRON IMPACT PHENOMENA

Field, F. H., and J. L. Franklin, *Electron Impact Phenomena and the Properties of Gaseous Ions*, Academic Press, New York, 1957.

Loeb, L. B., *Basic Processes of Gaseous Electronics*, University of California Press, Berkeley, 1955.

Massey, H. S. W., *Negative Ions*, 2nd ed., University Press, Cambridge, 1950.

Massey, H. S. W., and E. H. Burhop, *Electronic and Ionic Impact Phenomena*, Clarendon Press, Oxford, 1952.

McDaniel, E. W., *Collisional Phenomena in Ionized Gases*, Wiley, New York, 1964.

Mott, N. F., and H. S. W. Massey, *The Theory of Atomic Collisions*, Clarendon Press, Oxford, 1949.

Reed, R. I., *Ion Production by Electron Impact*, Academic Press, New York, 1962.

Stevenson, D. P., and D. O. Schissler, "Mass Spectrometry and Radiation Chemistry" in M. Haissinsky, Ed., *Actions Chimiques et Biologiques des Radiations*, Masson, Paris, 1961.

ISOTOPES—NUCLEAR CHEMISTRY—GEOCHEMISTRY

Birkenfeld, H., G. Haase, and H. Zahn, *Massenspektrometrische Isotopenanalyse*, VEB Deutscher Verlag der Wissenschaften, Berlin, 1962.

Duckworth, H. E., Ed., *Proceedings of the International Conference on Nuclidic Masses*, University of Toronto Press, Toronto, 1960.

Higatsberger, M. J., and F. Viehböck, *Electromagnetic Separation of Radioactive Isotopes*, Springer-Verlag, Vienna, 1961.

Hinterberger, H., Ed., *Nuclear Masses and Their Determination*, Pergamon Press, London, 1957.

Hintenberger, H., "High Sensitivity Mass Spectroscopy in Nuclear Studies," in E. Segré, G. Friedlander, and W. E. Meyerhof, Eds., *Annual Reviews of Nuclear Science*, Vol. 12, Annual Reviews, Palo Alto, Calif., 1962.

Mayne, K. I., "Stable Isotope Geochemistry and Mass Spectrometric Analysis," in A. A. Smales and L. R. Wager, Eds., *Methods in Geochemistry*, Interscience, New York, 1960.

Schaeffer, O. A., and J. Zähringer, *Potassium Argon Dating,*Springer-Verlag, Vienna, 1967.

Smith, M. L., Ed., *Electromagnetically Enriched Isotopes and Mass Spectrometry*, (Proceedings of Harwell Conference), Academic Press, New York, 1956.

Thode, H. G., C. C. McMullen, and K. Fritze, "Mass Spectrometry in Nuclear Chemistry," in H. J. Emeleus and A. G. Sharpe, Eds., *Advances in Inorganic Chemistry and Radiochemistry*, Vol. 2, Academic Press, New York, 1960.

Webster, R. K., "Mass Spectrometric Isotopic Analysis," in A. A. Smales and L. R. Wager, Eds., *Methods in Geochemistry*, Interscience, New York, 1960.

ANALYSIS OF SOLIDS

Ahearn, A. J., Ed., *Mass Spectrometric Analysis of Solids*, Elsevier, Amsterdam, 1966.

Herzog, L. F., D. J. Marshall, B. R. Kendall, and L. A. Cambey, "Ion Sources and Detectors for the Mass Spectroscopic Study of Solids," in C. N. Reilley, Ed., *Advances in Analytical Chemistry and Instrumentation*, Interscience, New York, 1964.

Honig, R. E., "Mass Spectrometry," in J. P. Cali, Ed., *Trace Analysis of Semiconductor Materials*, Macmillan, New York, 1964.

Roboz, J., "Mass Spectrometry," in G. H. Morrison, Ed., *Trace Analysis: Physical Methods*, Interscience, New York, 1965.

MISCELLANEOUS BOOKS AND REVIEWS

Ausloos, P. J., Ed., *Ion-Molecule Reactions in the Gas Phase (Advan. Chem. Ser.,* **58**), American Chemical Society, Washington, D.C., 1966.

"Applied Mass Spectrometry," Rept. Conf. Mass Spectrometry Panel Inst. Petroleum, Institute of Petroleum, London, 1954.

Barnard, G. P., *Mass Spectrometer Researches*, Her Majesty's Stationery Office, London, 1956.

Biemann, K., "Mass Spectrometry," in J. M. Luck, E. E. Snell, F. W. Allen, and G. McKinney, Eds., *Annual Reviews of Biochemistry*, Vol. 32, Annual Reviews, Stanford, Calif., 1963.

Bommer, P., and K. Biemann, "Mass Spectrometry," in H. Eyring, C. J. Christiensen, and H. S. Johnston, Eds., *Annual Review of Physical Chemistry*, Vol. 16, Annual Reviews, Stanford, Calif., 1965.

Cameron, A. E., "Electromagnetic Separations," in W. G. Berl, Ed., *Physical Methods in Chemical Analysis*, Vol. IV, Academic Press, New York, 1961.

Hutter, R. G., "The Deflection of Beams of Charged Particles," in L. Marton, Ed., *Advances in Electronics*, Vol. 1, Academic Press, New York, 1948.

Inghram, M. G., "Modern Mass Spectroscopy," in L. Marton, Ed., *Advances in Electronics*, Vol. 1, Academic Press, New York, 1948.

Kerwin, L., "Mass Spectroscopy," in L. Marton, Ed., *Advances in Electronics*, Vol. 8, Academic Press, New York, 1956.

Lampe, F. W., J. L. Franklin, and F. H. Field, "Kinetics of the Reactions of Ions with Molecules," in G. Porter, Ed., *Progress in Reaction Kinetics*, Pergamon Press, New York, 1961.

Mass Spectrometry (Report of a Conference Organized by the Mass Spectrometry Panel of the Institute of Petroleum, Manchester, April 20–21, 1950), The Institute of Petroleum, London, 1952.

Melpolder, F. W., and R. A. Brown, "Mass Spectrometry," in I. M. Kolthoff and P. J. Elving, Eds., *Treatise on Analytical Chemistry*, Part I, Vol. 4, Interscience, New York, 1963.

Meyerson, S., and J. D. McCollum, "Mass Spectra of Organic Molecules," in C. N. Reilley, Ed., *Advances in Analytical Chemistry and Instrumentation*, Vol. 2, Interscience, New York, 1963.

Mitchell, J. J., "Mass Spectroscopy in Hydrocarbon Analysis," in A. Farkas, Ed., *Physical Chemistry of the Hydrocarbons*, Vol. 1, Academic Press, New York, 1950.

Robinson, C. F., "Mass Spectrometry," in W. G. Berl, Ed., *Physical Methods in Chemical Analysis*, Vol. 1, 2nd ed., Academic Press, New York, 1960.

Septier, A., *Focusing of Charged Particles*, Vol. 1, Academic Press, New York, 1967.

Shumulovskiy, N. N., and R. I. Stakhovskiy, *Mass-spektral'nye Metody* (in Russian), Energiya, Moscow, 1966.

Washburn, H. W., "Mass Spectrometry," in W. G. Berl, Ed., *Physical Methods in Chemical Analysis*, Vol. 1, Academic Press, New York, 1950.

Young, W. S., "Mass Spectroscopy of Hydrocarbons," in B. T. Brooks, S. T. Kurtz, C. E. Boord, and L. Schmerling, Eds., *The Chemistry of Petroleum Hydrocarbons*, Reinhold, New York, 1954.

VACUUM TECHNIQUES

Only a few selected texts are listed here:

Brunner, W. F., Jr., and T. H. Batzer, *Practical Vacuum Techniques*, Reinhold, New York, 1965.

Dushman, S., in J. M. Lafferty, Ed., *Scientific Foundations of Vacuum Techniques*, 2nd ed., Wiley, New York, 1962.

Kohl, W. K., *Materials and Techniques for Electron Tubes*, 3rd ed., Reinhold, New York, 1967.

Leck, J. H., *Pressure Measurement in Vacuum Systems*, 2nd ed., Chapman & Hall, London, 1964.

ELECTRONICS

A review of the fundamentals of electronic circuits associated with mass spectrometers is given by G. D. Flesch and H. J. Svec in a report entitled *Electronics in Mass Spectrometry*. It is available from the Clearinghouse for Federal Scientific and Technical Information, U.S. Department of Commerce, Springfield, Virginia (Report IS-1193, listed in TID-4500, 43rd ed., July 1, 1965).

A popular text on elementary electronics is *Basic Electronics*, prepared by the Bureau of Naval Personnel. It was published in book form by Dover Publications, New York, in 1963.

III. DATA COMPILATIONS

The following is a list of the most important published data compilations presently available:

Beynon, J. H., and A. E. Williams, *Mass and Abundance Tables for Use in Mass Spectrometry*, Elsevier, Amsterdam, 1963.

Beynon, J. H., R. A. Saunders, and A. E. Williams, *Table of Metastable Transitions for Use in Mass Spectrometry*, Elsevier, Amsterdam, 1965.

Cornu, A., and R. Massot, *Compilation of Mass Spectral Data*, Heyden, London, 1966.

Cornu, A., and R. Massot, *Compilation of Exact Masses of Organic Ions for Use in High Resolution Mass Spectrometry*, Heyden, London, 1966.

Cornu, A., R. Massot, and J. Ternier, *Compilation of Exact Masses of Inorganic Ions for Use in High Resolution Spark Source Mass Spectrometry*, Heyden, London, 1965.

Cornu, A., and R. Massot, *List of Conversion Factors for Atomic Impurities to PPM by Weight*, Heyden, London, 1968.

Heath, R. L., *Table of Atomic Masses*, Monograph SCR-245, J. W. Guthrie, Ed., Sandia Corp., Albuquerque, N. M., 1961.

McLafferty, F. W., *Mass Spectral Correlations* (Advances in Chemistry Series, Vol. 40), American Chemical Society, Washington, D.C., 1963.

Owens, E. B., and A. M. Sherman, "Mass Spectrographic Lines of the Elements", Rept. Np. TR 265, Lincoln Laboratory, MIT, Lexington, Mass., 1962.

The American Petroleum Institute Research Project 44 (API Spectra, see Section 9-I) and the Manufacturing Chemists' Association Research Project have been relocated from Carnegie Institute of Technology to the Chemical Thermodynamic Properties Center of the Agricultural and Mechanical College of Texas (College Station, Texas). At the present time there are 2075 *certified* spectra available in the API Project 44 series, and 150 spectra in the MCA series. New certified spectra are issued periodically. Thousands of *uncertified* spectra are available in various private collections and in the literature. An *Index of Mass Spectral Data* incorporating 3200 spectra, both certified and uncertified, is available from the ASTM as Special Publication No. 356 (1963). These spectra are also available in punched card index form from the ASTM. It is recalled

(Section 9-I) that the Cornu-Massot compilation contains about 5000 spectra featuring the 10 most active peaks in various arrangements (by reference number, molecular weight, molecular formula, fragment ion values).

Two other useful publications should be mentioned at this point. The *Gas Data Handbook* of the Mathieson Co. (Rutherford, N. J., 1966) provides much valuable information on gases often dealt with in mass spectrometry. The second publication is the "Catalog and Price List of Standard Materials Issued by the National Bureau of Standards" (NBS Misc. Publ. 260, 1968). The Bureau has many active programs in mass spectrometry and several reference materials have recently become available for standardization and calibration.

IV. CURRENT INFORMATION SOURCES

Current information sources in mass spectrometry are the same as in other branches of the physical sciences: professional society meetings, publications in periodicals, reference and abstract journals, and the increasingly popular information and data centers.

The most important single source of information on current happenings in mass spectrometry, at least in the USA, is the so-called E-14 Meeting. Committee E-14 on Mass Spectrometry is a subcommittee of the American Society for Testing and Materials (ASTM), established to "promote the knowledge and advancement of the art of mass spectrometry." For 15 years the Committee has been sponsoring annual meetings which have become extremely popular. The week-long meetings are held in major cities in the United States and Canada. During the 1967 meeting 171 papers were presented and about 500 spectroscopists attended. The E-14 meetings are co-sponsored by all mass spectrometer manufacturers. In their hospitality suits mass spectroscopists congregate after the sessions and a great deal of information on current results, including failures, may be obtained in these informal discussions. Manufacturers also arrange "clinics" to discuss problems in instrumentation and announce new products and services. It may be said that the annual E-14 meetings have become an institution among mass spectroscopists in the USA. In recent years visitors from many countries have attended, and the international flavor should certainly further improve this already well-established institution. It has repeatedly been suggested that the formation of a more or less independent Society of Mass Spectrometry is desirable in the light of the swelling membership and expanding activity of the Committee E-14.

Since 1961 the *Proceedings* of the annual meetings have been printed in bound volumes. These volumes are, however, available only to those attending the meetings. This policy of limited circulation permits the contributors to publish in conventional journals. In addition to arranging the formal meetings, the E-14 Committee sponsors several subcommittees and task groups on various aspects of mass spectrometry (nomenclature, polymer analysis, solids analysis, fundamentals, etc.). An important activity of these groups is the establishment of standard analytical practices and methods. Table 14-1 lists the Recommended Practices and Methods of Test in the 1967 *Book of ASTM Standards*.

Table 14-1. ASTM Publications on Mass Spectrometry in *1967 Book of ASTM Standards*

Designation	Title
D1137-53	Analysis of Natural Gases and Related Types of Gaseous Mixtures by the Mass Spectrometer, Part 19, pp. 225–235.
D1302-61T	Analysis of Carburetted Water Gas by the Mass Spectrometer, Part 19, pp. 299–305.
D1658-63	Carbon Number Distribution of Aromatic Compounds in Naphthas by Mass Spectrometry, Part 17, pp. 602–605.
D2424-65T	Hydrocarbon Types in Propylene Polymer by Mass Spectrometry, Part 18, pp. 554–562.
D2425-65T	Hydrocarbon Types in Middle Distillates by Mass Spectrometery, Part 17, pp. 879–888.
D2498-66T	Isomer Distribution of Straight-Chain Detergent Alkylate by Mass Spectrometry, Part 18, pp. 601–606.
D2567-66T	Molecular Distribution Analysis of Monoalkylbenzenes by Mass Spectrometry, Part 18, pp. 729–733.
E137-65	Evaluation of Mass Spectrometers for Use in Chemical Analysis, Part 30, pp. 363–366.
E244-64T	Atom Per Cent Fission in Uranium Fuel, Mass Spectrometric Method, Part 30, pp. 767–772.
E304-66T	Use and Evaluation of Mass Spectrometers for Mass Spectrochemical Analysis of Solids, Part 30, pp. 979–985.
D2601-67T	Low-Voltage Mass Spectrometric Analysis of Propylene Tetramer. Published in the Annual Report of ASTM Committee D-2, 1967, pp. 93–96 and to appear in 1968 Book of ASTM Standards, Part 18.
D2650-67T	Chemical Composition of Gases by Mass Spectrometry, Published as information in 1966 Book of ASTM Standards, Part 18, pp. 679–686, and to appear in 1968 Book of ASTM Standards, Part 18.

The European counterpart of the annual E-14 meeting is the International Mass Spectrometry Conference, sponsored by a number of

British, French, German, and other European scientific societies. There have been four meetings thus far, one every two years since 1961. The most recent was held in September 1967 in Berlin. Proceedings of these meetings are officially published in the *Advances in Mass Spectrometry* series. Three volumes have been published to-date:

Waldron, J. D., Ed., *Advances in Mass Spectrometry*, Vol. 1, Pergamon Press, New York, 1959.

Elliott, R. M., Ed., *Advances in Mass Spectrometry*, Vol. 2, Macmillan Co., New York, 1963.

Mead, W. L., Ed., *Advances in Mass Spectrometry*, Vol. 3, The Institute of Petroleum, London, 1966.

A journal devoted entirely to mass spectroscopy has been published in Japan for many years (*Situryo Bunseki*). Since most articles appear in Japanese, circulation is obviously limited. Three new journals made their debut in 1968: *International Journal of Mass Spectrometry and Ion Physics* (Elsevier, Amsterdam), *Journal of Organic Mass Spectrometry* (Heyden & Sons, London), and *Archives of Mass Spectral Data* (Interscience, New York).

Articles on mass spectrometry appear in virtually all kinds of scientific papers ranging from food research through petroleum analysis to clinical techniques. The majority of publications are, naturally, concentrated in such journals as the *Review of Scientific Instruments*, *The Journal of Physical Chemistry*, *Analytical Chemistry*, the *Journal of the American Chemical Society*, and their counterparts abroad. Articles of interest to mass spectroscopists often appear in *Vacuum*, and in the *Journal of Vacuum Science and Technology*.

Mass spectrometry contributes a fair share to the recent explosion of the general scientific literature. A selected list of publications in the 1964 edition of the *Annual Reviews of Analytical Chemistry* (*Fundamental*) contains 614 references covering only a two-year period. In the 1966 edition of the *Annual Reviews of Analytical Chemistry* (*Fundamental*) it is stated that some 5000 new publications were collected in a two-year period, but only a selected 1171 of them have been reviewed. The problems involved in keeping up to date have been widely recognized and several attempts have been made to provide lists of publications. In addition to the above-mentioned reviews of *Analytical Chemistry*, the volumes in the *Advances in Mass Spectrometry* series also provide bibliographies. Several E-14 subcommittees have also been compiling references in their respective fields. An *Index and Bibliography of Mass Spectrometry*, covering the 1963–

1965 period, has been published by F. W. McLafferty and J. Pinzelik (Wiley, New York, 1967). J. Capellen, H. J. Svec, and C. R. Sage, compiled a bibliography by a computer method for the period of late 1964 to early 1966 (USAEC publication IS-1335).

In November 1966, the Mass Spectrometry Data Centre at Aldermaston (Berks, England) commenced the publication of a *Mass Spectrometry Bulletin*. This monthly guide to the current literature covers more than 150 journals together with important government reports and publications, and also provides information on new books, scientific meetings, new data compilations, etc. Literature references are grouped into eight sections: Instrument design and technique, isotopic analysis (including precision mass measurement, isotope separation and ageing), chemical analysis, organic chemistry, atomic and molecular processes, surface phenomena and solid state studies, thermodynamics and reaction kinetics, and other applications. The bibliography is thoroughly cross-referenced by subject index, compound classification, compounds, elements, ions, and materials indices, and author index. About 3500 articles have been referenced in the first twelve issues (Vol. 1). This bulletin appears to be the best source at the present time for current literature references. A similar type of service is offered by the *Current Awareness Service* of the Scientific Documentation Centre Ltd., Dunfermline, U.K. Weekly "Awareness Cards" lists titles, authors, and journal references.

In August 1965 the National Bureau of Standards (Washington, D.C.) has set up a Data Center for molecular ionization processes (Garrin, D., and H. M. Rosenstock, *J. Chem. Doc.*, **7**, 31 (1967)). The objective is to maintain a complete and current file on information on ionization and appearance potentials, electron affinities, heats of formations, and related data. In March 1966 the Mass Spectrometry Data Center of the Atomic Energy Authority (at Aldermaston) was established in the United Kingdom. Both data centers have as their long range goal the development of a central source of information. The current revolution in information storage and retrieval will undoubtedly result in automated information transfer networks. It will eventually only take a long distance telephone call to find out whether a complicated organic molecule has ever been analyzed in a mass spectrometer, or to obtain the possible metastable transitions for a particular fragmentation process.

Appendix I

Table of Nuclidic Masses and Relative Abundances for Naturally Occurring Isotopes

Atomic weights and nuclidic masses are based on the carbon-12 = 12.000000 scale. Most of the data on atomic weights and isotopic abundances have been taken from the 9th edition (1966) of the *Chart of Nuclides and Isotopes*, published by the General Electric Company. Data were also taken from Cameron, A. E., and E. Wickers, *J. Am. Chem. Soc.*, **84**, 4175 (1962) and from the 1967–68 edition of the *Handbook of Chemistry and Physics* (Chemical Rubber Publishing Co., Cleveland). The chief sources of nuclidic masses are: Everling, F., L. A. König, J. H. Mattauch, and A. H. Wapstra, *Nuclear Physics*, **18**, 529 (1960); Bhanot, V. B., W. H. Johnson, and A. O. Nier, *Phys. Rev.*, **120**, 235 (1960); Ries, R. R., R. A. Damerow, and W. H. Johnson, *Phys. Rev.*, **132**, 1662 and 1673 (1963). Also, several of the data compilations listed in Chapter 14 have been consulted.

Atomic no.	Element	Symbol	Mass no.	Nuclidic mass	Relative abundance	Atomic weight
1	Hydrogen	H	1	1.007825	99.9855	1.00797
	Deuterium	D	2	2.014102	0.0145	
2	Helium	He	3	3.016030	0.000137	4.0026
			4	4.002604	99.999863	
3	Lithium	Li	6	6.015126	7.50	6.939
			7	7.016005	92.50	
4	Beryllium	Be	9	9.012186	100.	9.0122
5	Boron	B	10	10.012939	19.78	10.811
			11	11.009305	80.22	

Atomic no.	Element	Sym-bol	Mass no.	Nuclidic mass	Relative abundance	Atomic weight
6	Carbon	C	12	12.000000	98.888	12.01115
			13	13.003354	1.112	
7	Nitrogen	N	14	14.003074	99.633	14.0067
			15	15.000108	0.367	
8	Oxygen	O	16	15.994915	99.759	15.9994
			17	16.999133	0.0374	
			18	17.999160	0.2039	
9	Fluorine	F	19	18.998405	100.	18.9984
10	Neon	Ne	20	19.992440	90.92	20.183
			21	20.993849	0.257	
			22	21.991384	8.82	
11	Sodium	Na	23	22.989773	100.	22.9898
12	Magnesium	Mg	24	23.985045	78.70	24.312
			25	24.985840	10.13	
			26	25.982591	11.17	
13	Aluminum	Al	27	29.981535	100.	26.9815
14	Silicon	Si	28	27.976927	92.21	28.086
			29	28.976491	4.70	
			30	29.973761	3.09	
15	Phosphorus	P	31	30.973763	100.	30.9738
16	Sulfur	S	32	31.972074	95.018	32.064
			33	32.971460	0.760	
			34	33.967864	4.215	
			36	35.967091	0.014	
17	Chlorine	Cl	35	34.968854	75.53	35.453
			37	36.965896	24.47	
18	Argon	Ar	36	35.967548	0.337	39.948
			38	37.962724	0.063	
			40	39.962384	99.600	
19	Potassium	K	39	38.963714	93.10	39.102
			40	39.964008	0.0118	
			41	40.961835	6.88	
20	Calcium	Ca	40	39.962589	96.97	40.08
			42	41.958628	0.64	
			43	42.958780	0.145	
			44	43.955490	2.06	
			46	45.953689	0.003	
			48	47.952363	0.185	
21	Scandium	Sc	45	44.955919	100.	44.956
22	Titanium	Ti	46	45.952633	7.93	47.90

Atomic no.	Element	Sym-bol	Mass no.	Nuclidic mass	Relative abundance	Atomic weight
	Titanium (*cont.*)					
			47	46.951758	7.28	
			48	47.947948	73.94	
			49	48.947867	5.51	
			50	49.944789	5.34	
23	Vanadium	V	50	49.947165	0.24	50.942
			51	50.943978	99.76	
24	Chromium	Cr	50	49.946051	4.31	51.996
			52	51.940514	83.76	
			53	52.940651	9.55	
			54	53.938879	2.38	
25	Manganese	Mn	55	54.938054	100.	54.9380
26	Iron	Fe	54	53.939621	5.82	55.847
			56	55.934932	91.66	
			57	56.935394	2.19	
			58	57.933272	0.33	
27	Cobalt	Co	59	58.933189	100.	58.9332
28	Nickel	Ni	58	57.935342	67.88	58.71
			60	59.930783	26.23	
			61	60.931049	1.19	
			62	61.928345	3.66	
			64	63.927959	1.08	
29	Copper	Cu	63	62.929594	69.09	63.54
			65	64.927786	30.91	
30	Zinc	Zn	64	63.929145	48.89	65.37
			66	65.926048	27.81	
			67	66.927149	4.11	
			68	67.924865	18.57	
			70	69.925348	0.62	
31	Gallium	Ga	69	68.925682	60.4	69.72
			71	70.924840	39.6	
32	Germanium	Ge	70	69.924277	20.52	72.59
			72	71.921740	27.43	
			73	72.923360	7.76	
			74	73.921150	36.54	
			76	75.921360	7.76	
33	Arsenic	As	75	74.921580	100.	74.9216
34	Selenium	Se	74	73.922450	0.87	78.96
			76	75.919229	9.02	
			77	76.919934	7.58	
			78	77.917348	23.52	
			80	79.916512	49.82	
			82	81.916660	9.19	

Atomic no.	Element	Symbol	Mass no.	Nuclidic mass	Relative abundance	Atomic weight
35	Bromine	Br	79	78.918348	50.537	79.909
			81	80.916344	49.463	
36	Krypton	Kr	78	77.920368	0.35	83.80
			80	79.916388	2.27	
			82	81.913483	11.56	
			83	82.914131	11.55	
			84	83.911504	56.90	
			86	85.910617	17.37	
37	Rubidium	Rb	85	84.911710	72.15	85.47
			87	86.909180	27.85	
38	Strontium	Sr	84	83.913376	0.56	87.62
			86	85.909260	9.86	
			87	86.908890	7.02	
			88	87.905610	82.56	
39	Yttrium	Y	89	88.905430	100.	88.906
40	Zirconium	Zr	90	89.904320	51.46	91.22
			91	90.905250	11.23	
			92	91.904590	17.11	
			94	93.906140	17.40	
			96	95.908200	2.80	
41	Niobium	Nb	93	92.906020	100.	92.906
42	Molybdenum	Mo	92	91.906290	15.84	95.94
			94	93.904740	9.04	
			95	94.905720	15.72	
			96	95.904550	16.53	
			97	96.905750	9.46	
			98	97.905510	23.78	
			100	99.907570	9.63	
44	Ruthenium	Ru	96	95.907600	5.51	101.07
			98	97.905500	1.87	
			99	98.906080	12.72	
			100	99.903020	12.62	
			101	100.904120	17.07	
			102	101.903720	31.63	
			104	103.905530	18.58	
45	Rhodium	Rh	103	102.904800	100.	102.905
46	Palladium	Pd	102	101.904940	0.96	106.4
			104	103.903560	10.97	
			105	104.904640	22.23	
			106	105.903200	27.33	
			108	107.903920	26.71	
			110	109.904500	11.81	

Atomic no.	Element	Symbol	Mass no.	Nuclidic mass	Relative abundance	Atomic weight
47	Silver	Ag	107	106.904970	51.817	107.870
			109	108.904700	48.183	
48	Cadmium	Cd	106	105.905950	1.22	112.40
			108	107.904000	0.88	
			110	109.902970	12.39	
			111	110.904150	12.75	
			112	111.902840	24.07	
			113	112.904610	12.26	
			114	113.903570	28.86	
			116	115.905010	7.58	
49	Indium	In	113	112.904280	4.28	114.82
			115	114.904070	95.72	
50	Tin	Sn	112	111.904940	0.96	118.69
			114	113.902960	0.66	
			115	114.903530	0.35	
			116	115.902110	14.30	
			117	116.903060	7.61	
			118	117.901790	24.03	
			119	118.903390	8.58	
			120	119.902130	32.85	
			122	121.903410	4.72	
			124	123.905240	5.94	
51	Antimony	Sb	121	120.903750	57.25	121.75
			123	122.904150	42.75	
52	Tellurium	Te	120	119.904510	0.089	127.60
			122	121.903000	2.46	
			123	122.904180	0.87	
			124	123.902760	4.61	
			125	124.904420	6.99	
			126	125.903242	18.71	
			128	127.904710	31.79	
			130	129.906700	34.48	
53	Iodine	I	127	126.904352	100.	126.9044
54	Xenon	Xe	124	123.906120	0.096	131.30
			126	125.904169	0.090	
			128	127.903538	1.919	
			129	128.904784	26.44	
			130	129.903510	4.08	
			131	130.905087	21.18	
			132	131.904162	26.89	
			134	133.905398	10.44	
			136	135.907221	8.87	

Atomic no.	Element	Sym-bol	Mass no.	Nuclidic mass	Relative abundance	Atomic weight
55	Cesium	Cs	133	132.905090	100.	132.905
56	Barium	Ba	130	129.906247	0.101	137.34
			132	131.905120	0.097	
			134	133.904310	2.42	
			135	134.905570	6.59	
			136	135.904360	7.81	
			137	136.905560	11.32	
			138	137.905010	71.66	
57	Lanthanum	La	138	137.906810	0.089	138.91
			139	138.906060	99.911	
58	Cerium	Ce	136	135.907100	0.193	140.12
			138	137.905720	0.250	
			140	139.905280	88.48	
			142	141.909040	11.07	
59	Praseodymium	Pr	141	140.907390	100.	140.907
60	Neodymium	Nd	142	141.907478	27.11	144.24
			143	142.909620	12.17	
			144	143.909900	23.85	
			145	144.912160	8.30	
			146	145.912690	17.22	
			148	147.916480	5.73	
			150	149.920710	5.62	
61	Promethium	Pm	—	—	—	—
62	Samarium	Sm	144	143.911650	3.09	150.35
			147	146.914620	14.97	
			148	147.914560	11.24	
			149	148.916930	13.83	
			150	149.917010	7.44	
			152	151.919490	26.72	
			154	153.922010	22.71	
63	Europium	Eu	151	150.919630	47.82	151.96
			153	152.920860	52.18	
64	Gadolinium	Gd	152	151.919530	0.20	157.25
			154	153.920720	2.15	
			155	154.922590	14.73	
			156	155.922100	20.47	
			157	156.923940	15.68	
			158	157.924100	24.87	
			160	159.927120	21.90	
65	Terbium	Tb	159	158.924950	100.	158.925

Atomic no.	Element	Symbol	Mass no.	Nuclidic mass	Relative abundance	Atomic weight
66	Dysprosium	Dy	156	155.923760	0.052	162.50
			158	157.923960	0.090	
			160	159.924830	2.29	
			161	160.926600	18.88	
			162	161.926470	25.53	
			163	162.928370	24.97	
			164	163.928830	28.18	
67	Holmium	Ho	165	164.930300	100.	164.930
68	Erbium	Er	162	161.928780	0.136	167.26
			164	163.929290	1.56	
			166	165.930400	33.41	
			167	166.932050	22.94	
			168	167.932380	27.07	
			170	169.935510	14.88	
69	Thulium	Tm	169	168.934350	100.	168.934
70	Ytterbium	Yb	168	167.933300	0.135	173.04
			170	169.934880	3.03	
			171	170.936460	14.31	
			172	171.936560	21.82	
			173	172.938300	16.13	
			174	173.939020	31.84	
			176	175.942740	12.73	
71	Lutetium	Lu	175	174.940890	97.41	174.97
			176	175.942740	2.59	
72	Hafnium	Hf	174	173.940260	0.18	178.49
			176	175.941650	5.20	
			177	176.943480	18.50	
			178	177.943870	27.14	
			179	178.946020	13.75	
			180	179.946810	35.24	
73	Tantalum	Ta	180	179.947520	0.012	180.948
			181	180.947980	99.988	
74	Tungsten	W	180	179.946980	0.14	183.85
			182	181.948270	26.41	
			183	182.950290	14.40	
			184	183.950990	30.64	
			186	185.954340	28.41	
75	Rhenium	Re	185	184.953020	37.07	186.2
			187	186.955960	62.93	

Atomic no.	Element	Symbol	Mass no.	Nuclidic mass	Relative abundance	Atomic weight
76	Osmium	Os	184	183.952560	0.02	190.2
			186	185.953940	1.59	
			187	186.955960	1.64	
			188	187.955970	13.3	
			189	188.958250	16.1	
			190	189.958600	26.4	
			192	191.961410	41.0	
77	Iridium	Ir	191	190.960850	37.3	192.2
			193	192.963280	62.7	
78	Platinum	Pt	190	189.959950	0.013	195.09
			192	191.961430	0.78	
			194	193.962810	32.9	
			195	194.964820	33.8	
			196	195.964381	25.3	
			198	197.967530	7.21	
79	Gold	Au	197	196.966552	100.	196.967
80	Mercury	Hg	196	195.965822	0.146	200.59
			198	197.966769	10.02	
			199	198.968256	16.84	
			200	199.968344	23.13	
			201	200.970315	13.22	
			202	201.970630	29.80	
			204	203.973482	6.85	
81	Thallium	Tl	203	202.972331	29.50	204.37
			205	204.974462	70.50	
82	Lead	Pb	204	204.973069	1.48	207.19
			206	205.974459	23.6	
			207	206.975898	22.6	
			208	207.976644	52.3	
83	Bismuth	Bi	209	208.980417	100.	208.980
84	Polonium	Po	—	—	—	—
85	Astatine	At	—	—	—	—
86	Radon	Rn	—	—	—	—
87	Francium	Fr	—	—	—	—
88	Radium	Ra	—	—	—	—
89	Actinium	Ac	—	—	—	—
90	Thorium	Th	232	232.038211	100.	232.038
91	Protactinium	Pa	—	—	—	—
92	Uranium	U	234	234.040900	0.0057	238.03
			235	235.043933	0.72	
			238	238.050760	99.27	

Appendix II

Atomic and Molecular Ionization Potentials

The ionization potentials presented in this Appendix are primarily for illustration. The data have been collected from many sources. Almost all atomic ionization potentials were determined by vacuum ultraviolet spectroscopy. The molecular ionization potentials were determined either by electron impact or by photoionization techniques. The former normally includes mass analysis; the latter usually does not. When the most recent and "best" value is desired for a particular compound, the Data Center at the National Bureau of Standards (Washington, D.C.) should be consulted, as described in Chapter 14.

A. ATOMIC IONIZATION POTENTIALS

Element	I	II	III	IV	V
Actinium	6.9	12.1	20.	—	—
Aluminum	5.96	18.74	28.31	119.37	153.4
Americium	6.0	—	—	—	—
Antimony	8.64	16.5	25.3	44.1	56.
Argon	15.755	27.62	40.90	59.79	75.0
Arsenic	9.81	18.63	28.34	50.1	62.6
Astatine	9.5	—	—	—	—
Barium	5.21	10.00	35.5	—	—
Beryllium	9.28	18.12	153.1	216.6	—
Bismuth	7.29	16.7	25.6	45.3	56.0
Boron	8.30	25.15	37.92	259.29	340.1
Bromine	11.84	21.6	35.9	47.3	59.7
Cadmium	8.99	16.90	37.47	—	—
Calcium	6.11	11.87	51.21	67.	84.4
Carbon	11.256	24.376	47.871	64.476	391.986

Element	I	II	III	IV	V
Cerium	5.6	12.3	20.	33.	—
Cesium	3.89	25.1	35.	—	—
Chlorine	13.01	23.80	39.90	53.5	67.80
Chromium	6.764	16.49	30.95	50.	73.
Cobalt	7.86	17.05	33.49	83.1	305.
Copper	7.72	20.29	36.8	671.	—
Dysprosium	6.8	—	—	—	—
Erbium	6.08	—	—	—	—
Europium	5.67	—	—	—	—
Fluorine	17.42	34.98	62.64	87.14	114.21
Francium	4.0	—	—	—	—
Gadolinium	6.2	12.	—	—	—
Gallium	6.00	20.57	30.70	64.2	—
Germanium	7.88	15.93	34.21	44.7	93.4
Gold	9.22	20.5	—	—	—
Hafnium	7.	14.9	23.2	—	—
Helium	24.481	54.403	—	—	—
Hydrogen	13.595	—	—	—	—
Indium	5.78	18.86	28.	54.4	—
Iodine	10.454	19.1	170.	—	—
Iridium	9.	—	—	—	—
Iron	7.87	16.18	30.64	56.8	151.
Krypton	13.996	24.56	36.9	43.5	63.
Lanthanum	5.61	11.43	19.17	—	—
Lead	7.42	15.03	31.93	42.3	68.8
Lithium	5.390	75.619	122.42	—	—
Lutetium	14.7	—	—	—	—
Magnesium	7.64	15.03	80.14	109.3	141.2
Manganese	7.43	15.64	33.69	52.	76.
Mercury	10.43	18.75	34.2	49.5	—
Molybdenum	7.10	16.15	27.13	46.4	61.
Neodymium	5.51	—	—	—	—
Neon	21.559	41.07	63.5	97.0	126.3
Nickel	7.633	18.15	35.16	350.	455.
Niobium	6.88	14.32	25.04	38.3	50.
Nitrogen	14.53	29.59	47.27	77.45	97.86
Osmium	8.5	17.	—	—	—
Oxygen	13.614	35.108	54.886	77.394	113.873

Element	I	II	III	IV	V
Palladium	8.33	19.42	32.92	—	—
Phosphorus	10.48	19.7	30.15	51.35	65.0
Platinum	9.0	18.5	—	—	—
Plutonium	5.1	—	—	—	—
Polonium	8.4	—	—	—	—
Potassium	4.4	31.8	46.	61.	82.6
Praesodymium	5.4	—	—	—	—
Radium	5.3	—	—	—	—
Radon	10.75	—	—	—	—
Rhenium	7.87	16.6	—	—	—
Rhodium	7.46	18.1	31.	—	—
Rubidium	4.176	27.5	40.	—	—
Ruthenium	7.36	16.7	28.	—	—
Samarium	5.6	11.2	—	—	—
Scandium	6.54	12.80	24.75	73.9	92.
Selenium	9.75	21.5	32.	43.	68.
Silicon	8.149	16.34	33.488	45.13	166.73
Silver	7.574	21.48	34.82	—	—
Sodium	5.138	47.29	71.71	98.88	138.37
Strontium	5.69	11.03	57.	324.	—
Sulfur	10.357	23.4	35.0	47.3	72.5
Tantalum	7.88	16.2	—	—	—
Technetium	7.28	15.26	29.54	—	—
Tellurium	9.01	18.6	31.	38.0	60.
Terbium	5.98	—	—	—	—
Thallium	6.11	20.42	29.8	50.7	—
Thorium	6.95	29.38	—	—	—
Thulium	5.81	—	—	—	—
Tin	7.34	14.63	30.49	40.72	72.3
Titanium	6.82	13.57	24.47	43.24	99.8
Tungsten	7.98	17.7	—	—	—
Uranium	6.08	—	—	—	—
Vanadium	6.71	14.1	29.3	48.	65.
Xenon	12.08	21.1	32.0	46.	53.
Ytterbium	6.2	12.1	—	—	—
Yttrium	6.4	—	—	—	—
Zinc	9.36	17.89	40.0	—	—
Zirconium	6.92	13.97	24.00	33.8	—

B. MOLECULAR IONIZATION POTENTIALS

1. Inorganic Compounds

Substance	Ionization potential, eV
Hydrogen	15.35
Nitrogen	15.60
Oxygen	12.50
Ozone	12.80
Chlorine	11.48
Bromine	10.92
Iodine	9.28
Carbon monoxide	14.01
Nitrogen monoxide	9.40
Hydrogen chloride	12.78
Water	12.83
Hydrogen sulfide	10.40
Carbon dioxide	13.78
Sulfur dioxide	12.42
Nitrogen dioxide	9.91
Ammonia	10.52
Hydrogen cyanide	13.91

2. Organic Compounds

Substance	Formula	Ionization potential
Hydrocarbons		
Methane	CH_4	13.12
Ethane	C_2H_6	11.65
Propane	C_3H_8	11.07
n-Butane	C_4H_{10}	10.80
Isobutane	C_4H_{10}	10.79
n-Octane	C_8H_{18}	10.24
n-Nonane	C_9H_{20}	10.21
n-Decane	$C_{10}H_{22}$	10.19
Deuteromethane	CH_3D	13.12
Dideuteromethane	CH_2D_2	13.14
Trideuteromethane	CHD_3	13.18
Tetradeuteromethane	CD_4	13.26
Ethylene	C_2H_4	10.56
Propylene	C_3H_6	9.80
1-Butene	C_4H_8	9.76
Isobutene	C_4H_8	9.35
cis-Butene	C_4H_8	9.29
trans-2- Butene	C_4H_8	9.27

Substance	Formula	Ionization potential
Hydrocarbons (cont.)		
Acetylene	$HC \equiv CH$	11.41
Diacetylene	C_4H_2	10.74
Propyne	$CH \equiv CCH_3$	10.30
1-Butyne	$CH \equiv C—CH_2CH_3$	10.34
2-Butyne	$CH_3C \equiv CCH_3$	9.85
Butadiene	C_4H_6	9.07
Cyclobutene	C_4H_6	8.90
Benzene	C_6H_6	9.44
Toluene	$C_6H_5—CH_3$	8.82
o-Xylene	$C_6H_4(CH_3)_2$	8.56
m-Xylene	$C_6H_4(CH_3)_2$	8.56
p-Xylene	$C_6H_4(CH_3)_2$	8.45
Styrene	$CH_2=CH—C_6H_5$	8.47
Diphenyl	$C_6H_5—C_6H_5$	8.27
Naphthalene	$C_{10}H_8$	8.12
Oxygen-Containing Compounds		
Methyl alcohol	CH_3OH	10.95
Ethyl alcohol	C_2H_5OH	10.60
n-Butyl alcohol	C_4H_9OH	10.30
Isobutyl alcohol	C_4H_9OH	10.17
t-Butyl alcohol	C_4H_9OH	9.92
Dimethyl ether	$(CH_3)_2O$	10.50
Diethyl ether	$(C_2H_5)_2O$	9.72
Dioxane	$C_4H_8O_2$	9.52
Ethylene oxide	C_2H_4O	10.56
Formaldehyde	$H—CHO$	10.83
Acetaldehyde	$CH_3—CHO$	10.18
Acetone	$(CH_3)_2CO$	9.92
Methyl ethyl ketone	$C_2H_5—CO—CH_3$	9.74
Ketene	H_2C_2O	9.61
Formic acid	$HCOOH$	11.05
Acetic acid	CH_3COOH	10.35
Methyl acetate	CH_3COOCH_3	10.51
Diacetyl	$CH_3CO—COCH_3$	9.54
Glyoxal	$CHO—CHO$	11.43
Acrolein	$CH_2=CHCHO$	10.34
Crotonaldehyde	$CH_3CH=CHCHO$	9.81
Phenol	C_6H_5OH	8.50
Benzaldehyde	$C_6H_5—CHO$	9.53
Anisol	$C_6H_5—O—CH_3$	8.22
Benzylmethyl ether	$C_6H_5CH_2—O—CH_3$	8.85

Substance	Formula	Ionization potential
Halogen Compounds		
Methyl chloride	CH_3Cl	11.28
Ethyl chloride	C_2H_5Cl	10.97
Methyl iodide	CH_3I	9.54
Ethyl iodide	C_2H_5I	9.33
Dichloromethane	CH_2Cl_2	11.35
Trichloromethane	$CHCl_3$	11.42
Tetrachloromethane	CCl_4	11.47
Vinyl chloride	$CH_2{=}CHCl$	10.00
Trichloroethylene	$ClCH{=}CCl_2$	9.47
Chlorobenzene	C_6H_5Cl	9.07
Iodobenzene	C_6H_5I	8.73
p-Chlorotoluene	$CH_3{-}C_6H_4{-}Cl$	8.69
m-Chlorotoluene	$CH_3{-}C_6H_4{-}Cl$	8.63
Heterocyclic Compounds		
Pyridine	C_5H_5N	9.26
Pyridazine	$C_4H_4N_2$	9.86
Pyrimidine	$C_4H_4N_2$	9.91
Pyrazine	$C_4H_4N_2$	10.00
Furan	C_4H_4O	8.89
Pyrrole	C_4H_5N	8.90
Thiophene	C_4H_4S	8.86
Tetrahydrofuran	C_4H_8O	9.45
Tetrahydropyrrole	C_4H_9N	8.60

Author Index

Numbers in parentheses are reference numbers and show that an author's work is referred to although his name is not mentioned in the text. Numbers in *italics* indicate the pages in which the full references appear. Note: Authors mentioned in Chapter 14 are *not* included in this Index.

Subject Index